BUILD YOUR OWN
LASER, PHASER, ION RAY GUN
& OTHER WORKING SPACE-AGE PROJECTS

BUILD YOUR OWN
LASER, PHASER, ION RAY GUN
& OTHER WORKING SPACE-AGE PROJECTS

BY ROBERT E. IANNINI

TAB BOOKS Inc.
BLUE RIDGE SUMMIT, PA 17214

FIRST EDITION

SECOND PRINTING

Printed in the United States of America

Library of Congress Cataloging in Publication Data

Iannini, Robert E.
Build your own laser, phaser, ion ray gun, and
other working space-age projects.

Includes index.
1. Electronics—Amateurs' manuals.   2. Lasers—
Amateurs' manuals.   I. Title.
TK9965.I2   1983   621.381   83-4872
ISBN 0-8306-0204-6
ISBN 0-8306-0604-1 (pbk.)

Contents

Introduction

Many fascinating and amazing scientific electronic developments have evolved during the past several decades. One of the most interesting and controversial is the "laser" (acronym for light amplification by stimulated emission of radiation). This device is often portrayed as the ultimate ray gun weapon of the future by many movie and TV presentations. The projection of destructive energy over a distance has always been an objective of man. Even in ancient times records show where man has taken advantage of heat beams to destroy the enemy by using the radiant energy from the sun. Often these rays would be used to set fire to the sails of an enemy fleet. Modern science has not only produced the means of projection but also can provide the source of energy to realistically destroy men and machines of war. Chemical lasers can convert the exothermic energy of a reaction into a powerful coherent beam of heat capable of penetrating and vaporizing the hardest of metals. X-ray and free electron lasers are now capable of producing high enough power levels where use in a vacuum such as space has definite advantages. Antipersonnel laser

weapons (now easily carried by man) have kill capability at distances up to miles but offer a definite cost disadvantage when compared to conventional weapons. Particle and molecular beams following evacuated channels in the atmosphere blasted by high-powered lasers offer a means of projecting this highly destructive energy to a target. It is obvious that the ability of projecting a beam of energy with extremely small divergence is a very positive asset to producing a weapon of destruction capable of being effective over considerable distances.

Lasers have many other practical uses owing also to these properties. These can range from surgical uses to metal fabricating and heat treating. The monochromaticity and coherence of the light allows many uses such as holography, optical ranging, detection analysis, radar, communication, special effects, etc. Equivalent line brightness cannot be duplicated by other sources of light.

This book describes in detail several laser projects that are capable of being constructed and used in many practical applications. These projects range from simple milliwatt-powered optical lasers to a multiwatt burning and cutting device. Unfortunately, lasers require high-voltage power supplies very similar to ham radio transmitters and consequently must be treated with total respect for all safety precautions. It is suggested that the hobbyist obtain *The Radio Amateur's Handbook* and study it. This excellent reference shows construction practices and procedures and many similar circuits that are used for laser power supplies.

This book also contains plans for many other useful scientific and electronic projects such as high voltage power supplies, Tesla coils, laboratory and science fair projects, ultrasonic pest control devices, personal protection and security devices.

The project text is divided into four sections, each with a general information type introduction briefly explaining pertinent points of interest along with any necessary safety precautions. All information including sources of individual parts as well as complete kits are provided so the builder may construct a useful working device. Reference is given throughout the text to the author's company, Information Unlimited, Inc., as a source of these parts and the builder may also call or write for assistance if needed. Some of the projects described in this book are patented and licensed by the author and are being manufactured and marketed to companies under different names and labels.

I wish to acknowledge the many contributions made by the following people. Rick Upham for his mechanical ability in laying out the circuits described in this book along with many of the assembled methods and designs; Keith Snow for the excellent artwork and illustrations; Dr. Harry E. Franks of Laser Nucleonics for use of his laser study course and help in solid-state lasers; Bill and Steve Sommer of IR Scientific for their preparation of the excellent chapter on constructing a See-In-The-Dark device; the entire staff of Information Unlimited, Karen Lavoie, John Devereaux, Ann Durocher, John Staiti, Constance and Clayton Ashly, Peter Riendeau, Janet Smith, Pat Upham, Jeff Vandeberghe, Margo Mack, my wife Cecile ("the Munchkin"), Jim Kennedy of Gazebo Advertising, and last, but not least, Jerry Vandeberghe and his wife Janet who typed the finished manuscript.

Section I
A Brief Description
of General Laser Theory

LASER is the acronym for "Light Amplification by Stimulated Emission of Radiation." Laser light differs from conventional light in several aspects. Even though both are electromagnetic in nature, laser light has certain properties that make it highly desirable in a multitude of fields and applications. Unlike conventional light, laser light is highly monochromatic, that is, pure light of almost a single wavelength or small line width (bandwidth). It is also temporally and spatially coherent. One may think of temporal coherence as a source of light where if the observer were to snap a picture of the wave, he would observe a group of peaks and valleys occurring with equal separations or wavelengths. This quality of the light contributes to the sharp line width approaching spectral purity. Temporal coherence is the time relation of the field intensity at a particular point in space. Spatial coherence of the wave now is the valley and peaks remaining stationary relative to space and position. It is this characteristic that gives the laser beam directionability due to the minimal constructional and destructional interferences. We shall refer to

laser light as being both temporal and spatially coherent and spontaneous light as being random and unsynchronized.

Atoms absorb or emit energy in the form of electromagnetic waves as a result of electrons changing from one energy level to another. These energy levels are numerous and well defined being characteristic of the particular atom in question. They also have an equivalent frequency response defined as resonance that is functional of the difference in energy of these levels as stated by the familiar relation $E = hfc$.

Absorption is the result of electrons jumping from a lower state to a higher state (excitation) taking energy at the resonant frequency from the system. Typical are the spectral lines of absorption in spectrographic analysis. An excited atom may now contribute this energy to a system in the form of spontaneous or stimulated emission. Spontaneous emission is the random natural decay of these higher energy levels dropping to their normal lower rest levels and emitting energy equivalent in wavelength to that particular transition. This

1

energy equivalent bandwidth is broad and incoherent. Energy must obviously be externally supplied to the atom for it to achieve a higher energy state.

Stimulated emission is the reciprocal process of resonant absorption and it only occurs when a population inversion is produced. A population inversion is when a higher energy level contains more population than its lower counterpart. When this occurs stimulated emission is easily obtained by photons of energy near to that of the inverted transition now triggering other identical electrons to drop to a lower level producing the in-phase coherent emission energy characteristic of the laser beam.

The trick is to create this necessary population inversion between these levels in question, by forcing, or pumping externally. Achieving this unnatural phenomena is accomplished by several different means, usually depending on the particular laser in question. The usual means is to excite a medium containing three or more involved energy levels. These levels are usually pumped either electrically or optically to their respective excited states. If part of this decay from the highest level drops down to the intermediate level and adds to the population, temporarily causing it to have more population than the ground level, this creates the population inversion necessary for stimulated emission between the intermediate and ground levels. This emission when directed between two mirrors such as a Fabry-Perot resonator is reflected back through the laser medium many times stimulating more photons and exiting the device as a powerful beam of energy.

An excellent home or classroom study course is now available to the laser enthusiasts desiring further technical subject material on these fascinating and interesting devices. The course consists of a 55 minute tape and thirty-three 35 mm slides for classroom use or full sized photographs for the home student. The course clearly explains laser theory and applications starting from basic fundamentals and reaching an intermediate level. The student will achieve an excellent understanding of these devices plus obtain a good foundation as a prerequisite for taking advantage of more advanced and expensive courses. Dr. Harry E. Franks of Laser Nucleonics and a well known pioneer in high-powered laser technology has prepared this excellent study course. Also contained in this program are four laser construction plans that include all the required information for constructing and using a Helium-Neon Visible light, Ruby or Yag Rod Pulsed, Carbon dioxide continuous beam, and galium arsenide injection pulsed infrared laser system. Write Information Unlimited, Inc., P.O. Box 716, Amherst, N.H. 03031, or call 1-603-673-4730 for further information.

LASERS AND GENERAL SAFETY PRECAUTIONS AND LABELING

Most laser projects that involve sufficient power levels to burn or cut usually require power supplies capable of producing severe electrical shock hazards. It must be stressed to the builder to exercise necessary safety precautions involving proper handling, building and labeling of potential danger spots. Some of the power supplies used here are very similar to those used in Ham Radio transmitters therefore it is suggested that the builder consult the Amateur Radio Handbook on safety and construction practices.

Lasers not only can be an electrical hazard but some also generate a beam of energy that can be optically dangerous. This beam of energy can start fires, burn flesh, ignite combustible materials and cause eye damage from unsuspecting reflection.

The following section is dedicated to the use of proper labeling for several of the projects in this book to comply with normal safety procedures as well as compliance with the Bureau of Radiological Health and Welfare.

LRG3 Plans—Invisible Laser Rifle Project

This device does not burn or cut but can be optically hazardous if viewed directly on axis.

2

Voltages used can produce a moderate shock. (Black, white, and red label.)

LGU3 Plans—Visible Laser Light Gun Project

This device generates a visible beam of light that will not burn or cut. It is moderately hazardous and should not be viewed on axis. Voltage used can produce a moderate shock. (Black and yellow label.)

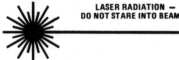

RUB3 Plans—Ruby/Yag Laser Ray Gun Project

This device generates an intense burst of light capable of blasting holes through metal. It can be optically hazardous and electrically dangerous. Use both labels. (Black, white, and red labels.)

LC5 Plans—Carbon Dioxide Burning Laser Project

This device generates a powerful continuous beam of heat that can cause fires, burn flesh, and is obviously optically dangerous. Use both labels. (Black, white, and red labels.)

Laser safety goggles are available from the Glendale Optical Co., 130 Crossways Park Dr., Woodbury, L.I., New York 11797. All projects using high voltages should be properly labeled as instructions indicate.

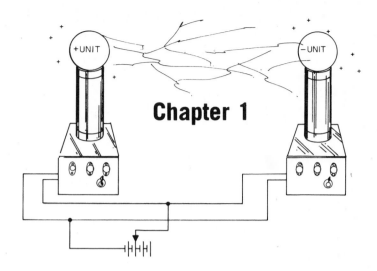

Chapter 1

Beginner's Pulsed Visible Simulated Laser (LHP2)

The following laser project is intended for beginners and younger students. It is not dangerous or hazardous and therefore, is an excellent classroom demonstrator. It is not a true lasing device, yet it produces a highly monochromatic visible source of red or infrared light that can be optically or electronically detected for several kilometers using a suitable detector. It can also be used for spot illumination, short range target sighting and designation, optical communications, optical aligning of more complicated systems, long range intrusion alarms, stroboscopic uses and many more. The finished unit is completely self-contained in a circular housing with batteries and necessary optics.

CIRCUIT DESCRIPTION

Refer to Fig. 1-1 and Table 1-1. The circuit pulses a monochromatic emitter of an adjustable rate of one to 20 pulses per second at a peak current of 0.5 amps. Transistor Q1 and Q2 comprise an adjustable astable multivibrator whose repetition rate is made adjustable by the setting of R3. Note that the collector of Q2 goes positive during the separation time between pulses and approaches zero volts for the duration time of these pulses. These transitions are applied to the base of Q3 that is a pnp connected with its collector lead resistor R7 in series with the base emitter junction of Q4. When the collector of Q2 conducts, Q3 base goes negative, "pulse on time" (respect to emitter) and Q3 conducts causing current to flow. Q4 now conducts heavily through R8 and the diode emitter LA1. R8 limits the current to 500 mA. It is this time duration that the device emits. Now when the base of Q3 goes positive, "pulse time off", Q3 and Q4 now does not conduct. Therefore, LA1 ceases emission.

The optical output of the diode LA1 is centered to the bore of the tube TU1 and must be at the focal point of the collimating lens LE2 for best results.

CONSTRUCTION STEPS

1. Separate and identify all parts and pieces.
2. Round corners of PB1 so it will match end

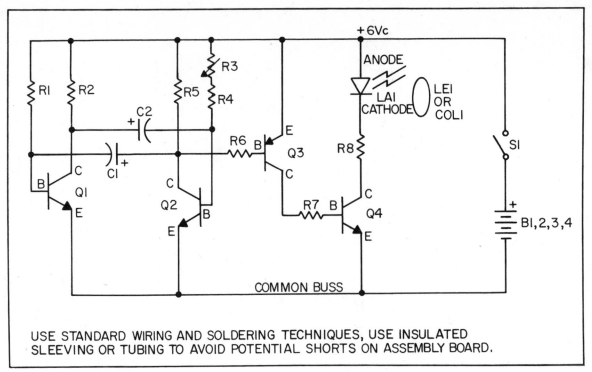

USE STANDARD WIRING AND SOLDERING TECHNIQUES, USE INSULATED SLEEVING OR TUBING TO AVOID POTENTIAL SHORTS ON ASSEMBLY BOARD.

Fig. 1-1. Circuit schematic.

Table 1-1. Beginner's Pulsed Visible Simulated Laser Parts List (LHP2).

R1	(1)	68 k ¼ watt resistor
R2, 5	(2)	2.2 k ¼ watt resistor
R3/S1	(1)	500 k pot and switch combination
R4	(1)	15 k ¼ watt resistor
R6	(1)	6.8 k ¼ watt resistor
R7	(1)	3.9 k ¼ watt resistor
R8	(1)	27 ohm 1/2 watt resistor
C1,2	(2)	.47/35 volt tantalum or electrolytic capacitor
Q1,2,4	(3)	PN2222 npn transistor
Q3	(1)	PN2907 pnp transistor
LA1	(1)	FLV-104 emitter diode for visible red* or FPE-104 for infrared*
BH1	(1)	4-"AA" battery holder
CL1	(1)	Battery clip for holder
B1,2,3,4	(4)	1.5 volt "AA" batteries
PB1	(1)	1.2" × 1.2" of .1 × .1 grid perfboard
TU1	(1)	10" × 1.5" OD × .035 wall alum tube
TU2	(1)	12" × 1.625" × .058" wall alum tube
LE2 Collimating Lens	(1)	38 × 250 mm simple lens
LC5,6	(2)	Plastic cap 1⅝"
LRT1,2	(2)	Lens retainer 1½" reworked plastic plug
CA1	(1)	Plastic plug 1 7/16"

*If FLV104 or FPE104 are unavailable use Hewlett Packard #HLMP3750 for red output. Use #HLMP3850 for yellow, #3950 for green, and XCITON XC880-A for infrared.

Complete kit of the above including printed circuit board is available through Information Unlimited, Inc., P.O. Box 716, Amherst, N.H. 03031. Write or call 1-603-673-4730 for price and availability.

An excellent detector for controlling anything electrical via the light output of this device is the "LLD1 Laser Light Detector and Controller", also available through Information Unlimited, Inc.

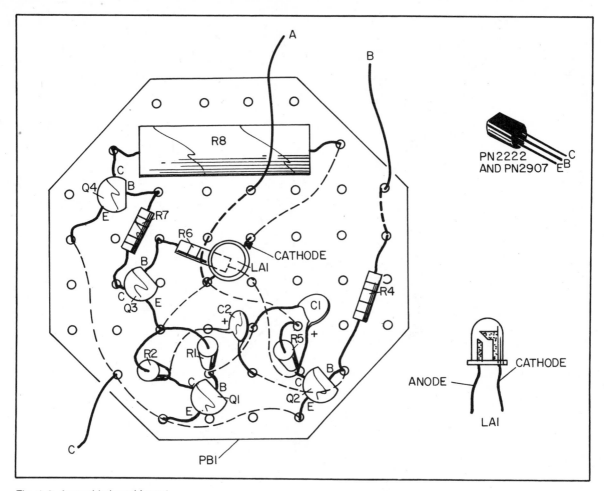

Fig. 1-2. Assembly board layout.

contour of BH1 for snug sliding fit inside TU1 (Figs. 1-2, 1-3).

3. Assemble board as shown either using perfboard (Fig. 1-2) or printed circuit board (Fig. 1-4). Observe position of transistors Q1-Q4. Note Q3 is a pnp type. Observe polarity of C1, C2, and LA1.

4. Connect 4″ wire A from switch section of R3/S1 to assembly board (Fig. 1-2). Connect small jump from outer pin of R3 to the pin on switch section (Fig. 1-3.)

5. Connect 4″ wire B from arm of R3 to point on assembly board (Figs. 1-2 and 1-3).

6. Connect battery clip BH1 as shown in Fig.

1-3. Note wire C from BH1 going to assembly board (Fig. 1-2).

7. Check wiring for accuracy and shorts.

8. Turn S1 to "off" position and insert four "AA" cells correctly in BH1 holder. Clip CL1 to BH1.

9. Turn S1 to "on" and note a bright flashing in LA1 that increases to about 20 per second as R3 is adjusted. Note the voltage points in Fig. 1-1.

10. Tape end of BH1 holder to cover contacts. Glue PB1 to BH1 or use a piece of two-sided tape, etc. (Fig. 1-3).

11. Deburr and polish TU1 for a smooth, sliding fit into TU2. Fit must be even and smooth or

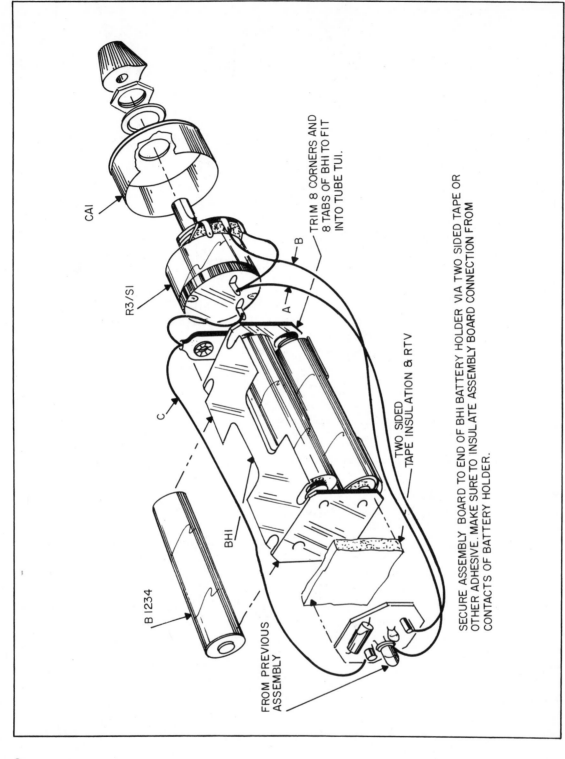

CAI

R3/SI

B

TRIM 8 CORNERS AND
8 TABS OF BHI TO FIT
INTO TUBE TUI.

A

C

BHI

B 1234

FROM PREVIOUS
ASSEMBLY

TWO SIDED
TAPE INSULATION & RTV

SECURE ASSEMBLY BOARD TO END OF BHI BATTERY HOLDER VIA TWO SIDED TAPE OR
OTHER ADHESIVE. MAKE SURE TO INSULATE ASSEMBLY BOARD CONNECTION FROM
CONTACTS OF BATTERY HOLDER.

Fig. 1-3. Electronic assembly.

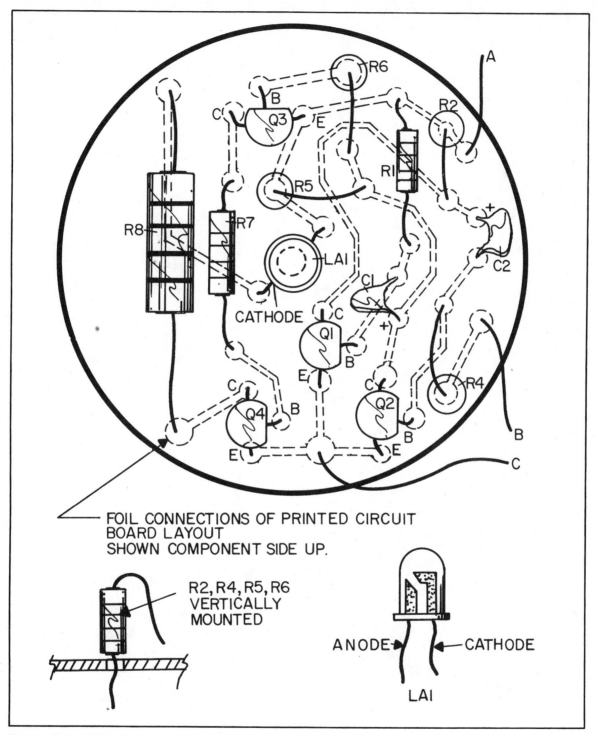

FOIL CONNECTIONS OF PRINTED CIRCUIT
BOARD LAYOUT
SHOWN COMPONENT SIDE UP.

R2, R4, R5, R6
VERTICALLY
MOUNTED

ANODE CATHODE

LAI

Fig. 1-4. Pc board layout.

FABRICATE AND CUT ENDS OUT OF THE TAPERED PLASTIC PLUGS SO AS TO FORM TWO CYLINDRICAL RINGS ABOUT 3/8" TO 1/2" WIDE DESIGNATED LRTI. RINGS MUST BE NEAR EQUAL IN WIDTH FOR PROPER SEATING OF LENS (LE2). INSERT FIRST RING (LRTI) AND RECESS ABOUT 1/2" IN FROM END OF TU2 TUBE. CHECK FOR EVENNESS WITH END OF TUBE TU2. CLEAN LENS (LE2) AND CAREFULLY PLACE IN TUBE ABUTTING ON FIRST RING LRTI. INSERT SECOND RING ABUTTING EVENLY AGAINST LE2 AND FLUSHING WITH END OF TUBE TU2. THIS ARRANGEMENT SANDWICHES THE LENS BETWEEN THE TWO PLASTIC REWORKED RINGS AND MAY REQUIRE ADHESIVE UNDER NORMAL CONDITIONS. LENS IS EVENLY POSITIONED FOR MAXIMUM OUTPUT.

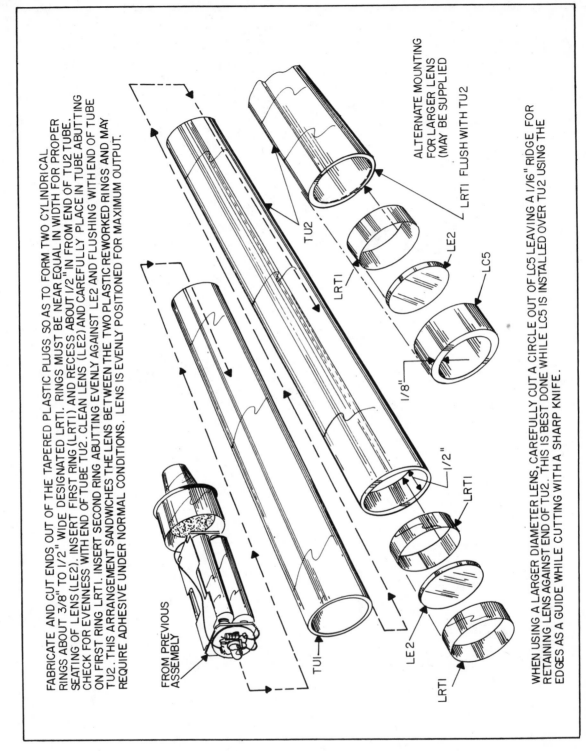

FROM PREVIOUS ASSEMBLY

TUI

TU2

LRTI

LE2

LC5

1/8"

1/2"

LRTI

LE2

LRTI

ALTERNATE MOUNTING FOR LARGER LENS (MAY BE SUPPLIED

LRTI FLUSH WITH TU2

WHEN USING A LARGER DIAMETER LENS, CAREFULLY CUT A CIRCLE OUT OF LC5 LEAVING A 1/16" RIDGE FOR RETAINING LENS AGAINST END OF TU2. THIS IS BEST DONE WHILE LC5 IS INSTALLED OVER TU2 USING THE EDGES AS A GUIDE WHILE CUTTING WITH A SHARP KNIFE.

Fig. 1-5. Blowup.

10

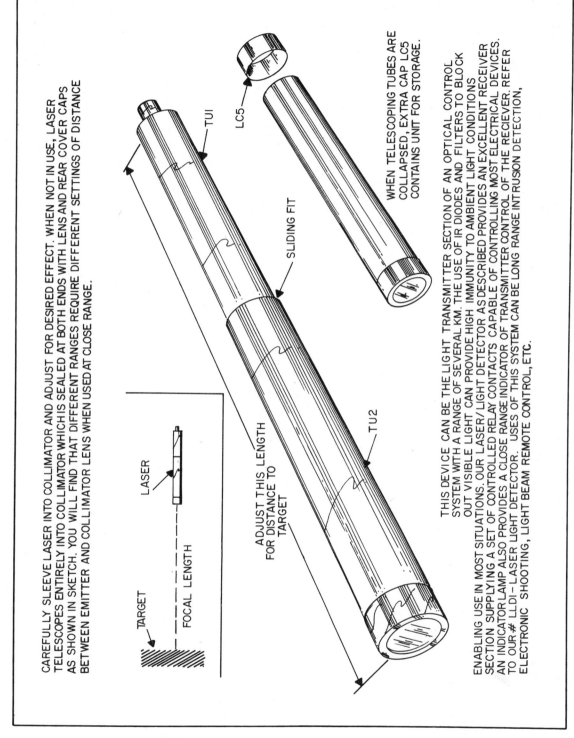

CAREFULLY SLEEVE LASER INTO COLLIMATOR AND ADJUST FOR DESIRED EFFECT. WHEN NOT IN USE, LASER TELESCOPES ENTIRELY INTO COLLIMATOR WHICH IS SEALED AT BOTH ENDS WITH LENS AND REAR COVER CAPS AS SHOWN IN SKETCH. YOU WILL FIND THAT DIFFERENT RANGES REQUIRE DIFFERENT SETTINGS OF DISTANCE BETWEEN EMITTER AND COLLIMATOR LENS WHEN USED AT CLOSE RANGE.

TARGET

LASER

FOCAL LENGTH

ADJUST THIS LENGTH FOR DISTANCE TO TARGET

TU1

LC5

SLIDING FIT

TU2

WHEN TELESCOPING TUBES ARE COLLAPSED, EXTRA CAP LC5 CONTAINS UNIT FOR STORAGE.

THIS DEVICE CAN BE THE LIGHT TRANSMITTER SECTION OF AN OPTICAL CONTROL SYSTEM WITH A RANGE OF SEVERAL KM. THE USE OF IR DIODES AND FILTERS TO BLOCK OUT VISIBLE LIGHT CAN PROVIDE HIGH IMMUNITY TO AMBIENT LIGHT CONDITIONS ENABLING USE IN MOST SITUATIONS. OUR LASER/LIGHT DETECTOR AS DESCRIBED PROVIDES AN EXCELLENT RECEIVER SECTION SUPPLYING A SET OF CONTROLLED RELAY CONTACTS CAPABLE OF CONTROLLING MOST ELECTRICAL DEVICES. AN INDICATOR LAMP ALSO PROVIDES A CLOSE RANGE INDICATOR OF TRANSMITTER CONTROL OF THE RECIEVER. REFER TO OUR # LLDI-LASER LIGHT DETECTOR. USES OF THIS SYSTEM CAN BE LONG RANGE INTRUSION DETECTION, ELECTRONIC SHOOTING, LIGHT BEAM REMOTE CONTROL, ETC.

Fig. 1-6. Final assembly.

problems will result. Also burns of aluminum can cause scratching that will eventually completely disable the sliding action of the two tubes. Assemble lens LE2 as shown in Fig. 1-5.

12. Mount R3/S1 to CA1. The hole in plastic cap can easily be done with an exacto knife for securing the shaft with a nut (Fig. 1-3).

13. Shim BH1 with tape until moderate friction is developed to secure inside of TU1. Insert assembly into rear of TU1 (Fig. 1-5). Place a piece of thin white paper over exit end of TU1 and observe that the red emitter beam is centered in tube.

Adjust LA1 for centered radiation and glue into position with a small dab of RTV or equivalent.

14. Slide on collimator and adjust for correct focal length. Point assembly to a white paper target about 3 to 5 feet away and adjust assembly position (Fig. 1-6).

Note when using wide angle emitting diodes such as the Hewlett-Packard devices specified it may be desirable to paint the insides of the enclosure flat black to prevent side reflections from deteriorating the beam spot size.

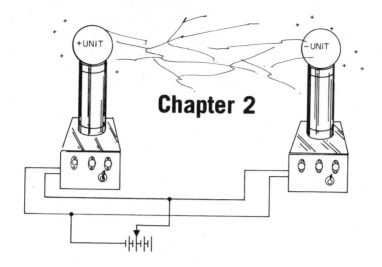

Chapter 2

Visible Red
Continuous Laser Gun (LGU3)

The following plans show how to construct a device capable of projecting a visible red laser light beam at 6328 Å. The property of this light is very high spatial and temporal coherence with a color temperature equivalent of many millions of degrees. This property allows projection of this energy for great distances with little beam divergence. As an example, one could easily note a brilliant spot on a reflective surface such as a road sign for distances of one mile or more. The unit described here is built in a rifle/pistol configuration. It contains its own rechargeable batteries and power supply and weighs about one pound. When used without the collimator, the unit is only about 15" long, hence a pistol configuration, when used with the long range collimator it is about 19" long, hence a rifle configuration.

The unit is shown constructed around an easily available helium-neon laser tube whose output is 1 mW at 6328 Å with a beam divergence of 1.3 milliradians. The beam width is about 0.65 mm. When collimated the divergence decreases and the beam width increases by the magnification factor of the optical system, used in these plans about 10 times. A shorter version of this unit can be built housing the batteries in an enclosure and attaching them to your belt, etc. The unit also works well on any vehicle 12 volt system.

THEORY OF OPERATION

This device, in principle, is not complicated in either theory or operation. (It would be very tough and critical in adjustment, construction and operation if the tube was to be made from scratch.) The basic device is very simple, that of a gas discharge tube that is highly evacuated and then filled with gas, placed between two mirrors forming a resonant optical cavity. When the gas is excited via an external energy source such as a current discharge, photons are produced and due to the amplifying action of the cavity and mirrors, laser radiation is produced. This sounds simple, however, certain limiting parameters and obstacles are present and must be reduced or eliminated to obtain any decent performance.

A helium-neon laser is a poor amplifier as lasers go, consequently all the following efforts must be made if lasing is to take place. The gas discharge tube must be properly made and sealed for a high vacuum, impurities cannot exist, gas mixture must be pure and applied at the correct pressure. Mirrors must be of the dielectric type for maximum optical efficiency (approaching 100%). Optical alignment must be nearly perfect. Once these demands are fulfilled the following takes place; atoms of gas are excited by an electric discharge to an energy state where photons of coherent light are produced upon these atoms returning to a stabler state. Much of this external energy is wasted in exciting atoms whose photons do not contribute to the laser beam. Some of these photons, however, manage to be reflected back and forth by the mirror and consequently stimulates other excited atoms to do the same thus producing the necessary synchronized wavelets increasing the beams intensity several percent on each pass between the mirrors. Placement of these mirrors must be in the same plane for near perfect transmissions. They must be of the dielectric type which are formed from multiple layers of a nonconducting film. This film is formed from transparent material such as the chlorides and fluorides. The thickness of these layers determines attenuation or enhancement of certain frequencies by creating destructive or constructive reinforcing waves. It is this effect that when repeated in many layers develops nearly 100% reflection at the desired frequency.

Note: For those who may consider making their own tubes, care must be taken to mention that regular silver mirrors seldom exceed 95% and tarnish quickly in air. Aluminum coated mirrors are only around 99% reflective. These losses are sufficient to prevent the device from performing.

CIRCUIT THEORY

The circuitry (Fig. 2-1 and Table 2-1) consists of a high-voltage transistor inverter. Q1 and Q2 switch the primary windings of transformer (T1) via a square wave at a frequency determined by its magnetic properties. Diodes (D1 and D2) provide base return paths for the feedback current of Q1 and Q2. Resistor (R2) limits this base current while R1 provides the necessary electrical imbalance to commence oscillation. The output winding of T1 is connected to a multiple section voltage multiplier. The multiplier consists of capacitors (C1-C5) and diodes (D3-6). This circuit quadruples the 400 volt ac output of T1 to approximately 1600 volts dc across C6-8. Resistors (R3 and R4) divide the 800 volts taken off at the junction of C3, and C5 for charging the dump capacitor C9 in the ignitor circuit. This voltage should be approximately 400 volts. The ignitor, consisting of the (T2) pulse transformer and capacitor discharge circuitry, provides the high-voltage dc pulse to ignite the laser tube LT1. This high-voltage pulse is the result of the SCR1 dumping the energy of capacitor C9 into the primary of T2. The high-voltage pulses in the secondary are now integrated onto high-voltage capacitor C11 through rectifier diode D9.

The SCR switching rate is determined by unijunction transistor (Q3) connected as a relaxation oscillator whose frequency is determined by the values of R8 and C10. When C11 is charged to a value sufficient to break down the gas in laser tube LT1, ignition takes place and a current now flows that is sufficient to sustain itself at the lower voltage output of the voltage multiplier section (usually about 1100 volts). The path for this sustaining current is through the secondary of T2 and ballast resistors (R11 and R12). These two resistors limit the current through LT1 and may be adjusted for a current flow of up to 5.5 mA to allow maximum laser output. The ignitor circuit is now deactivated by the clamping of Q3 emitter via Q4 being turned on by the voltage drop occurring across R10. This voltage drop will only occur when the laser tube is ignited and causes the SCR1 to cease firing as the ignitor circuit would continue to operate unnecessarily drawing on the limited power available.

The unit can be powered either by internal or external batteries or any source of 12 to 14 volts at 1 amp. Batteries must be of the ni-cad type if the small "AA" cells are used as these provide the high current necessary. Larger sized batteries will inherently have the current capacity necessary. A small battery is shown and is seriously suggested if

Table 2-1. Visible Red Continuous Laser Gun Parts List (LGU3).

R1	(1)	2.2 k ¼ watt resistor
R2	(1)	220 ohm 1 watt resistor
R3,4	(2)	1 meg ¼ watt resistor
R5,7	(2)	100 ohm ¼ watt resistor
R8	(1)	100 k ¼ watt resistor
R9	(1)	1 k ¼ watt resistor
R10	(1)	220 ohm ¼ watt resistor
R11,12	(2)	47 k 1 watt resistor
R13	(1)	470 Ω ¼ watt resistor
C1	(1)	10 µfd 25 electrolytic capacitor
C2,3,4,5 6,7,8	(7)	.01 @ 1.6 kV disc capacitor
C9	(1)	.1 µF/400 volt paper capacitor
C10	(1)	1 µF/50 volt electrolytic capacitor
C11	(1)	.001 µF-10 kV ceramic capacitor
D1,2	(2)	1N4001 50 volt 1 amp diode
D3,4,5,6 7,8	(6)	1N4007 1 kV 1 amp diode
D9	(1)	10 kV diode
D10	(1)	Red LED 20 mA for indicator
Q1,2	(2)	D40D5 silicon npn power tab transistor
Q3	(1)	2N2646 UJT transistor
Q4	(1)	PN2222 npn silicon transistor
SCR1	(1)	2N4443 SCR transistor
T1	(1)	Type I, 400 volt inverter transformer Figs. 2-2 and 2-3
T2	(1)	Reworked trigger transformer Fig. 2-4
FER2	(1)	Ferrite core 1¼" × .250 dia.
PR5	(12")	Primary winding 15 turns #24 wire
S1	(1)	Push button switch
CL1,2	(2)	10" × 20 ga battery clips
BH1	(1)	4 "AA" holder
BH2	(1)	8 "AA" holder
PB1	(1)	4½" × 1½" perfboard
EN1	(1)	15" × 1.9" PVC tube
TU1	(1)	4" × 1.5" × .035 al tube
TU2	(1)	4" × 1.625" × .058 al tube
HA1	(1)	6" × 1.5" al tube
BK1	(1)	7" × ½" × 24 ga al
SM1	(1)	PVC corona shield .875 × .6 tubing
CA1,2	(2)	1⅞ plastic caps (used for system covers) Fig. 2-8
CA3	(1)	1⅝ plastic caps (used for handle cover) Fig. 2-8
CA4	(1)	Reworked 1½ plastic cap used for retaining LE1 Fig. 2-9.
CA5	(1)	1-⅝" metal bottle cap reworked (Fig. 2-9)
CA6,8	(2)	1-½ plastic cap reworked by removing end leaving a ⅜" ring for shimming to correct diameter Fig. 2-7
CA7,9	(2)	Reworked plastic cap #Q3 Fig. 3-7
BU1,2	(2)	Reworked 1" × ½" × ⅛" neoprene bushing Fig. 2-7
WR1	(18")	#18 wire
WR3	(6")	#24 wire
LE1	(1)	38 × 100 mm convex lens
LE2	(1)	+14 × −9 mm concave lens
SW8	(2)	#6 × ¼ sheet metal screws
LAB1	(1)	Label Class II Laser warning
LAB2	(1)	Aperture label

Note on Batteries

B1-12 (12) AA ni-cad batteries
DUM1 Dummy battery fabbed from equivalent length of aluminum rod, etc. May not be required (see text). Suggested batteries are GE #GC1 nickel cadium along with our optional charger. #BCM1. Fully charged batteries should give at least 30 minutes operation. For short period operation (2) GE #GC9 7.2 volt ni-cad batteries may be used.

The Visible Red Laser Gun model LGU3BK is available as a complete kit through Information Unlimited, P.O. Box 716, Amherst, N.H. 03031. Write or call 603-673-4730 for price and delivery.

Note on Laser Tube

The laser tube LT1 shown in these plans is not to exceed one milliwatt in order to comply with the class II compliance as per LAB1. More powerful tubes will work but places the unit into a class III B compliance. Laser tubes are also available INFORMATION UNLIMITED.

A special detector suitable for use as a targe is available through Information Unlimited, Inc. This device generates an audible tone whenever the target is illuminated by the laser.

LDT1 Plans	$ 5.00
LDT1K Kit	$34.50
LDT10 Assembled Electronics	$49.50

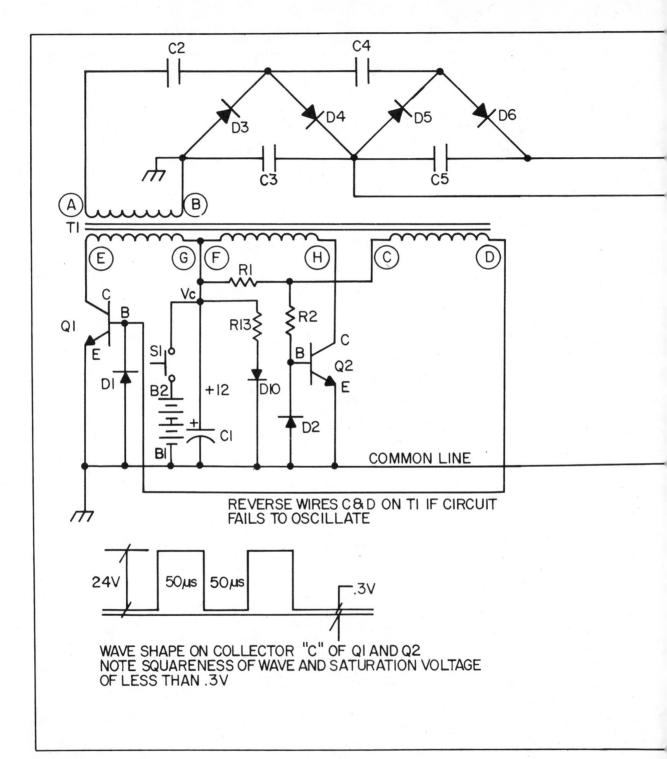

Fig. 2-1. Circuit schematic.

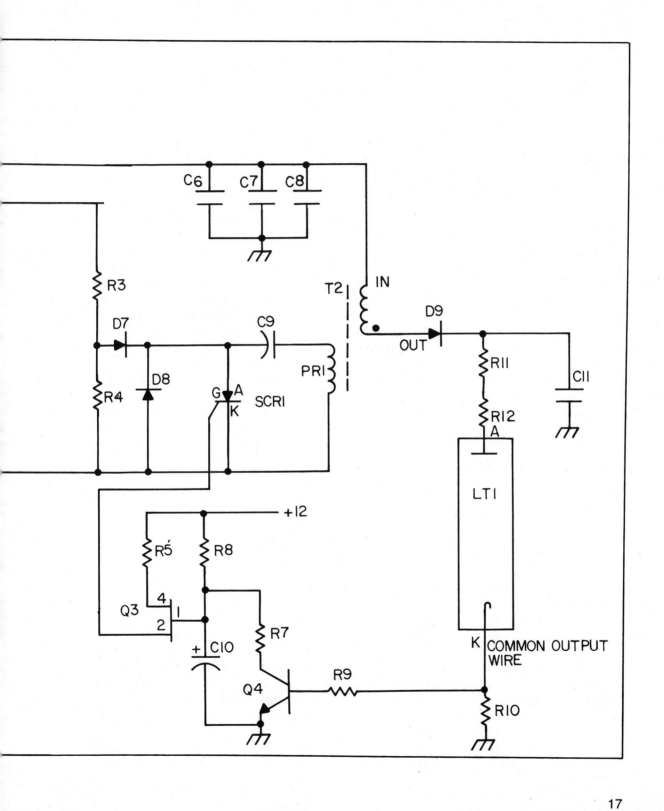

the small ni-cads are to be used as they can all be charged simultaneously.

CONSTRUCTION STEPS

1. Construct T1 transformer as shown in Figs. 2-2 and 2-3.

2. Construct T2 trigger transformer as shown in Fig. 2-4.

3. Layout the assembly board as shown in Figs. 2-5A and 2-5B. Use Figs. 2-6 and 2-7 as aids. Observe the correct polarity of all diodes and capacitors C1 and C10. Bend over dissipation tabs of transistors Q1 and Q2 as shown. Observe chamfered edge indicating collector pin. Wire in T1 and secure to assembly board using silicon RTV. Tabs of Q1 and Q2 are also secured to T1 via RTV as these help dissipate any collector heating. Pay particular attention to the location of T2, D9, C11, and R11 and R12. These points are at a potential of 10 kV during triggering of the laser tube and must be positioned so as not to allow high voltage breakdown. This section is a critical area and must be laid out as shown. A layer of paraffin or several coats of varnish on this section is strongly advisable. RTV silicon rubber properly applied will also suffice.

4. Wire in switch S1, common output lead and battery clips CL1, 2 for battery packs. Do not connect the laser tube at this time.

5. Verify accuracy and quality of wiring. Check for shorts, dangerous proximity points, and miswiring.

PRELIMINARY TEST OF ASSEMBLY BOARD

1. Position "common" output wire and lead of R12 so there is about 1/16″ to 1/8″ spacing. These are the leads that would normally be connected to the laser tube.

2. Connect 10 volts dc to $+V_c$ and common line. Note current draw of about 250 mA and a spark jumping between output approximately one to two times a second. Gap may sustain a dc arc once the spark jumps causing breakdown. This is normal and should cause an input current draw of about 750 mA. The arc may burn the leads once initiated since it is drawing the normal tube operating current. Do not continue to operate in this state but remove power

and spread output leads further apart in order to eliminate the sustaining of the dc arc. Spark should still jump up to ¼″ or more. *Do not exceed ¼″ gap or breakdown of ignition components T2, D9, and C22 may occur.*

3. Connect a sensitive voltmeter to the output wires and measure approximately 2500 volts. Note that the meter load will extinguish ignition spark and will only indicate a light jitter. The above tests should only be done to verify measurements. Do not maintain operation in this state.

4. Once these are verified you can connect the laser tube into the circuit. Use the actual lead of R12 to the anode pin of the tube and temporarily connect a current meter in series with the "common" output lead and the cathode pin of the laser tube. The anode lead must be as short as possible.

5. Apply 10 V power and note the laser tube flashing as it attempts to operate. Increase output voltage to a point where the tube remains ignited. Note series tube current should be around 5 mA (optimum value) with input voltage at 12 volts and current around 750 mA. You will note that the laser tube operates between 4.5 and 5.5 mA for best performance. Varying the input voltage *will greatly change this current value.*

Battery life and circuit performance are obtained when the input battery voltage is selected to produce reliable sustaining of the laser plasma at the tube current of about 5 mA. Further touching up of this current can be obtained by trimming resistors R11 and R12.

The plans show using 12 AA ni-cads which produce a voltage of 14.4 volts as ni-cads are only 1.2 volts each and are necessary to supply the current necessary. The recommended voltage in our prototype was 13.2 and it was necessary to use a dummy battery in one of the holders. You also can use forward biased diodes such as 1N4001 rated at 1 amp, these provide approximately 0.6 volts drop per diode and can reduce the voltage when connected in series with the battery pack. Use as many to obtain proper operating tube current as per the above step 5.

Ni-cad AA cells will usually provide about ½ to ¾ hours total operation. Longer battery life re-

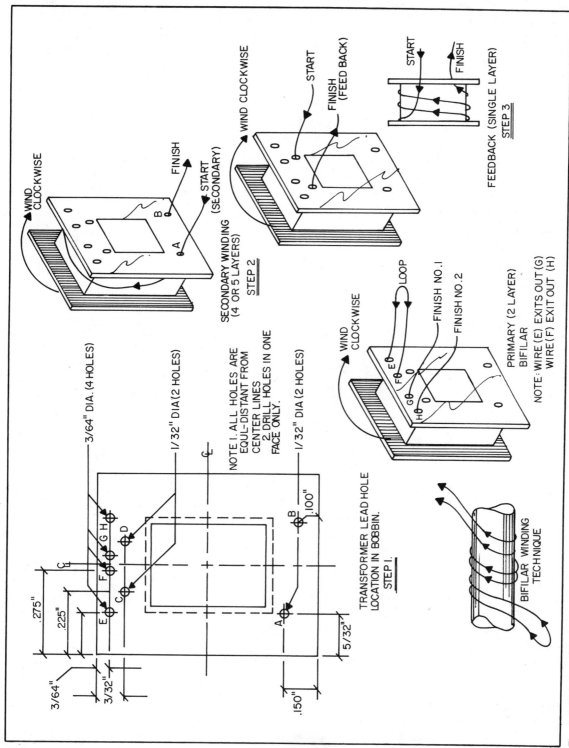

Fig. 2-2. TI inverter transformer type I.

19

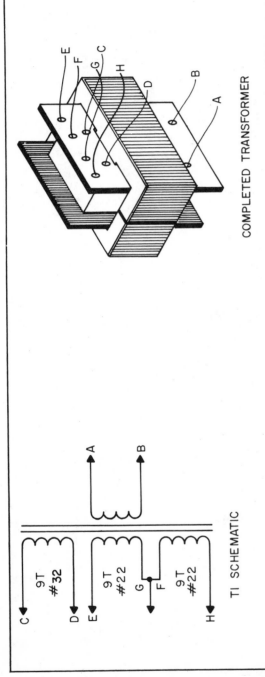

COMPLETED TRANSFORMER

PLEASE NOTE THAT TRANSFORMER IS SYMMETRICAL AND THAT LEADS MAY BE REVERSE IMAGED.

TI SCHEMATIC

PARTS LIST

(2) "E" CORES STACKPOLE #57-0048
(1) BOBBIN FERROX CUBE 990-012-01
6' #22 ENAMELED WIRE
100' #32 ENAMELED WIRE
MYLAR TAPE .001 THICK

CONSTRUCTION STEPS

1 DRILL BOBBIN AS SHOWN
2 SECONDARY 340 VOLTS—250 TURNS OF #32 WIRE A & B WIND SECONDARY IN TIGHT, EVEN LAYERS. INSULATE EVERY TWO LAYERS WITH A LAYER OF 1 MIL MYLAR TAPE TO PREVENT BREAKDOWN. NOTE SKETCH 2. TAPE COMPLETED SECONDARY
3 WIND 9 TURNS OF #32 WIRE C & D
4 TAKE A LENGTH OF #22 WIRE ABOUT SIX FEET LONG AND DOUBLE IT. INSERT THE TWO CUT ENDS THROUGH HOLES E & F LEAVING ABOUT 6" FOR LEADS. WIND 9 TURNS OF THESE PARALLEL WIRES NEATLY AND TIGHTLY AND EXIT AT HOLES G & H AS SHOWN.
5 CUT WIRE LOOP AND JOIN WIRES FROM HOLES F & G. MEASURE CONTINUITY BETWEEN WIRES FROM HOLES E & H.
6 TAPE FINISHED WINDINGS AND PLACE ON E CORES. TAPE THESE CORES TIGHTLY TOGETHER.

9T #32

9T #22

9T #22

Fig. 2-3. TI inverter transformer type I.

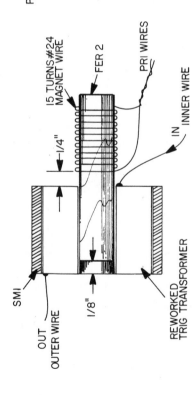

PARTS LIST:
1-SMI PVC CORONA SHIELD .875 X .6 TUBING
1-REWORKED TRIG TRANSFORMER
1-FER 2 FERRITE CORE 1-1/4" X .250 DIA.
15 TURNS #24 ENAMELED MAGNET WIRE

15 TURNS #24
MAGNET WIRE

FER 2

-1/4"-

PRI WIRES

IN
INNER WIRE

SMI

OUT
OUTER WIRE

1/8"

REWORKED
TRIG TRANSFORMER

T2 ASSEMBLY INSTRUCTIONS

1. FABRICATE SMI CORONA SHIELD AS SHOWN.

2. VERY CAREFULLY REMOVE THE SMALL FERRITE CORE AND PRIMARY WIRES OF THE TRIGGER TRANSFORMER. BE
   EXTRA CAREFUL NOT TO BREAK ANY OF THE INNER SECONDARY WIRES AS THESE MUST BE INTACT FOR PROPER
   OPERATION. THIS IS BEST ACCOMPLISHED BY CLIPPING THE WIRES TO THE BASE SECTION AND APPLYING PRESSURE
   TO THE CENTER OF THE FERRITE CORE AND PUSHING OUT THIS SECTION ALONG WITH THE PRIMARY WIRES ALSO.
   LOCATE AND CAREFULLY BRING OUT INNER LEAD OF COIL TO A USABLE LENGTH. REMOVE ANY INNER TURNS NECESSARY
   TO MAKE SURE THIS LEAD IS NOT BROKEN . YOU MAY REMOVE UP TO ONE INNER LAYER WITHOUT IMPAIRING
   PERFORMANCE.

3. LOCATE AND BRING OUT OUTER SECONDARY WIRE DIRECTLY OPPOSITE OF LOCATION OF INNER WIRE. THIS HELPS
   PREVENT ANY BREAKDOWN FROM OCCURING. PLACE A SMALL DAB OF ADHESIVE OR WAX TO STRAIN RELIEVE
   AND HOLD IN PLACE.

4. WIND 15 TURNS OF # 24 ENAMELED WIRE ON FERRITE CORE FER2 AS SHOWN. THIS IS PRIMARY. WRAP ENTIRE
   CORE WITH A LAYER OF ELECTRICAL TAPE. THIS HELPS INSULATE PRIMARY FROM SECONDARY WIRES.

5. PLACE REWORKED TRIGGER TRANSFORMER INTO SMI CORONA SHIELD AND INSERT TAPED PRIMARY SECTION.
   APPLY HOT PARAFFIN WAX AND POSITION PRIMARY CORE 1/8" FROM END AS SHOWN BEFORE WAX SETS. NOTE
   THAT OUTER WIRE IS AT 10KV WHEN TRIGGERING AND SHOULD BREAKOUT OPPOSITE OF INNER WIRE TO PROVIDE
   THE LARGEST BREAKDOWN PATH POSSIBLE.

Fig. 2-4. T2 reworked trigger transformer.

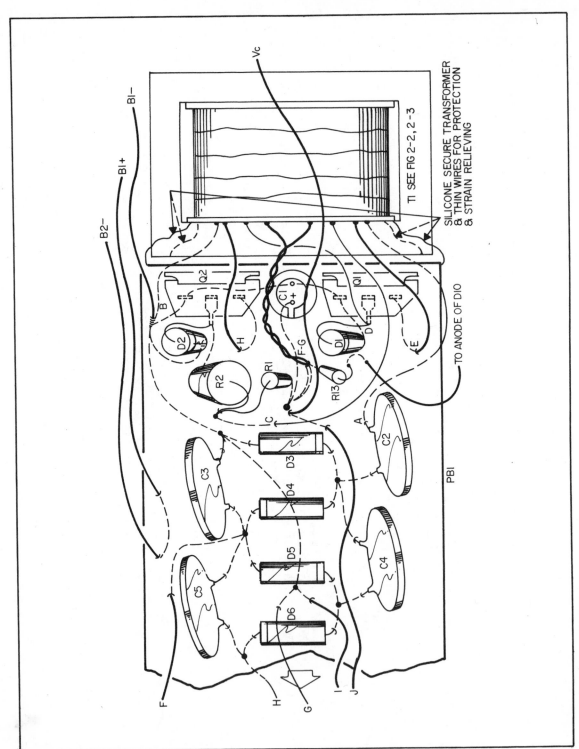

Fig. 2-5A. Board assembly (½).

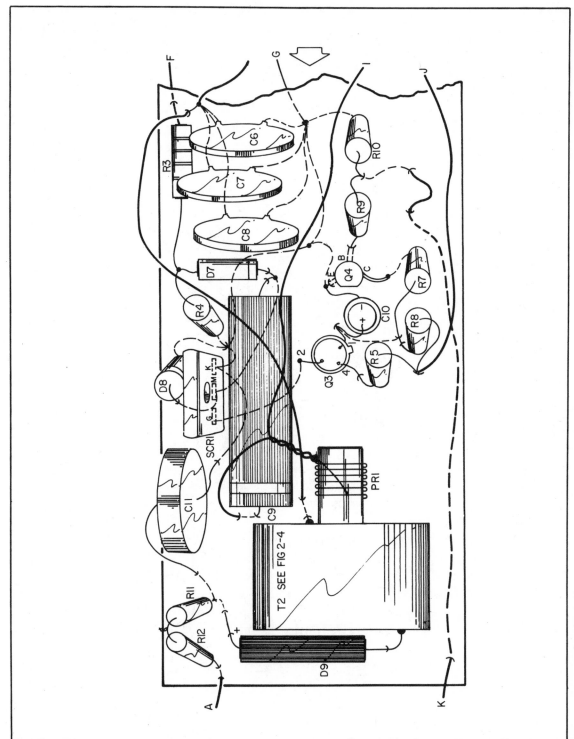

Fig. 2-5B. Board assembly (½).

23

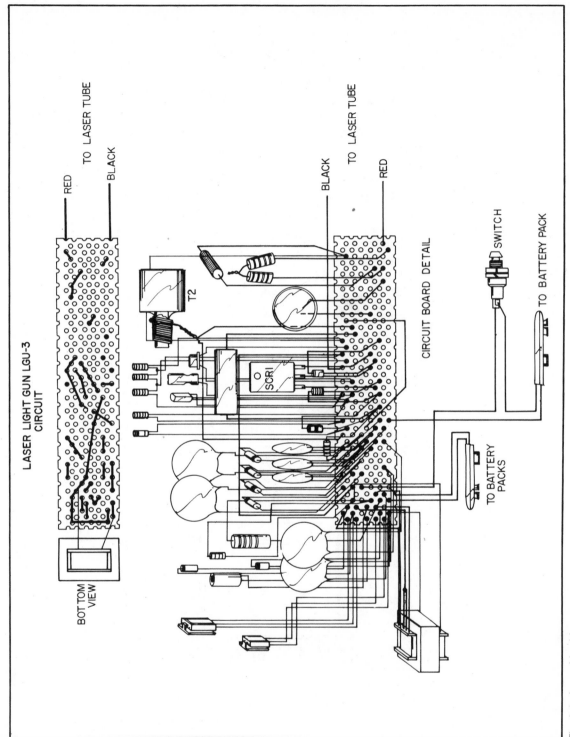

Fig. 2-6. Assembly aid.

24

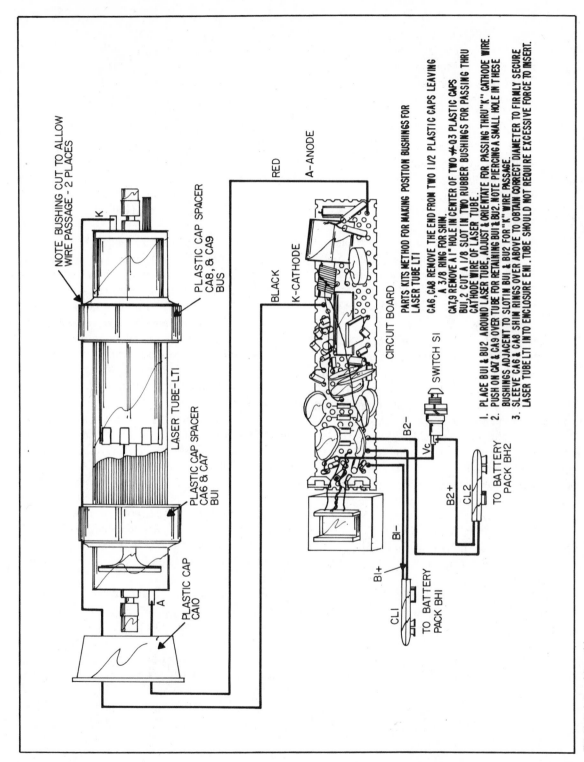

NOTE BUSHING CUT TO ALLOW WIRE PASSAGE - 2 PLACES

PLASTIC CAP SPACER CA8, & CA9 BU5

LASER TUBE-LT1

PLASTIC CAP SPACER CA6 & CA7 BU1

PLASTIC CAP CA10

RED

A-ANODE

BLACK

K-CATHODE

CIRCUIT BOARD

SWITCH S1

Vc

B2−

B2+

CL2

TO BATTERY PACK BH2

B1−

B1+

CL1

TO BATTERY PACK BH1

PARTS KITS METHOD FOR MAKING POSITION BUSHINGS FOR LASER TUBE LT1

CA6, CA8 REMOVE THE END FROM TWO 1 1/2 PLASTIC CAPS LEAVING A 3/8 RING FOR SHIM.
CA7,9 REMOVE A 1" HOLE IN CENTER OF TWO #Q3 PLASTIC CAPS
BU1,2 CUT A 1/8 SLOT IN TWO RUBBER BUSHINGS FOR PASSING THRU CATHODE WIRE OF LASER TUBE.

1. PLACE BU1 & BU2 AROUND LASER TUBE, ADJUST & ORIENTATE FOR PASSING THRU "K" CATHODE WIRE.
2. PUSH ON CA7 & CA9 OVER TUBE FOR RETAINING BU1 & BU2. NOTE PIERCING A SMALL HOLE IN THESE BUSHINGS ADJACENT TO SLOT IN BU1 & BU2 FOR "K" WIRE PASSAGE.
3. SLEEVE CA6 & CA8 SHIM RINGS OVER ABOVE TO OBTAIN CORRECT DIAMETER TO FIRMLY SECURE LASER TUBE LT1 INTO ENCLOSURE EMI. TUBE SHOULD NOT REQUIRE EXCESSIVE FORCE TO INSERT.

Fig. 2-7. Interwiring aid.

25

quires an external battery of a physically larger size or a rechargeable gel cell or other appropriate energy pack. Short time operation for gun sighting and pointing can be obtained by using two series connected 7.2 V #GC9 ni-cads. This approach allows the electronics to be placed in the handle and the batteries in the barrel thus reducing the overall length considerably. However, continuous operation will quickly discharge these batteries allowing only 10-15 minutes of operating time. For troubleshooting refer to the section on circuit theory and waveshapes shown in Fig. 2-1.

FINAL ASSEMBLY

It is assumed that the electronic section and laser tube are all functioning properly and are ready to be placed into the cylindrical enclosure. Note that if the two small 7.2 volt ni-cads are used that the electronics may also be assembled in the main housing (see Figs. 2-8 and 2-9).

1. EN1—Cut a 16.5" length of 1.9" OD Schedule 40 PVC tube. Polish with steel wool for clean finish. Drill ⅝" hole for wires to S1 and BH1 as shown. Hone out end of EN1, 1⅛" deep to 1⅝" diameter to accept TU2. Use a hole cutting saw and drill press for this step to firmly secure EN1.

2. TU1—Cut a 4" length of 1.5" OD × 0.035" wall aluminum tubing. Polish and remove all burrs.

3. TU2—Cut a 4" length from 1.625" OD × 0.058" wall aluminum tubing. Polish and remove all burrs. Check for a smooth sliding fit with TU1.

4. HA1—Cut a 6" length from 1.5" OD × 0.035" aluminum tubing. File groove to fit contour of EN1 at desired angle for pistol grip configuration. Drill holes for S1 and mounting bracket BK1. Polish and remove burrs.

5. BK1—Fabricate bracket from a 7" × ½" × 24 gauge aluminum as shown for attaching handle HA1 to enclosure EN1.

6. CA5—Punch or drill a ½" hole in the center of a 1⅝" metal bottle cap. This must be done accurately and carefully for alignment reasons.

7. CA4—Remove the center of a 1½" plastic cap leaving approximately ⅛" lip to retain lens LE1. Use exacto knife or equivalent. Use wall of TU1 as a gauge when making cut.

8. Center and glue lens LE2 to center of CA5 as near the center as possible. Insert this lens and cap assembly into TU2 and glue in place.

9. Place lens LE1 over TU1 and retain via reworked plastic cap CA4.

10. Insert TU1 lens assembly into TU2 and adjust using like a telescope. Note approximate ×10 magnification when viewing objects. The image will only be clear in the center of the lens. This is due to the chromatic and spherical aberrations using the simple uncorrected lens. The laser beam is monochromatic and only uses the center so these effects are not a detriment when used as a collimator.

11. Shim laser tube LTO5R for a firm sliding fit into enclosure EN1. This can be done in several ways. The simplest is to wrap around successive layers of tape building up the diameter of the tube to the proper dimension. You will note that the cathode common wire must be snaked back to the circuit board and this may be done by carefully cutting slot-like passages in the shims with the final layer of tape being unslotted retaining the lead.

12. Insert the assembly into the enclosure from the front. Start with battery clips, S1 end first, and laser tube last. Gently position the laser tube so that the output mirror is flush with the honed out section of the enclosure. The laser tube should be firmly positioned yet not requiring excessive force. Fish out battery clips and S1 leads through ⅝" hole and attach to HA1 handle as shown. Note battery holder BH2 in EN1 enclosure section. Note eight-cell battery pack may be placed in handle HA1 with four cells in enclosure. This will reduce the length requirement of EN1. The opposite is shown in the drawings.

13. Attach handle. Insert batteries and retainer caps CA1 and CA3. Apply proper "aperture" and "laser class" decals for safety and identification reasons.

14. Point laser and press S1 noting a bright red dot. Beam is clearly visible when looking up or down optical axis in heavy air or fog.

15. Insert collimator section firmly seating into honed out section of closure EN1. *Make sure laser tube is recessed so that lens mount CA5 of*

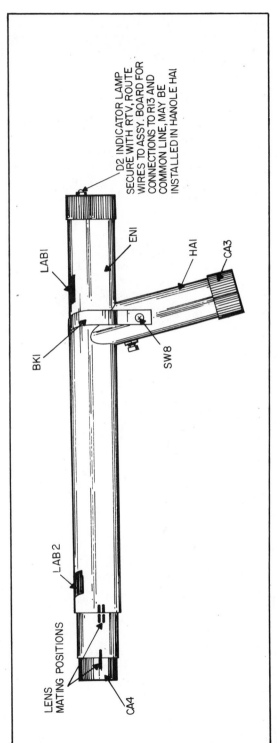

LENS
MATING POSITIONS

LAB 2

CA4

BK1

LAB1

ENI

SW8

HA1

CA3

D2 INDICATOR LAMP
SECURE WITH RTV, ROUTE
WIRES TO ASSY. BOARD FOR
CONNECTIONS TO RI3 AND
COMMON LINE, MAY BE
INSTALLED IN HANOLE HAI

Fig. 2-8. Finished assembly.

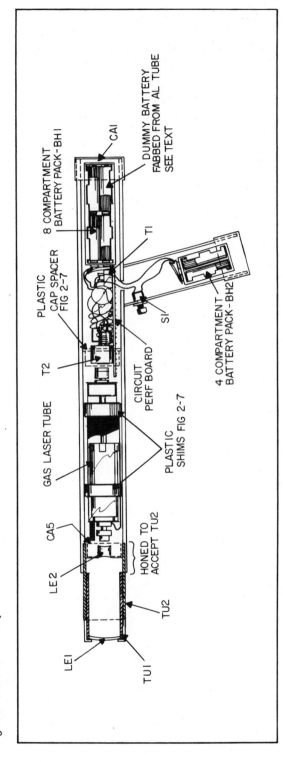

8 COMPARTMENT
BATTERY PACK-BH1

CAI

DUMMY BATTERY
FABBED FROM AL TUBE
SEE TEXT

TI

PLASTIC
CAP SPACER
FIG 2-7

SI

4 COMPARTMENT
BATTERY PACK-BH2

T2

GAS LASER TUBE

CIRCUIT
PERF BOARD

PLASTIC
SHIMS FIG 2-7

CA5

LE2

HONED TO
ACCEPT TU2

TU2

LE1

TU1

Fig. 2-9. X-ray view.

27

collimator properly clears its output mirror.

16. Point laser at a distant object over 100 meters and carefully adjust telescoping collimator for the sharpest and smallest spot size. Check at close range to make sure that the optics are properly adjusted. This is indicated by the main part of the beam centered within the fringe regions of the light that is reflected from the inner surfaces of the tube. You may have to readjust or realign the lenses to obtain this effect. The collimator increases the beam width by approximately ×10 and decreases the divergence by the same factors. Secure position with tape, etc. The lenses and output mirror of the system may be cleaned using a Q-tip and glass cleaner such as Windex, etc. *Be gentle with the laser mirror.*

APPLICATIONS

Applications of this device can be numerous such as *special effects, spotting, pointing, leveling, pipe laying, construction, gun sight*—wherever a straight line and transit may be required. The device can make a great steadiness indicator for those who shoot handguns. Try to hold the beam spot on a target.

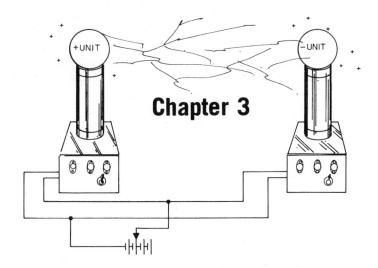

Chapter 3

Invisible Infrared
Pulsed Laser Rifle (LRG3)

Caution—Use of controls or adjustments or performance of procedures other than those specified herein may result in hazardous radiation exposure.

This project shows how to construct a low-powered portable, solid-state gallium arsenide laser operating from batteries, being completely self-contained, hand-held and built into a rifle configuration. The device is intended as a source of adjustable frequency pulses of infrared energy at 9000 A. Peak power can be from 7 to 35 watts depending on the diode used and the pulse rate is adjustable from 50 to 2000 pps.

The electronics are shown built on a perfboard assembly that fits into a tubular enclosure fitted with a lens for preliminary collimating. This tubular enclosure is mounted onto a circular housing via screws and nuts. Housing serves as the enclosure for the batteries and contains the control panel at its rear. It is fitted with a handle that houses the trigger switches. A conventional sighting system is easily adapted to the device. It is intended as a rifle-type weapon whose range can be several thousand yards.

It can easily be used with a detection type device such as the *Laser Light Detection* project as described.

The device generates an adjustable frequency of low to medium powered IR pulses of invisible energy and must be treated with care. At no time should it be pointed at anyone or anything that could reflect these pulses. Never look into the unit when the power is on. It is intended to be used for ranging, simulated weapons practice, intrusion detection, communications and signaling and a variety of related scientific, optical experiments and uses.

THEORY OF OPERATION

A laser diode is nothing more than a three layer device consisting of a pn junction of n-type silicon, and p-type of gallium arsenide and a third p layer of doped *gallium arsenide* with aluminum. The n-type material contains electrons that readily migrate across the pn junction and fill the holes of the p-type material, conversely holes in the p-type migrate to the n-type and join with electrons. This migration causes a potential hill or barrier consisting of negative charges in the p-type material and positive charges in the n-type material that eventually ceases growing when a charge equilibrium exists. In order for current to flow in this device, it must be supplied at a voltage to overcome this potential barrier. This is the forward voltage drop across a common diode. If this voltage polarity is reversed, the potential barrier is simply increased assuring no current flow. This is the reversed bias condition of a common diode.

A diode with no external voltage applied to it contains electrons that move and wander through the lattice structure at a low, lazy average velocity as a function of temperature. When an external current at a voltage exceeding barrier potential is applied, these lazy electrons now increase their velocity to where some by colliding acquire a discrete amount of energy and become unstable eventually emitting this acquired energy in the form of a photon upon returning to a lower energy state. These photons of energy are random both in time and direction, hence any radiation produced is incoherent such as that of a light-emitting diode.

The requirements for coherent radiation are that these discrete packets of radiation be in the form of lockstep phase and in a definite direction. The above demands two essential requirements: (A) Sufficient electrons at the necessary excited energy levels and (B) An optical resonant cavity capable of trapping these energized electrons for stimulating more and directing them. The number of energized electrons is determined by the forward diode current. A definite threshold condition exists where the device emits laser light rather than incoherent as a light-emitting diode. This is why the device must be pulsed with high current. The radiation from these energized electrons is reflected back and forth between the square cut edges of the crystal which form reflecting surfaces due to the index of refraction of the material and air. The electrons are initially energized in the region of the pn junction. When these energized electrons drift into the p-type transparent region, they spontaneously liberate other photons that travel back and forth in the optical cavity interacting with other electrons commencing laser action. A portion of the radiation traveling back and forth between the reflecting surfaces of these mirrors escapes and constitutes the output of the device.

CIRCUIT THEORY

The portable battery pack is stepped up to 200 to 300 volts by the inverter circuit consisting of Q1, Q2, and T1 (Fig. 3-1 and Table 3-1). Q1 conducts until saturated, at which time, the base no longer can sustain it in an "on" state and Q1 turns "off", causing the magnetic field in its collector winding to collapse thus producing a voltage or proper phase in the base drive winding that turns on Q2 until saturated, repeating the above sequence of events in an "on/off" action. The diodes connected at the bases provide a return path for the base drive current. The stepped up squarewave voltage on the secondary of T1 is rectified and integrated on C2.

TRIGGER CIRCUIT

This circuit determines the pulse repetition rate of the laser and uses a unijunction (Q3) whose pulse rate is determined by R6 and C5. You will note that the maximum permissible pulse rate is determined by the laser diode rating, the RC time constant of the charging circuit and the current capability of the power supply.

DISCHARGE CIRCUIT

The discharge circuit generates the current pulse in the laser and consequently is the most important section of the pulser. The basic configuration of the pulse power supply is shown in the system schematic. The current pulse is generated by the charging storage capacitor (C3) through SCR1 and laser diode (LA1). The rise time of the current pulse is usually determined by the SCR1 while the fall time is determined by the capacitor value and the total resistance in the discharge circuit. Figure 3-2A shows typical anode voltage and current waveforms of the SCR during the current pulse through the diode laser.

The peak current, pulse width and voltage of the capacitor discharge circuit are interrelated for various load and capacitance values. The peak laser current and charged capacitor voltage relationships are given in Fig. 3-2B for several different capacitor values and typical laser types. The voltage and current limits of the SCR are also shown. Short pulse widths provide less time for the SCR to turn on than longer pulse widths; therefore, the SCR impedence is higher and more voltage is required to generate the same current. Figure 3-2C shows the current pulse waveforms for the three different values of capacitance. The capacitor is charged to the same voltage in all three cases, i.e., 400 volts.

SCR SWITCH

In conventional operation of an SCR, the anode current, initiated by a gate pulse, rises to its maximum value in about 1 microsecond. During this time the anode-to-cathode impedance drops from open circuit to a fraction of an ohm. In injection laser pulsers, however, the duration of the anode-cathode pulse is much less than the time required for the SCR to turn on completely. Therefore, the anode-to-cathode impedance is at the level of 1 to 10 ohms throughout most of the conduction period. The major disadvantage of the high SCR impedance is that it causes low circuit efficiency. For example, at a current of 40 amps, maximum voltage would be across the SCR while only 9 volts would be across LA1. These values represent a very low circuit efficiency. The efficiency of a laser "array" is greater due to its circuit impedance being more significant.

Because the SCR is used unconventionally, many of the standard specifications such as peak current reverse voltage, on-state forward voltage, and turn-off time are not applicable. In fact, it is difficult to select an SCR for a pulsing circuit on the basis of normally specified characteristics. The specifications important to laser pulser applications are forward blocking voltage and current rise time. A "use test" is the best and many times, the only practical method of determining the suitability of a particular SCR.

STORAGE CAPACITOR (C3)

The voltage rating of the storage capacitor

Table 3-1. Invisible IR Pulsed Laser Rifle Parts List (LRG3)

R1	(1)	1 k ¼ watt resistor
R2	(1)	220 ohm 1 watt resistor
R3	(1)	33 k 1 watt resistor
R4	(1)	3.3 k 1 watt resistor
R5	(1)	1 ohm ½ watt resistor
R6/S1	(1)	100 k pot and switch
R7,9	(2)	10 k ¼ watt resistor
R10	(1)	100 k 1 watt resistor
R8,11	(2)	100 ohm ¼ watt resistor
C1	(1)	10 μF @ 25 volt electrolytic capacitor
C2 A & B	(2)	5 μF @ 150 volt (two connected in series)
C3	(1)	.03 μF @ 400 volt paper capacitor (or other for pulse discharge)
C4,5	(2)	.05μF @ 25 volt disc capacitor
S2,3	(2)	Push Button switch
S4	(1)	Spst toggle switch
Q1,2	(2)	npn power tab transistor D40D5
Q3	(1)	2N2646 UJT
Q4	(1)	2N3439 HV transistor
SCR1	(1)	2N4443 SCR 400 volt
D1,2	(2)	1N4001 50 volt 1 amp rectifier
D3,4,5, 6,9	(5)	1N4007 600 volt 1 amp rectifier
D7,8	(2)	1N914 signal diode
T1BOB	(1)	Small bobbin Ferroxcube #570048 (not needed if T1 is purchased assembled)
T1CORES	(2)	Small "E" core stackpole #990012
CA1,2	(2)	1.3" 33 mm bottle cap
CA3	(1)	Plastic cap Niagara #988
CA4	(1)	1⅝" ID plastic cap Alliance #A1-⅝
CA5	(1)	1.5" bottle cap
CA6	(1)	1⅞" plastic cap Alliance #A1-⅞
CA7,8	(2)	3½" plastic cap Alliance
BH1A,B	(2)	4 "C" cell battery holder
MP1	(1)	3¼" × 3¼" × 22 ga sheet metal fab
PB1	(1)	4" × .9" × .1 grid perfboard
PL1	(1)	¾" × 6½" × .035" copper
TU1	(1)	10" × 1.5" × 0.35 al tube
TU2	(1)	6" 1.625" × .058 al tube
TU3	(1)	8" × 3½" sked 40 PVC
HA2	(1)	6" × 2" sked 40 PVC
BU1,2	(1)	Plastic bushing ⅜" hole
KNI	(1)	Small knob ¼"
NU1	(1)	6 32 kep nut
SW3	(2)	6 32 × 1" screw
SW5	(1)	6 32 × ⅜ nylon screw
WR4	(10")	#24 black hook-up wire
WR5	(5')	#20 red hook-up wire
MIW1	(1)	Small mica insulating washer
LE1	(1)	29 × 43 mm convex lens
SW8	(1)	6 × ¼" sheet metal screws
PC1	(1)	Printed circuit board optional
T1	(1)	Assembled transformer Type II
LAB1	(1)	Class III B Laser warning label
LAB2	(1)	Aperture Label
LRG3B		Kit of all the above items
BAT	(8)	1.2 volt C sized ni-cads (not included in kit)

Use one of the following diodes:

LD740	(1)	7 watt laser diode for LA1
LD660	(1)	14 watt laser diode for LA1
LD1630	(1)	35 watt laser diode for LA1

Useful items with this project

LLD1	Laser Light Detector and Control unit as described.
IRP1	Infrared indicator—this useful device produces visible light from infrared and is a must for anyone experimenting with IR. It is truly a super handy tool to have and is virtually indestructible for it never wears out.
GPV1	General Purpose Viewer. Features high resolution and sensitivity of type 6032 image converter (500 to 1200 nanometer spectral response, 50 lines/mm resolution). Built in 142 mm, f3, 3, 10.5° field infrared telephoto lens and 7X eyepiece, gives 2X magnification over range from 3 feet to maximum range of invisible source. Approximately 150 feet. SIZE: 13" long × 3½" wide × 6½" high (less handle).

Above items and kits available through Information Unlimited, Inc. P.O. Box 716, Amherst, N.H. 03031. Write or call 603-673-4730 for price and delivery.

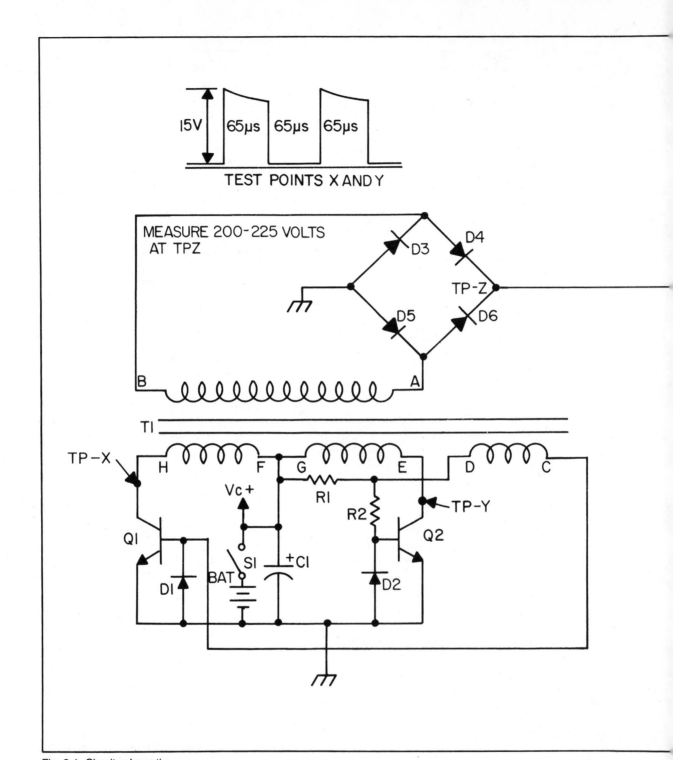

Fig. 3-1. Circuit schematic.

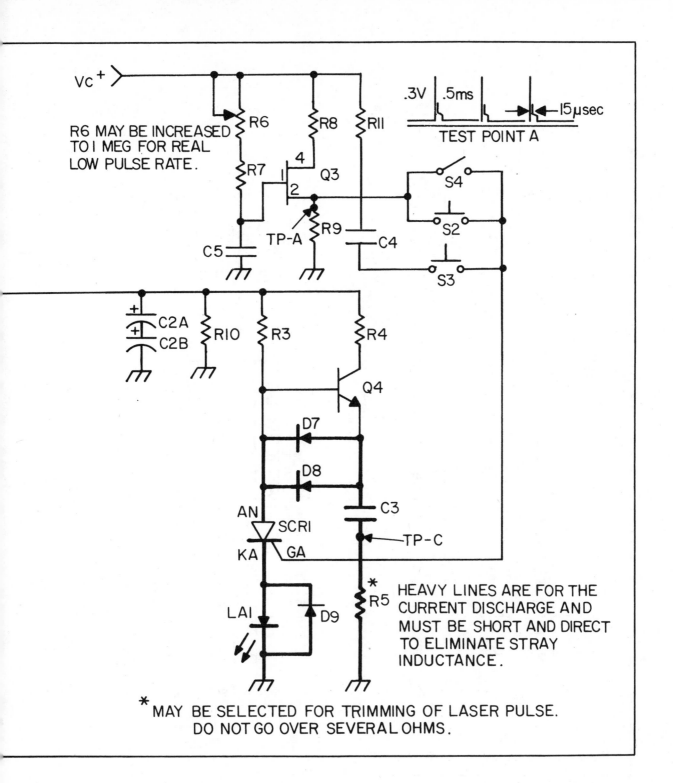

Vc +

R6 MAY BE INCREASED
TO 1 MEG FOR REAL
LOW PULSE RATE.

R6
R7
R8
Q3
R11
C5
TP-A
R9
C4

.3V   .5ms   15μsec
TEST POINT A

S4
S2
S3

C2A
C2B
R10
R3
R4
Q4
D7
D8
AN
SCR1
KA   GA
C3
TP-C
LA1
D9
R5 *

HEAVY LINES ARE FOR THE
CURRENT DISCHARGE AND
MUST BE SHORT AND DIRECT
TO ELIMINATE STRAY
INDUCTANCE.

*MAY BE SELECTED FOR TRIMMING OF LASER PULSE.
DO NOT GO OVER SEVERAL OHMS.

33

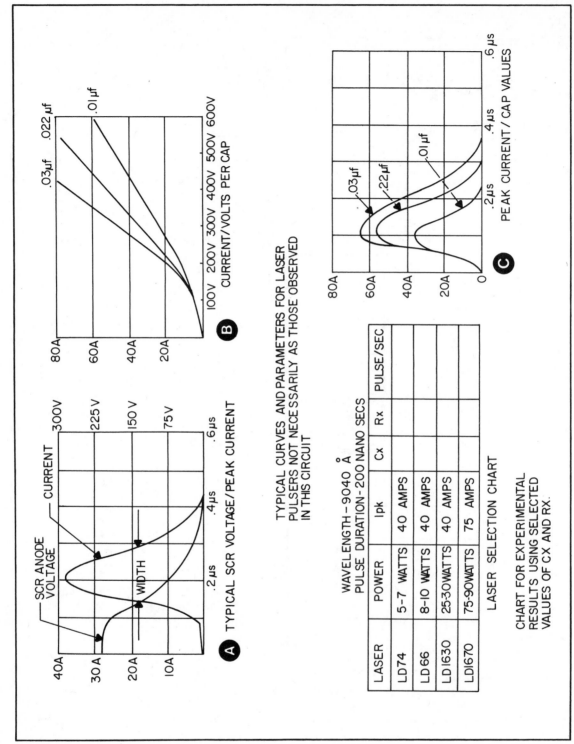

A TYPICAL SCR VOLTAGE/PEAK CURRENT

B CURRENT/VOLTS PER CAP

C PEAK CURRENT / CAP VALUES

TYPICAL CURVES AND PARAMETERS FOR LASER
PULSERS NOT NECESSARILY AS THOSE OBSERVED
IN THIS CIRCUIT

WAVELENGTH - 9040 Å
PULSE DURATION - 200 NANO SECS

LASER	POWER	Ipk	Cx	Rx	PULSE/SEC
LD74	5-7 WATTS	40 AMPS			
LD66	8-10 WATTS	40 AMPS			
LDI630	25-30 WATTS	40 AMPS			
LDI670	75-90 WATTS	75 AMPS			

LASER SELECTION CHART

CHART FOR EXPERIMENTAL
RESULTS USING SELECTED
VALUES OF CX AND RX.

Fig. 3-2. Circuit parameters.

34

must be at least as high as the supply voltage. With the exception of ceramic types, most capacitors (metallized paper, mica, etc.) will perform well in this circuit. Ceramic capacitors have noticeably greater series resistance, but are usable in slower speed pulsing circuits.

LAYOUT WIRING

Lead lengths and circuit layout are very important to the performance of the discharge circuit. Lead inductance affects the rise time and peak value of the current and can also produce ringing and an undershoot in the current waveform that can destroy the laser.

A well built discharge circuit might have a total lead length of only one inch and therefore an inductance of approximately 20 nanohenries. If the current rises to 75 amperes in 100 nanoseconds, the inductive voltage drop will be approximately 15 volts, or 1.6 times the voltage drop across a typical laser. If proper care is not taken in wiring the discharge circuits, high inductive voltage drops will result.

A two ohm resistor in the discharge circuit will greatly reduce the current undershoot in single diode lasers. Laser "arrays" usually have sufficient resistance to eliminate undershoot. The small resistance in the discharge circuit is also useful in monitoring the laser current, as described in the following section.

A clamping diode (D5) is added in parallel with the laser to reduce the current undershoot. Its polarity should be opposite to that of the laser. Although the clamping diode is operated above its usual maximum current rating, the current undershoot caused by ringing is very short and the operating life of the diodes is satisfactory.

CURRENT MONITOR

The current monitor in the discharge circuit provides a means of observing the laser current waveform with an oscilloscope. A resistive type monitor (R5), reduces circuit ringing and current undershoot, but the lead inductance of the resistor may cause a higher than actual current reading. A current transformer such as the Tektronix CT-2 can also be used to monitor the current and is not affected by lead inductance. Because the transformer does not respond to low-frequency signals, it should be used with fast time, short pulse width, fast-fall-time waveforms.

CHARGING CIRCUITS

The second major section of the pulser is the charging circuit. The circuit charges the capacitor to the supply voltage during the time interval between laser current pulses, and isolates the supply voltage from the discharge circuit during the laser current pulse, thereby allowing the SCR to recover to the blocking state. Because the response times of the charging circuit are relatively long, lead lengths are not important and the circuit can be remotely located from the discharge circuit.

The simplest charging circuit is a resistor/capacitor combination. The resistor must limit the current to a value less than the SCR holding current, but should be as low as practical because this resistance also determines the charging time of the capacitor C3. For example, a resistance of 40 kilohms limits the current to 10 milliamperes from a 400 volt supply. This current value is just at the holding level of an average 2N2443 SCR. A time of almost 3 milliseconds is required to charge a 0.022 microfarad capacitor to the supply voltage in three time constants through a 40 kilohm resistor. Therefore the pulse repetition rate (PRR) of the pulsing circuit is limited to about 375 Hz. If the PRR exceeds this value, the capacitor does not completely recharge between pulses and the peak laser current decreases with increasing PRR.

The peak current in the discharge circuit is controlled by varying the supply voltage, provided the PRR is low enough to allow the capacitor to recharge fully between current pulses. Both the supply voltage and the PRR determines the peak laser current. There is considerable risk in increasing the supply voltage to compensate for insufficient recharge time. However, modifying the number of secondary turns in T1 will change the voltage. If the PRR is decreased while the supply voltage is high, the capacitor again recharges completely and the laser pulse current increases to a value that may damage or destroy the laser diode.

The need for a variable voltage supply and the low PRR limit are the major disadvantages of the resistive type charging network. Therefore the limitations of the simple resistor drivers is that a resistor large enough to keep the current below the holding current of the SCR also limits the pulse repetition rate. The frequency capability of the pulse power supply can be improved with the charging circuits used in these plans. Capacitor C3 is charged to the supply voltage when the SCR is not conducting. Diodes D7 and D8 are also in the off state because zero current flows through them.

At the onset of a pulse trigger, the impedance of the SCR drops rapidly and capacitor C3 discharges toward ground potential. During the period when current is surging through diodes D7 and D8, transistor Q4 is reverse biased by the diode voltage drop. The SCR turns off when the current drops below its holding current, which determines the value of R3. The voltage across capacitor C3 then charges back to the supply voltage. During the capacitor charge cycle, diodes D7 and D8 pass no current and Q4 is forward biased into the saturation region. Obviously, the charging time of C3 is now shorter due to being charged through the small resistor R4.

CONSTRUCTION STEPS

1. *Read all indicated notes on plans before starting (especially those pertaining to short and direct leads).*

2. Assemble T1 as per the instructions given in Figs. 3-3A and 3-3B.

3. Drill extra holes for base "B" of Q1 and Q2, and holes for terminals of SCR1. Trial position the parts to guarantee correct fit.

4. Position and mark perforated board at reference corner as shown in Fig. 3-4.

5. Assemble chassis and end caps as shown in Fig. 3-5.

6. Assemble and solder as per the schematic (Fig. 3-1) and board layout (Fig. 3-4). Attach leads from T1 as shown and wires designated as shown in Fig. 3-6, to controls shown in Fig. 3-7. Assemble board to chassis and solder ground connections as

indicated by triangles shown in Fig. 3-4. Note polarity of semiconductors, etc. Please note—do not connect LA1 laser diode at this point. Verify accuracy of wiring and quality of solder joints.

TEST PROCEDURE FOR ELECTRONIC ASSEMBLY

This device using the small rectangular ni-cad battery and the enclosure shown is not intended for continuous use. Larger amp capacity batteries and ventilation holes will be necessary if prolonged use is anticipated.

1. Obtain an oscilloscope capable of measuring up to 35 MHz such as a Tektronix T932/935, a volt-ohm-mulliammeter such as a Simpson 260, a source of dc voltage preferably adjustable from 6-12 volts capable of producing up to 1 amp, an infrared indicating device, such as the IR-sensitive paper offered by Kodak Eastman, and finally a laser light beam detector such as that described, if needed.

2. Apply 6 volts to CL1, turn on S1 and measure 200-225 volts at point "Z". Ripple should be less than 0.2 volts. Observe waveforms at point "X" and "Y". See Fig. 3-1. Current input should be between 250-300 mA. Allow to run for several minutes and finger touch tabs of Q1 and Q2 (should only be warm). If above does not occur, try reversing feedback wires on T1, C and D. Note that the remaining circuit is not operating, due to S2 being normally "off".

3. Observe waveform at point "A" (3 volts/15 microsecs). See Fig. 3-1. Rotate R6 and note pulse changing from 200 Hz (5 ms) to 2500 Hz (0.4 ms).

4. Connect a dummy diode using short leads where LA1 laser diode is intended to go. Short out contacts of S2.

5. Adjust R6 for a minimum pulse rate of 200-300 for the remaining test. The reason we do this is the following: the lower pulse rate places less of a demand on the power supply and any noted measurements will be highest in this mode. It may be advantageous to use a variable-voltage power supply to compensate for this effect and determine the input necessary for the correct current pulse.

Damage to the laser diode could result, for

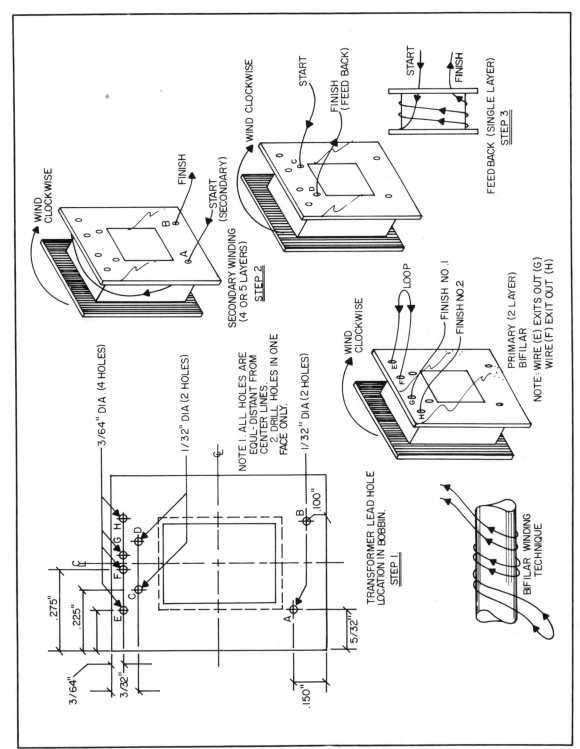

WIND CLOCKWISE

FINISH

START (SECONDARY)

B

A

SECONDARY WINDING (4 OR 5 LAYERS)
STEP 2

WIND CLOCKWISE

START

FINISH (FEED BACK)

C

D

START

FINISH

FEED BACK (SINGLE LAYER)
STEP 3

LOOP

FINISH NO.1

FINISH NO.2

WIND CLOCKWISE

E

F

G

H

PRIMARY (2 LAYER) BIFILAR

NOTE: WIRE (E) EXITS OUT (G)
WIRE (F) EXIT OUT (H)

3/64" DIA. (4 HOLES)

1/32" DIA (2 HOLES)

NOTE 1. ALL HOLES ARE EQUI-DISTANT FROM CENTER LINES.
2. DRILL HOLES IN ONE FACE ONLY.

1/32" DIA (2 HOLES)

.100"

.275"

.225"

3/64"

3/32"

E

F G H

C

D

B

A

5/32"

.150"

TRANSFORMER LEAD HOLE LOCATION IN BOBBIN.
STEP 1.

BIFILAR WINDING TECHNIQUE

Fig. 3-3A. T1 inverter transformer type II.

37

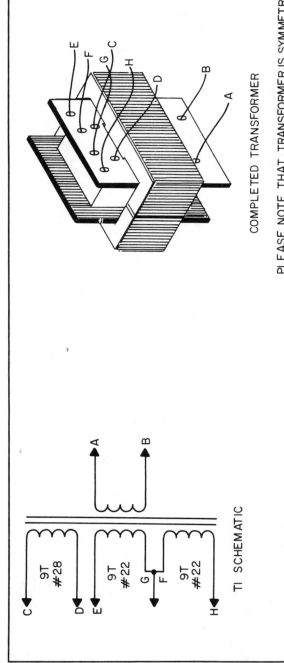

COMPLETED TRANSFORMER

PLEASE NOTE THAT TRANSFORMER IS SYMMETRICAL AND THAT LEADS MAY BE REVERSE IMAGED.

T1 SCHEMATIC

9T #28

9T #22

9T #22

G F

PARTS LIST

(2) "E" CORES STACKPOLE #57-0048
(1) BOBBIN FERROX CUBE 990-012-01
6' #22 ENAMELED WIRE
50' #20 ENAMELED WIRE
MYLAR TAPE .001 THICK

CONSTRUCTION STEPS

1 DRILL BOBBIN AS SHOWN
2 SECONDARY 220 VOLTS–16.5 TURNS OF #28 WIRE A & B WIND SECONDARY IN TIGHT, EVEN LAYERS. INSULATE EVERY TWO LAYERS WITH A LAYER OF IMIL MYLAR TAPE TO PREVENT BREAKDOWN. NOTE SKETCH 2. TAPE COMPLETED SECONDARY
3 WIND 9 TURNS OF #28 WIRE C & D
4 TAKE A LENGTH OF #22 WIRE ABOUT SIX FEET LONG AND DOUBLE IT. INSERT THE TWO CUT ENDS THROUGH HOLES E & F LEAVING ABOUT 6" FOR LEADS. WIND 9 TURNS OF THESE PARALLEL WIRES NEATLY AND TIGHTLY AND EXIT AT HOLES G & H AS SHOWN.
5 CUT WIRE LOOP AND JOIN WIRES FROM HOLES F & G. MEASURE CONTINUITY BETWEEN WIRES FROM HOLES E & H.
6 TAPE FINISHED WINDINGS AND PLACE ON E CORES. TAPE THESE CORES TIGHTLY TOGETHER.

Fig. 3-3B. T1 inverter transformer type II.

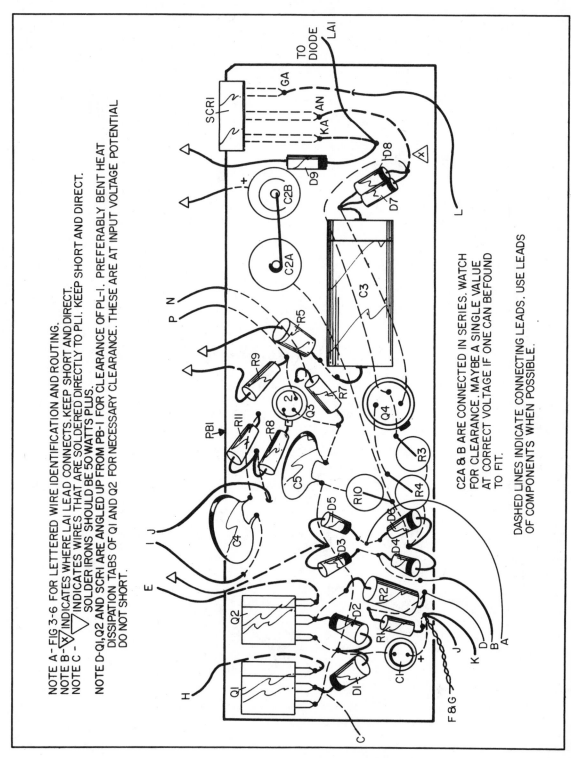

NOTE A - FIG 3-6 FOR LETTERED WIRE IDENTIFICATION AND ROUTING.
NOTE B - ⟨W⟩ INDICATES WHERE LAI LEAD CONNECTS. KEEP SHORT AND DIRECT.
NOTE C - INDICATES WIRES THAT ARE SOLDERED DIRECTLY TO PLI. KEEP SHORT AND DIRECT. SOLDER IRONS SHOULD BE 50 WATTS PLUS.
NOTE D - QI, Q2 AND SCRI ARE ANGLED UP FROM PB-I FOR CLEARANCE OF PL-I. PREFERABLY BENT HEAT DISSIPATION TABS OF QI AND Q2 FOR NECESSARY CLEARANCE. THESE ARE AT INPUT VOLTAGE POTENTIAL DO NOT SHORT.

C2A & B ARE CONNECTED IN SERIES. WATCH FOR CLEARANCE. MAYBE A SINGLE VALUE AT CORRECT VOLTAGE IF ONE CAN BE FOUND TO FIT.

DASHED LINES INDICATE CONNECTING LEADS. USE LEADS OF COMPONENTS WHEN POSSIBLE.

Fig. 3-4. Assembly board wiring.

39

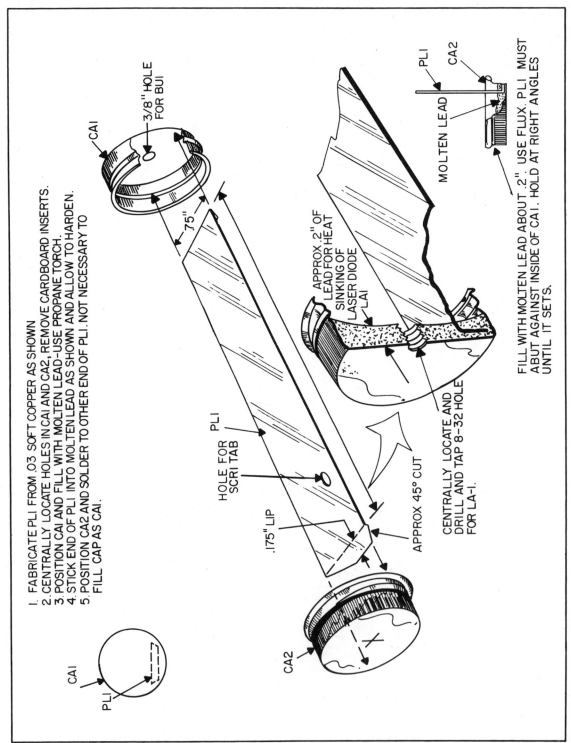

1. FABRICATE PLI FROM .03 SOFT COPPER AS SHOWN
2. CENTRALLY LOCATE HOLES IN CAI AND CA2, REMOVE CARDBOARD INSERTS.
3. POSITION CAI AND FILL WITH MOLTEN LEAD–USE PROPANE TORCH.
4. STICK END OF PLI INTO MOLTEN LEAD AS SHOWN AND ALLOW TO HARDEN.
5. POSITION CA2 AND SOLDER TO OTHER END OF PLI. NOT NECESSARY TO FILL CAP AS CAI.

CAI

PLI

3/8" HOLE FOR BUI

CAI

.75"

HOLE FOR SCRI TAB

PLI

.175" LIP

APPROX 45° CUT

CA2

APPROX .2" OF LEAD FOR HEAT SINKING OF LASER DIODE LAI

CENTRALLY LOCATE AND DRILL AND TAP 8-32 HOLE FOR LA-I.

PLI

CA2

MOLTEN LEAD

FILL WITH MOLTEN LEAD ABOUT .2". USE FLUX. PLI MUST ABUT AGAINST INSIDE OF CAI. HOLD AT RIGHT ANGLES UNTIL IT SETS.

Fig. 3-5. Mechanical assembly.

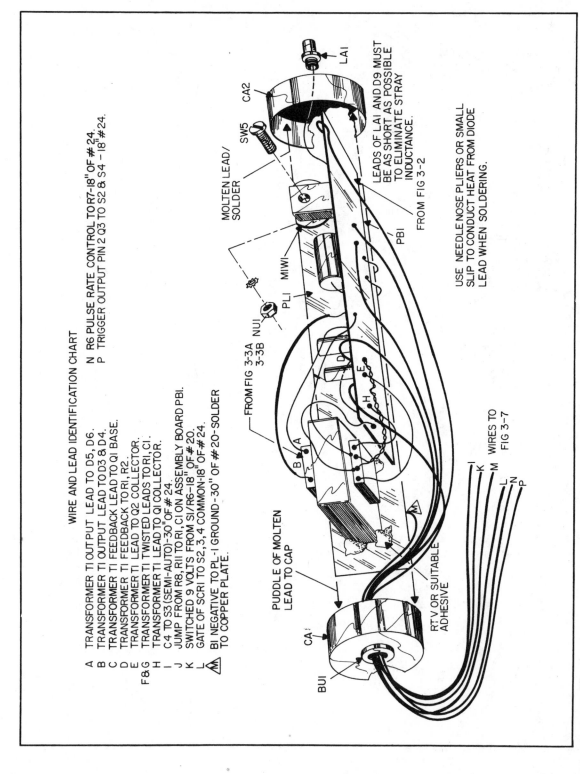

WIRE AND LEAD IDENTIFICATION CHART

A   TRANSFORMER T1 OUTPUT LEAD TO D5, D6.
B   TRANSFORMER T1 OUTPUT LEAD TO D3 & D4.
C   TRANSFORMER T1 FEEDBACK LEAD TO Q1 BASE.
D   TRANSFORMER T1 FEEDBACK TO R1, R2.
E   TRANSFORMER T1 LEAD TO Q2 COLLECTOR.
F&G TRANSFORMER T1 TWISTED LEADS TO R1, C1.
H   TRANSFORMER T1 LEAD TO Q1 COLLECTOR.
I   C4 TO S3 (SEMI-AUTO)-30" OF #24.
J   JUMP FROM R8, R11 TO R1, C1 ON ASSEMBLY BOARD PB1.
K   SWITCHED 9 VOLTS FROM S1/R6-18" OF #20.
L   GATE OF SCR1 TO S2, 3, 4 COMMON-18" OF #24.
⚠   B1 NEGATIVE TO PL-1 GROUND-30" OF #20-SOLDER
     TO COPPER PLATE.

N   R6 PULSE RATE CONTROL TO R7-18" OF #24.
P   TRIGGER OUTPUT PIN 2 Q3 TO S2 & S4 -18" #24.

LEADS OF LA1 AND D9 MUST
BE AS SHORT AS POSSIBLE
TO ELIMINATE STRAY
INDUCTANCE.

FROM FIG 3-2

USE NEEDLE NOSE PLIERS OR SMALL
SLIP TO CONDUCT HEAT FROM DIODE
LEAD WHEN SOLDERING.

MOLTEN LEAD/
SOLDER

LA1

SW5

CA2

PB1

MIW1

PL1

NU1

FROM FIG
3-3A
3-3B

WIRES TO
FIG 3-7

PUDDLE OF MOLTEN
LEAD TO CAP

RTV OR SUITABLE
ADHESIVE

CA

BU1

Fig. 3-6. Electronic assembly.

41

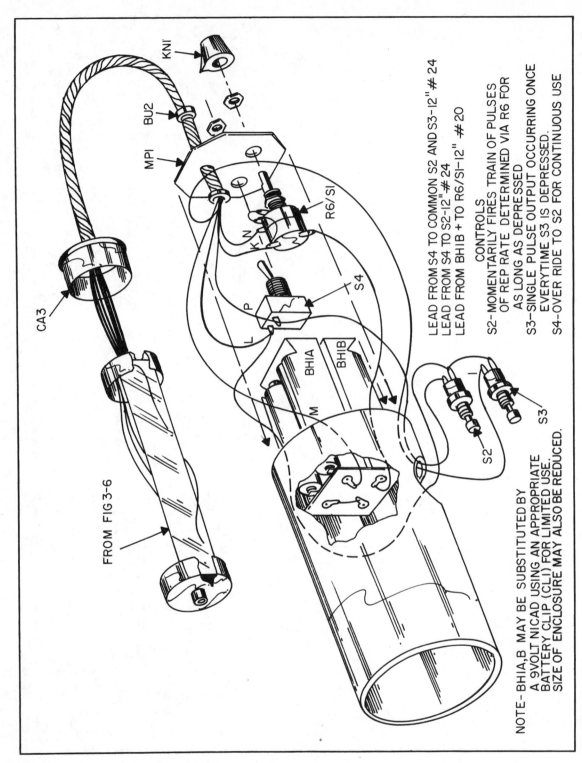

KNI

BU2

MPI

CA3

FROM FIG 3-6

R6/SI

S4

BHIA

BHIB

M

S2

S3

LEAD FROM S4 TO COMMON S2 AND S3-12" # 24
LEAD FROM S4 TO S2-12" # 24
LEAD FROM BHIB + TO R6/SI-12" # 20

CONTROLS
S2-MOMENTARILY FIRES TRAIN OF PULSES
OF REP RATE DETERMINED VIA R6 FOR
AS LONG AS DEPRESSED
S3-SINGLE PULSE OUTPUT OCCURRING ONCE
EVERYTIME S3 IS DEPRESSED.
S4-OVER RIDE TO S2 FOR CONTINUOUS USE

NOTE— BHIA,B MAY BE SUBSTITUTED BY
A 9VOLT NICAD USING AN APPROPRIATE
BATTERY CLIP (CLI) FOR LIMITED USE.
SIZE OF ENCLOSURE MAY ALSO BE REDUCED.

Fig. 3-7. Final wiring.

instance, if we were to adjust the input voltage for a 40 amp pulse at the high pulse repetition and then decrease our pulse repetition rate, it could cause the current pulse to increase to a dangerous value. This effect can be greatly reduced by using good solid connections to the power supply with short leads helping to decrease feed voltage differences. Another point of measurement is to monitor the input voltage right at the device, i.e. between PL1 plate and point "FG" on T1. You may be surprised to see this voltage differ as a result of the current demand between low and high pulse rates. Keeping this voltage adjusted to a constant value and taking a measurement at point "C" as the pulse repetition rate is varied will determine the circuit capability to maintain fairly good regulation throughout the pulse repetition range. Please note the regulation curve in Fig. 3-8.

Caution—Caution—pay particular attention to correct pulse amplitude especially when using the LD1630 diode.

It is assumed at this point that we are familiar with our circuit regulation versus pulse repetition rate along with the proper current pulses for our particular diode. We now can carefully install the laser diode LA1 in place of our test diode, by securing into the threaded can and heatsink. Thighten and solder using the shortest direct lead possible to connection point. Energize the system and note a visible indication of laser output via the infrared indicator or other suitable detection device.

FINAL ASSEMBLY

It is assumed that the electronics section is performing correctly, the correct battery combination has been selected and the unit checked for

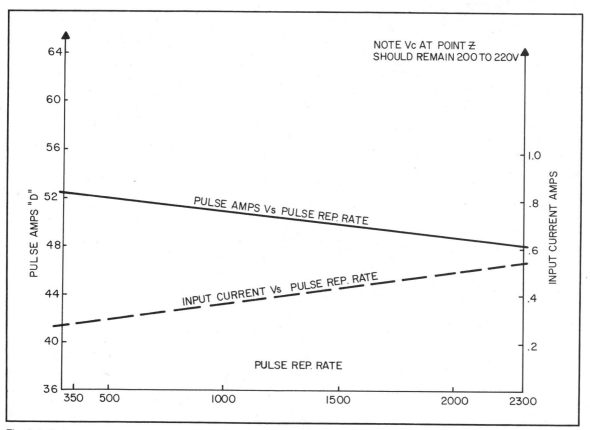

Fig. 3-8. Pulse rate vs pulse amps, pulse rate vs input current.

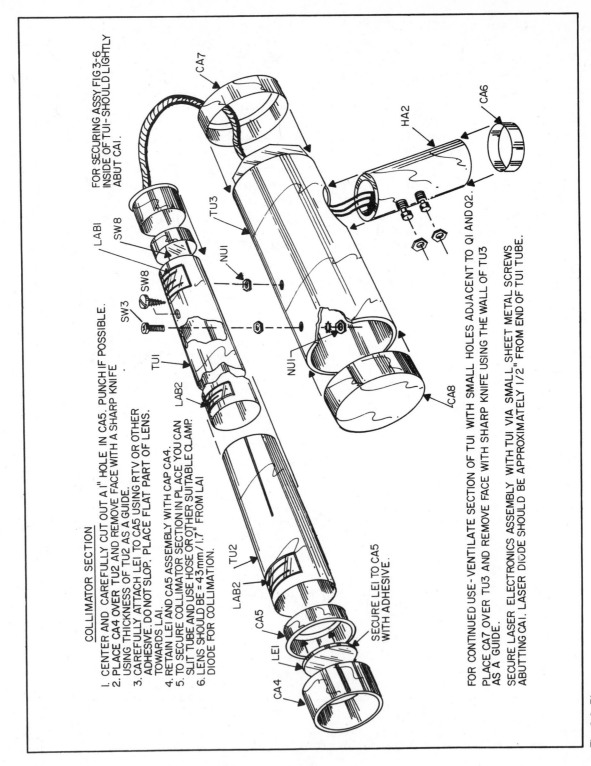

COLLIMATOR SECTION
1. CENTER AND CAREFULLY CUT OUT A 1" HOLE IN CA5. PUNCH IF POSSIBLE.
2. PLACE CA4 OVER TU2 AND REMOVE FACE WITH A SHARP KNIFE USING THICKNESS OF TU2 AS A GUIDE.
3. CAREFULLY ATTACH LEI TO CA5 USING RTV OR OTHER ADHESIVE. DO NOT SLOP. PLACE FLAT PART OF LENS. TOWARDS LAI.
4. RETAIN LEI AND CA5 ASSEMBLY WITH CAP CA4.
5. TO SECURE COLLIMATOR SECTION IN PLACE YOU CAN SLIT TUBE AND USE HOSE OR OTHER SUITABLE CLAMP.
6. LENS SHOULD BE ≈ 43mm/1.7" FROM LAI DIODE FOR COLLIMATION.

SECURE LEI TO CA5 WITH ADHESIVE.

FOR SECURING ASSY FIG 3-6. INSIDE OF TUI - SHOULD LIGHTLY ABUT CAI.

FOR CONTINUED USE - VENTILATE SECTION OF TUI WITH SMALL HOLES ADJACENT TO QI AND Q2. PLACE CA7 OVER TU3 AND REMOVE FACE WITH SHARP KNIFE USING THE WALL OF TU3 AS A GUIDE.

SECURE LASER ELECTRONICS ASSEMBLY WITH TUI VIA SMALL, SHEET METAL SCREWS ABUTTING CAI. LASER DIODE SHOULD BE APPROXIMATELY 1/2" FROM END OF TUI TUBE.

Fig. 3-9. Blowup.

44

quality of wire and solder joints.

1. Fabricate TU1 tube as shown in Fig. 3-9. Drill all holes and carefully deburr. Polish tube with steel wool to a dull finish for aesthetics. Fabricate TU2 in a likewise fashion.

2. Check for smooth sliding fit. Burrs must be completely removed or you will surely have a problem.

3. Fabricate caps CA4, CA5, and CA7 as shown in Fig. 3-9.

4. Attach and secure lens LE1 as shown in Fig. 3-9.

5. Fabricate HA2 as shown in Fig. 3-9. Fabricate TU3. Cut hole for handle HA2. Use a 1⅞" circle cutter, fabricate MP1 as shown in Fig. 3-7. Remove corners for proper fit to TU3, retained via CA7.

6. Carefully insert the electronics assembly into TU1 as shown in Fig. 3-7. Make sure to check for components that are dangerously close to shorting out on inner surface of TU1. It may be necessary to wrap a thin layer of plastic or similar material around this assembly to prevent shorting. Mount controls to MP1 and insert battery holders into TU3. Wire as shown in Fig. 3-7.

7. Connect battery to unit and retest to verify output using an infrared indicator.

8. Sleeve TU2 collimator section and set lens to approximately 43 mm (1.7") from face of LA1.

9. Secure laser and aim at IR-indicator paper starting at two to three feet. Adjust collimator for sharpest point. Increase distance and readjust. Collimation will usually occur at 5 feet to infinity or more from the device. Closer range work will require a resetting of the collimator tube resulting in increasing the distance between LE1 and LA1.

10. Assemble grips, install battery, etc. Unit is ready for use.

For medium range detection (several Km) it is suggested to use the *laser light detector* as described. For maximum range a special fast rise time, low capacity *pin photo diode* such as RCA C30807 should be used.

TROUBLESHOOTING

For protection of the laser diode, the pulse should be tested initially with the "dummy" load. Laser diodes can also be replaced by a one-ohm to two-ohm resistor. In tests of laser pulsers, the difficulty encountered most frequently is synchronization of the test oscilloscope. Most commercial pulse generators provide a trigger output which should be used if practical. The Hewlett-Packard Model 214 provides versatile controls.

The SCR anode is a good test point for pulser testing and troubleshooting. The voltage waveform at this point, as well as the current waveform of the current monitor, should be similar to those in Fig. 3-2A. For display of the discharge current waveform, the oscilloscope must have a high sweep speed.

If the voltage on the SCR anode remains at the value of the power supply, the discharge circuit is inoperative. The discharge circuit may be open, or the SCR trigger signal may be inadequate or of reversed polarity. If the anode voltage is constant at one or two volts, the SCR is holding, The charging circuit may be malfunctioning or the SCR may be overheating and locking on. If the anode voltage is zero, the SCR or capacitor is shorted or the charging circuit is malfunctioning.

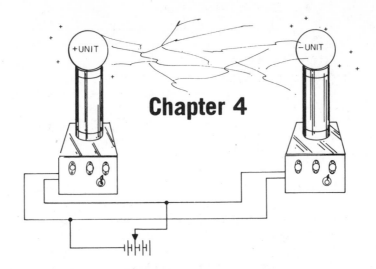

Chapter 4

High-Power
Pulsed Red Ruby Laser Gun (RUB3)

Caution—this laser project involves the use of high voltages and capacitances that can be lethal. Use the same safety procedures that you would use assembling a high-powered ham radio transmitter. When firing laser always wear safety glasses and direct beam into a light absorbing target. This project should not be attempted without experience in handling high voltage circuits. It is an advanced project.

All light, conventional or laser, is the result of emission between energy levels of atoms and molecules. These energy levels are characterized by certain quantum levels inherent of the particular atom. These particular levels have the property or resonance both in the absorption and emissions of radiant energy. This energy when emitted as radiation (light) is equal to E joules = hf where h is Planck's constant and f is the frequency of radiation. Conversely, the frequency of the emitted radiation is functional to E higher − E lower and applies to both laser and conventional systems.

Energy (radiation) is emitted (emission) when the atoms make a transition (change) from a higher level (excited state) to lower level (relaxed or ground state). See Fig. 4-1. An atom, however, must first absorb energy (Fig. 4-2). All light requires a certain amount of absorption of energy by an atom to a higher level with conventional light being the result of spontaneous emission and laser light the result of stimulated emission. See the electromagnetic spectrum sheet (Table 4-1).

It is obvious that spontaneous emission never produces light of the quality of laser light. Conventional light is the result of a system of thermal equilibrium where energy levels are always populated with the least in the highest and the most in the lowest. This condition always allows the atoms to be absorptive in nature only becoming saturated at infinite temperatures (all energy levels filled). The basic character of conventional light is wide spectral distribution, random polarization, circular and irregular wavefronts and relatively low resultant color temperatures.

A laser differs from conventional light in that the radiation produced is not by the spontaneous emission of energy levels, but by stimulated emis-

Table 4-1. Electromagnetic Energy Spectrum.

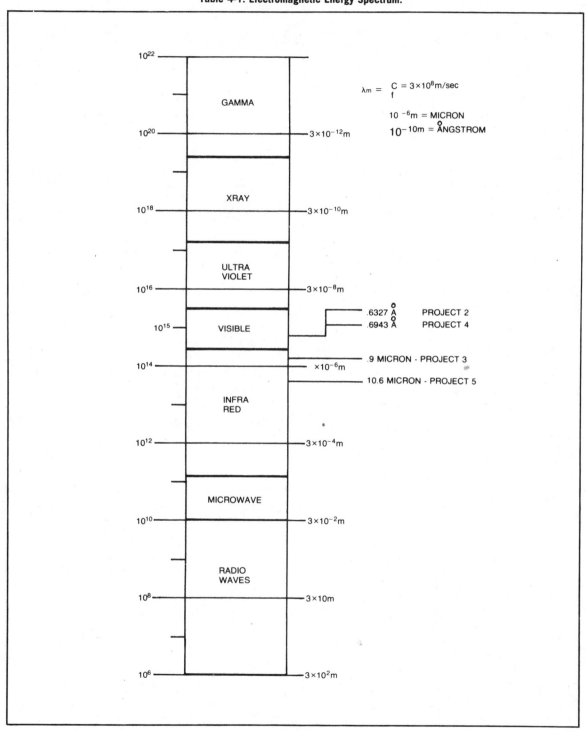

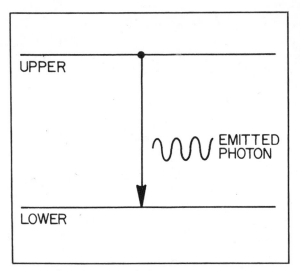

Fig. 4-1. Spontaneous emission.

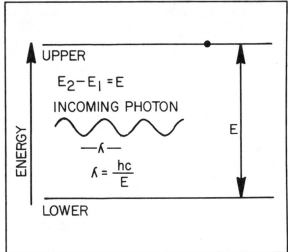

Fig. 4-3. Stimulated emission (initial state).

sion. This emission is the result of nearly equal levels of energy stimulating other levels into emitting photons of radiation of a nearly pure single frequency when relaxing to the lower energy states of the atom (Figs. 4-3 and 4-4). However, to obtain this stimulated emission effect a population inversion consisting now of a maximum occupancy of the higher energy levels rather than lower levels must occur and can only be achieved by forcing or pumping the system with external energy. This pumping or exciting is accomplished via flash lamps for pulsed solid-state devices, arc lamps for continuous solid-state, electrical discharge for gas systems, other laser optically pumping an active medium, chemical reaction, etc.

Many materials will lase, however, only certain materials are worthwhile and produce usable output. Laser materials must be of an atomic structure where energy levels are able to be excited by practical means to achieve the necessary population inversion. Suitable materials are usually crystalline for solid devices, and gaseous or liquid for others.

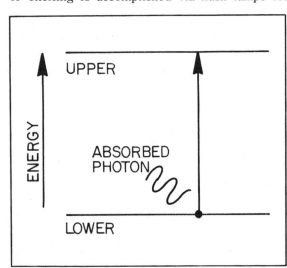

Fig. 4-2. Absorption.

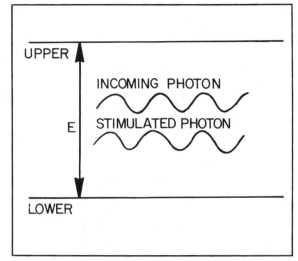

Fig. 4-4. Stimulated emission (final state).

These materials are usually placed between two mirrors where the radiation can make multiple passes through the materials stimulating more excited atoms into producing more radiation and so until the radiation exits a powerful beam of nearly pure light.

Of all the many laser devices and systems possible, the ruby laser takes precedence as being the first optical laser developed. (Schawlow and Townes early 1960.) The ruby laser while being extremely simple to construct also produces one of the most powerful sources of light when Q-switched. This type of device is referred to as a solid-state laser along with its neodymium yag and neodymium glass cousins.

The ruby laser while being only fractions of a percent efficient (0.1 percent) produces these high powered optical pulses in the visible red range at 6943 Å. See Table 4-1. These pulses are capable of blasting holes through steel, bouncing off the moon, and special military systems capable of being used for antipersonnel weaponry when utilizing the more efficient neodymium glass devices.

A ruby laser is a three level device (Fig. 4-5), consisting of an active medium containing about 0.3 to 0.4% chromium Cr^{+3} atoms to the main aluminum oxide (AL_2O_3) crystal host. The geometry of the medium is usually a cylindrical rod of from 1 cm to 2 cm in diameter and 5 to 20 cm long. It is usually pink to red in color depending on the concentration of Cr^{+3} atoms present in the host. The ends of the rod are usually optically coated with one being totally reflective and the other being partially reflective. The partial end is the output end. Light is reflected back and forth between these ends and is amplified by further stimulation of other atoms and eventually exits as a visible beam of laser light. Laser action is the result of the energy levels of the Cr^{+3} atoms.

Note that the 2E level is pumped with at least ½ the ions of the 4A ground state before laser action commences. The absorption of the ruby rod medium consists of two regions, one in the violet T1 and one in the green T2. These regions are about 1000 Å wide and are comparatively efficient when pumped by white light. Upon reaching the T1 and T2 absorption levels the excited ions quickly drop down to the 2E level creating the necessary population inversion required for laser action. The transition from this 2E level to the ground level produces the output wavelength of 6943 Å. It is this transition from the 2E level to ground state that is the stimulated emission, i.e., atoms of nearly equal energy levels producing photons of energy that stimulate other atoms to produce more photons, etc. It should be noted that the diagram (Fig. 4-5) is greatly simplified and fails to show the closely separated defined lines of both the 2E and the 4A levels. These levels only serve to increase the spectral distribution of this output and can be "resolved" by cooling the rod to about 75° K, when the line widths become about 10 to 15 GHz.

One now sees that the first step is to provide a pulse of light rich in green and violet to pump the ruby rod into the population inverted state. This is done via a xenon flash lamp that is physically placed near the ruby rod (Fig. 4-6). There are several accepted methods of accomplishing this, one being the use of a helical flash lamp placed around the ruby rod. This method is not quite as efficient as using linear flash lamps that are placed inside of a reflective cylindrical or elliptical reflector. The laser

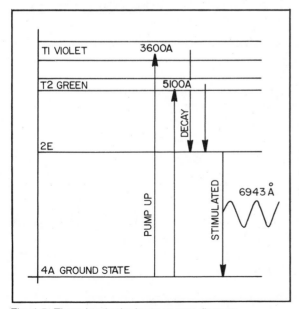

Fig. 4-5. Three-level ruby laser energy diagram.

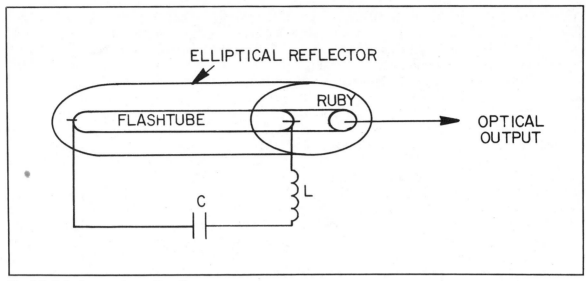

Fig. 4-6. Ruby laser system schematic.

and lamps are located at the respective foci of these enclosures.

The laser pulse starts about 500 μsec after the pumping flash and lasts for another 500 μsec (Fig. 4-7). Depopulation of the upper levels occurs many times faster than the pumping rate can replenish them. Consequently, the system must rest until population inversion is again achieved, consequently, the output pulse consists of many spikes lasting for about 1 μsec each for each optical flash of the lamp.

The flashlamp of the system is usually energized by a conventional dc power supply charging a bank of low loss capacitors. Voltages usually are from 1 to 4 kV with energy from 20 to 2000 joules. Joules = $\frac{1}{2}$ CV^2. C = capacitance in farads, V = voltage. This corresponds to about 50 to 1000 μF of capacity. You remember from high school physics

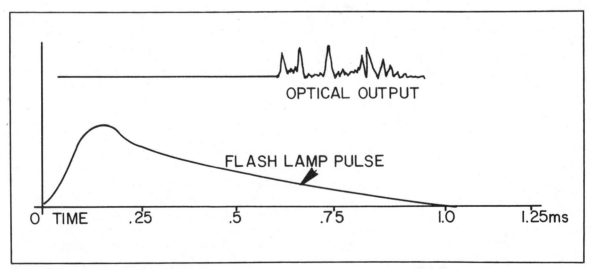

Fig. 4-7. Flashlamp pulse/optical output versus time.

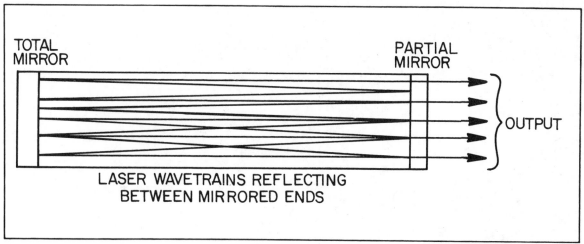

Fig. 4-8. Optical resonator.

that joules = energy = watts – sec. Therefore, watts = joules/sec. A typical moderately powered system may consist of the following: 500 μF at 2000 volts = 1000 joules. If the light pump pulse occurs in 5×10^{-4} secs the pulse power = 2 megawatts. A conversion efficiency of about 0.1 % optical output will be a 2000 watt pulse lasting almost a millisecond. If, however, we were to Q-switch the above system at a pulse width of 10 nanoseconds, we would produce peak power of 1000 joules/10^{-8} sec = 100 GW optical power at an efficiency of 0.1% now produces an output pulse of 1000 megawatts. Quite an impressive amount of optical intensity consisting of an equivalent color temperature when considering conventional sources of this bandwidth of astronomical proportions.

In order to obtain useful output from the ruby rod, it must be optically prepared to be totally reflective at one end and partially on the other (Fig. 4-8). This allows the optical energy to pass through the active medium of the rod on axis many times stimulating other transition and increasing the (optical) energy on each successive pass and eventually exiting out of the partial end as the useful output beam. (This is termed a Fabry-Perot optical resonator.) Q-switching is the spoiling of the above that allows a heavy population of the system creating a super high extremely short pulse of energy. Q-switching is done mechanically, optically, etc.

CIRCUIT THEORY

The laser described in these plans (Fig. 4-9 and Tables 4-2 through 4-6) utilizes a portable energy storage charging power supply using a highly efficient transistor switching oscillator circuit and ferrite transformer (T1). The switching frequency of these transistors, (Q1 and Q2) is determined by the magnetics of T1 and is about 20 kHz for the values shown in these plans. Resistor (R1) limits the base drive to Q1 and Q2 obtained via the feedback winding on T1. Diodes (D1 and D2) form the return path for the base drive of the opposite transistor. Resistor (R2) provides the dc unbalance necessary to initiate oscillator switching. Power to the circuit is obtained from a 12-volt gel cell (B1) and is controlled via switch (S1). Resistor (R3) limits the current to the "power on" lamp (LA1). High voltage square-wave pulses are obtained from the secondary of T1 and are rectified and voltage doubled via diodes (D3-D8). Resistors (R4-R9) divide the reverse voltage across the diodes during their nonconducting period. The rectified dc pulses are now integrated into the energy storage capacitors (C2-C9).

The circuit shows the dual capacitor bank option where two flash tubes are utilized. The builder may decide only to use one section for the single lamp configuration, eliminating (Bank B) C7-C9 and surge separating resistors (R10 and R11). These

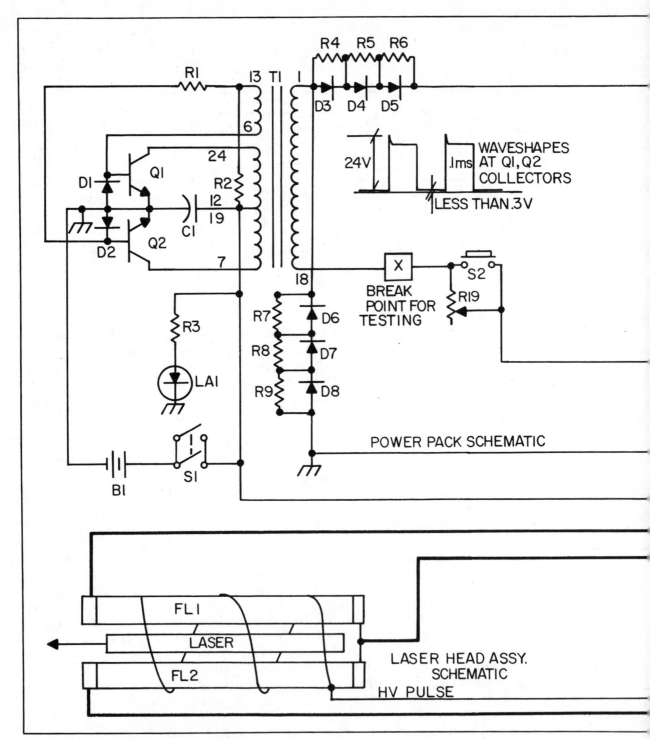

Fig. 4-9. Power pack and laser head schematic.

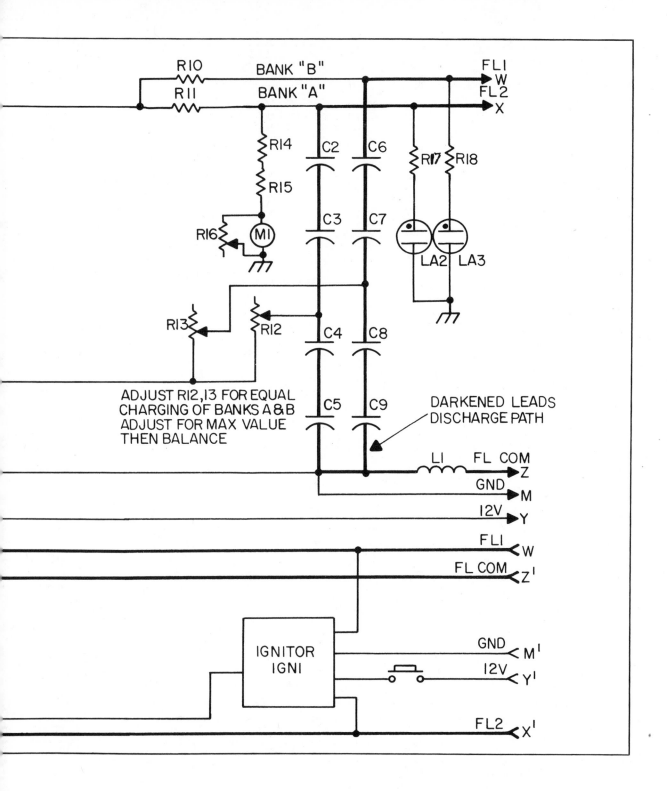

ADJUST R12,13 FOR EQUAL
CHARGING OF BANKS A & B
ADJUST FOR MAX VALUE
THEN BALANCE

DARKENED LEADS
DISCHARGE PATH

Table 4-2. Wire and Leads Designation Chart

Wire ID#	Length	Gauge	Type	Routing from - to
A	10″	#20	vinyl red	Junction D3,6 - T1 pin 1
B	4″	#24	vinyl black	common assembly board-M1 neg
C	5″	#20	vinyl black	common-C5 Neg
D	10″	#20	vinyl black	common-TE1 grd lug
E	12″	#24	vinyl black	common-LA2,3 common
F	4″	#24	vinyl red	junction R15,16-M1 pos.
G	12″	#24	vinyl red	R17-LA2
H	18″	#20	vinyl red	R11-C2 pos.
I	12″	#20	vinyl red	R10-C6 pos.
J	12″	#24	vinyl red	R18-LA3
K	4″	#20	vinyl red	Q2 collector to T1 pin 7
L	10″	#18	vinyl black	TE1 grd lug B1 neg
M	Umbilical	#20	vinyl black	TE1 grd lug laser head grd
N	6″	#20	vinyl black	Junction R1,R2-T1 pin 13
P	6″	#20	vinyl black	R2 to T1 pin 19
Q	6″	#20	vinyl black	Q1 base to T1 pin 6
R	6″	#18	vinyl black	Q1 collector to T1 pin 24
S	6″	#18	vinyl black	S1 to T1 pin 19
T	10″	#18	vinyl black	S1 to B1 pos.
V	10″	#20	vinyl black	S2 to R12
U	6″	#20	vinyl black	S2 to T1 pin 18
W	Umbilical	#12	vinyl black	C6 pos. to laser head FL1
X	Umbilical	#12	vinyl black	C2 pos. to laser head FL2
Y	Umbilical	#20	vinyl black	T1 #19 to laser head +12
Z	Umbilical	#12	vinyl black	L1 to laser head FL com.
AA	4″	#20	vinyl black	R12 to R13
BB	4″	#20	vinyl black	R13 arm to C7,C8 buss
CC	12″	#20	vinyl black	R12 arm to C3,C4 buss
DD	6″	#20	vinyl black	R12 to T1 pin 18

resistors are only required for even peak power balancing between the two flash lamps (FL1 and FL2). Metering of the energy bank voltage is done by monitoring only (Bank A) C2-C5. Resistors (R14, R15) determine the meter current while (R16) shunts this current for meter calibration. LA2 and LA3 are neon lamps with current limiting resistors (R17 and R18) and give the operator indication that the energy bank still maintains a potentially hazardous energy charge at the time.

Whenever charging any capacitor bank in this manner a 50% loss of energy is always encountered due to the lumped value equivalent of a series current-limiting charge resistor. The circuit shown contains several circuit resistances such as transformer secondary resistance, forward diode resistances, etc. These are too low to properly limit this initial charging current to a value that can be safely handled by the transistor switching circuit. Hence, resistors R12, R13, and R19 provide this necessary resistance and may be adjustable slider types of 10 watts or more. These are set to a value for obtaining the fastest charge possible without circuit overloading. S2 shorts out R19 speeding up the charging time only after some value of initial charge is already accumulated. These values and parameters will vary from circuit to circuit and may require final adjustment for optimum performance.

When the charges on both Banks A and B are sufficient (usually somewhere between 1500 and 1800 volts) the laser is ready to be fired. When ignition occurs and the flash lamps FL1 and FL2 conduct, a pulse of current occurs that must be shaped in reference to duration and value. This is accomplished via inductor (L1) forming a resonant discharge circuit with a period determined by and depending upon the type of laser medium used. Example: when pumping a ruby rod, the flash lamps should be more favorable in the green/blue region of color temperatures. This involves a high voltage and usually lower capacitance and consequently a different value of inductance of L1. A Yag laser requires a lower voltage and larger capacity for a reddening of pumping color.

The ignitor (Fig. 4-10) is powered by leeching a small amount of power from both energy banks simultaneously to maintain charge balance. These are lines FL1 and FL2 as shown and charge the ignitor (IGN) through a high resistance.

The ignitor circuit IGN consists of a simple

Table 4-3. Power Pack Section Parts List.

R1	(6)	82 Ω ohm 1 watt or use (1) 15 Ω 10 watts power resistor
R2	(1)	1 k ½ W
R3	(1)	470 ohm ¼ watt
R4-9	(6)	1 meg. ¼ watt
R10,11	(2)	220 ohm 2 watts
R12,13,19	(3)	10 k at 25 watts adjustable
R14	(1)	2.7 M ¼ watt
R15	(1)	1.8 M ¼ watt
R16	(1)	2 k vert trimpot
R17,18	(2)	22 meg. ½ watt
C1	(1)	8000 μF/16 V electrolytic
C2-9	(8)	1100 μF/450 electrolytic capacitors (must have insulating sleeve around metal case)
D1,2	(2)	1N4002 silicon diode 100 PIV 1 amp
D3-8	(6)	1N4007 silicon diode 1000 PIV 1 amp
Q1,2	(2)	2N3055 power silicon npn transmitter
LA1	(1)	LED FLV106
LA2,3	(2)	Neon lamp leads/RTV to secure lamps into bushings
M1	(1)	Small 100 ua panel meter
S1	(1)	10 amp dpdt switch
S2,3	(2)	Push button switch
T1	(1)	See test Fig. 4-11
L1	(1)	See text
MK1,2	(2)	Mtg kits for Q1,2
PB1	(1)	Perfboard 3 × 1¾
TE1	(1)	7 lug terminal strip
BU1	(1)	⅜" plastic bushing
BU2,3	(2)	Red lamp bushing
BU4	(1)	Heyman clamp bushing
WN1,2	(2)	4 wire nuts
CA1	(1)	12 × 8 × 3½ fabbed case Fig. 4-12
BK1 A & B	(2)	Capacitor bracket fabbed Fig. 4-12
STR1 A & B	(2)	Capacitor retaining straps fabbed Fig. 4-12
BK3	(1)	Battery retaining bracket 1" wide fabbed Fig. 4-12
WR7	(3')	#16 buss wire
WR3	(4')	#24 vinyl stranded wire
WR20	(18')	#20 vinyl stranded wire
WR1	(3')	#18 vinyl stranded wire
WR12	(50')	#12-14 wire for winding L1
SW12	(12)	6 32 × ⅜ type F sheet metal screws
SW3	(4)	Screws 6 32 × 1" for straps STR1
SW2	(10)	Screws 6 32 × ½ general assembly
NU1	(14)	6 32 kep nuts
LUG1	(1)	Solder lug for C1 ground

Table 4-4. Laser Head Section Parts List.

IGN1	(1)	Ignitor assembly Fig. 4-10, 4-21
TU1	(1)	8″ length of 3 1/2″ OD sked 40 PVC fab Fig. 4-24
TU2	(1)	6″ length of 1/9″ OD sked 40 PVC fab Fig. 4-24
CA3,4	(2)	3 1/2″ plastic caps fab Fig. 4-24
CA5	(1)	1 7/8″ plastic cap for handle
HC1	(1)	Chassis head fab Fig. 4-22
RP1	(1)	Rear plate fab Fig. 4-22
BAR1	(1)	Barrel fab from a 3″ to 6″ piece of 1/2 to 3/4 copper tube
BU5,6	(2)	1/2″ plastic bushing
TA1	(1)	1″ square piece of double sided tape
BLK1,2	(2)	1 1/4 × 1 × 3/4 teflon block fab as shown Fig. 4-23
PRC1,2	(2)	1/2 × 9/32 OD × .014 wall brass
SW1	(2)	6 32 × 1/4 screws
SW12	(4)	6 32 × 3/8 self tapping screws
NU1	(2)	6 32 kep nuts
WN1	(2)	Small wire nuts or crimp lugs
SW40	(4)	#4 × 1/4 sheet metal screws
CL1	(1)	1/2″ Nylon clamp
FCL1,2,3,4	(4)	Small 1/4″ spring contact clip
LAB1,2	(2)	Aperture label
LAB3	(1)	Class IV laser warning label
LAB4,5	(2)	Danger HV label
FOIL	(8″)	2 1/2″ width of 1 to 3 mil reflective al foil Fig. 4-23
FL1,2	(2)	Flashlamps EGG FX103C2 shown, see text
LASER ROD	(1)	Select type and vendor, see text

capacitor discharge system. A dump capacitor (C1) is charged via divider resistors (R1), (R2), (R3) and (R4) and divided down by R5. Diodes (D1) and (D2) are isolation diodes. A voltage developed across R5 now charges C1 to 300 to 400 volts through the primary of the HV pulse transformer (T1). SCR1 is triggered "ON" and shorts capacitor C1 across the primary of T1 and dumps its energy producing a high voltage pulse at the output of the secondary winding. When C1 is dumped in the primary inductance of T1 a ringing wave is produced whose frequency is equal to the resonance of these components. Diode (D3) recovers the negative part of this energy in this waveform. SCR1 is triggered "ON"

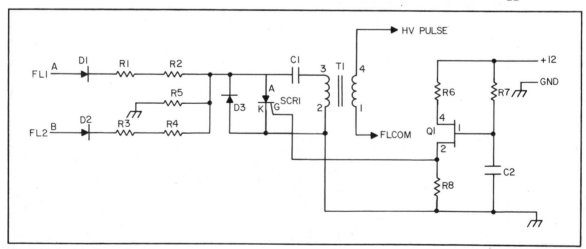

Fig. 4-10. Ign1 ignitor schematic.

Table 4-5. Ignitor Parts List.

R1,2,3, 4,5	(5)	1 meg ¼ watt resistor
R6,8	(2)	100 ohm ¼ watt resistor
R7	(1)	100 K ¼ watt resistor
D1,2,3	(3)	1N4007 diode
C1	(1)	.1M/400 paper capacitor
C2	(2)	10 μF/25 electrolytic capacitor
SCR1	(1)	SCR400 volt
Q1	(1)	2N2646 UJT transistor
T1	(1)	10KV trigger transformer
PB1	(1)	1 1/4″ × 1 1/4″ perfboard
WR3	(60″)	#24 plastic wire
EN1	(1)	1 7/8″ plastic cap

by the UJT (Q1 connected as a free-running relaxation oscillator whose repetition is determined by R7 and C2). This circuit is activated via trigger switch S3 that energizes the UJT circuit. C2 now charges up and eventually fires the UJT turning "ON" SCR1 producing the HV trigger pulse. This pulse is 6 to 20 kV and triggers the flash tube of the laser system that supplies the necessary optical pump pulse for the Ruby or Yag laser.

CONSTRUCTION STEPS

The Ruby Laser project is divided into two sections: *power pack* and *laser head*.

1. Assemble T1 transformer as shown in Fig. 4-11A and B.

Table 4-6. The Following are Manufacturers of Laser Rods Valid in 1983.

LASER NUCLEONICS
1-617-891-7880
Dr. Franke

COMMERCIAL CRYSTAL LABS
111 Chevalier Ave.
Samboy, NJ 08879
1-201-727-1055
Yag

SAPHIRWERK USA
17 Barstow Rd.
Great Neck, NY 11021
1-516-466-8275
(YLF) - Yttrium Lithium Fluoride Rods, Ruby

DR STEEG & REUTER
Berner Strabe 109
6000 Frankfurt 56 Germany
0611-507-1024/25
Neodynium Glass, Yag, Ruby

LASER CRYSTAL
154 Edison Rd.
Lake Hopatcong, NJ 07849
1-201-663-1322
Yag

KIGRE
5333 Secor Rd.
Toledo, OH 43623
1-810-442-1653
Neodynium Glass

ALLIED SYNTHETIC CRYSTALS
1201 Continental Blvd.
Charlotte, NC 28231
1-704-588-2340
Yag

ADOLPH MELLER
P.O. Box 6001
Providence, RI 02904
1-401-331-3717
Yag

SCHOTT
400 York Ave.
Duryea, PA 18642
1-717-457-7485
Neodynium Glass

ATOMERGIC CHEMICALS
100 Fairchild Ave.
Plainview, NY 11803
1-516-349-8800
Yag, Ruby

UNION CARBIDE
8888 Balboa Ave.
San Diego, CA 92123
1-714-279-4500
Yag, Ruby

AIRTRON
200 E. Hanover Ave.
Morris Plains, NJ 07950
1-201-539-5500
Yag, YLF

A source of low to medium quality rods are available through Jack Ford Scientific. We do not guarantee the operation of these rods.

An excellent source of Ruby Rods, mirrors, assorted accessories and controls may be obtained from:

Dennis Meredith
6517 West Eva
Glendale, AZ 85302
1-602-934-9387

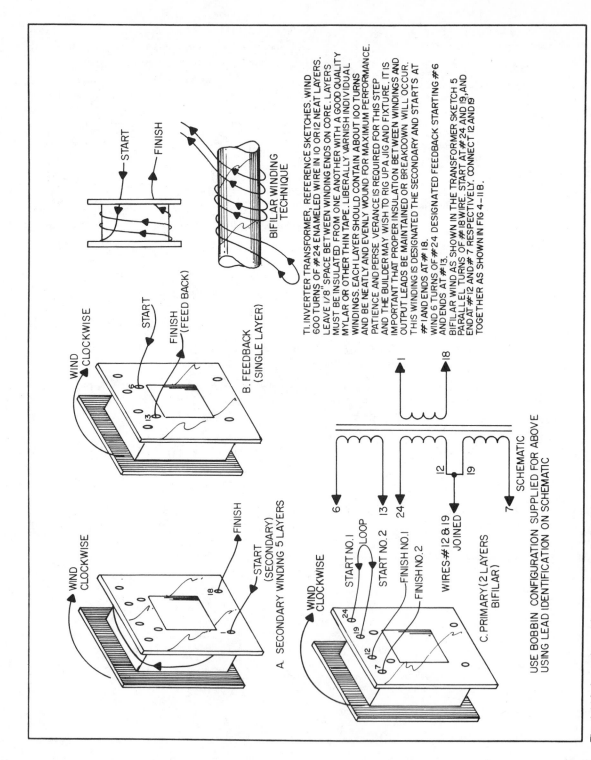

Fig. 4-11A. Inverter transformer winding instructions.

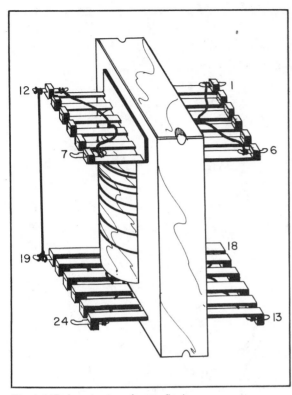

Fig. 4-11B. Inverter transformer final.

2. Fabricate CA1 as shown in Figs. 4-12 and Fig. 4-13. Case inner dimensions are 12″ × 8″ × 3½″. Use 22 gauge galvanized or 20 gauge aluminum. Form from a single sheet or fabricate from individual pieces and sheet metal screwed together. The holes for Q1 and Q2 can be laid out using the mica insulating washers as templates. These are supplied with the mounting kits MK1,2. These holes must be completely deburred or breakdown may occur between Q1, Q2 and the case. Note layout of inverter and control circuitry.

3. Fabricate brackets and straps BK1 and STR1. Use 1″ width of galvanized #24 or #20 aluminum. Remove any burrs and wrap with a layer of tape.

4. Layout and assemble rectifier board Fig. 4-14. Connect wires A-J. See wire and lead chart.

5. Wind inductor L1 using either #12 solid or stranded wire. The stranded wire may also be used

for the umbilical wire where the solid may be too stiff and brittle. Fourteen gauge wire may also be used but offers slightly more I^2R loss. Wind on 1½″ form in 5 to 6 even layers of 10 to 11 turns for a total of 50 to 60 turns. Hold tightly together with tye wraps as assembly must be kept intact to properly fit. See text on the trimming and calibration of this part for proper flash pulse shape. Secure to case via several nylon tye wraps.

6. Mount components shown in Figs. 4-12, 4-13, and 4-15.

7. Mount electrolytic charging capacitors C2-9 as shown. Make sure that metal cases are completely insulated as these may become charged and breakdown if allowed to make electrical contact with each other. Also contact with the retaining brackets BK1 or straps ST1 may cause breakdown if not free of burrs and sharp edges and wrapped with insulating tape to prevent any possible contact with the capacitor cases. Do not over tighten retaining brackets and straps. **Caution: Capacitors should be encased in plastic to prevent accidental contact should case failure occur. Remember there is in excess of 1500 volts at unlimited peak current across these capacitors.**

8. Wire in components as shown in Fig. 4-15. Watch polarity of diodes and C1.

9. Connect wires per wiring aid Fig. 4-16 to assembly board Fig. 4-14.

10. Connect wires per wiring aid Fig. 4-17 and Fig. 4-15 from inverter and control panel section. Note wire description chart.

11. Connect wires as per wiring aids Fig. 4-18 and Fig. 4-19. Note #14 buss wire jumps between capacitors C2-C9. See Fig. 4-12. Use screws and double loops. Connect the heavy wires of inductor L1. Note junction going to C5 negative Bank A and C9 Negative Bank B.

12. Connect the umbilical wires. These are the interconnecting wires that connect to the laser head. The length should not exceed 6 feet.

13. Strain relieve umbilical wires with clamp bushing BU4. Twist leads and tape or use tye wraps every 12 inches. See Fig. 4-20.

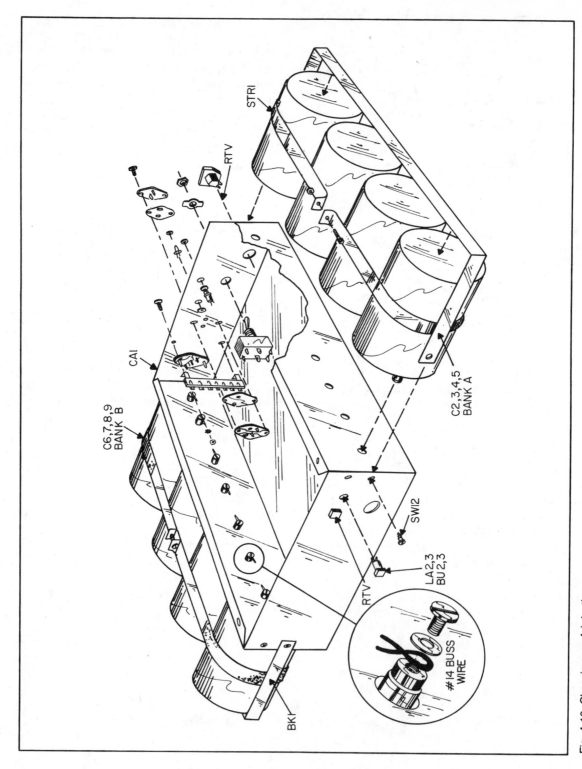

STRI

RTV

CAI

C6,7,8,9
BANK B

C2,3,4,5
BANK A

SWI2

RTV

LA2,3
BU 2,3

#14 BUSS
WIRE

BKI

Fig. 4-12. Chassis case fabrication,

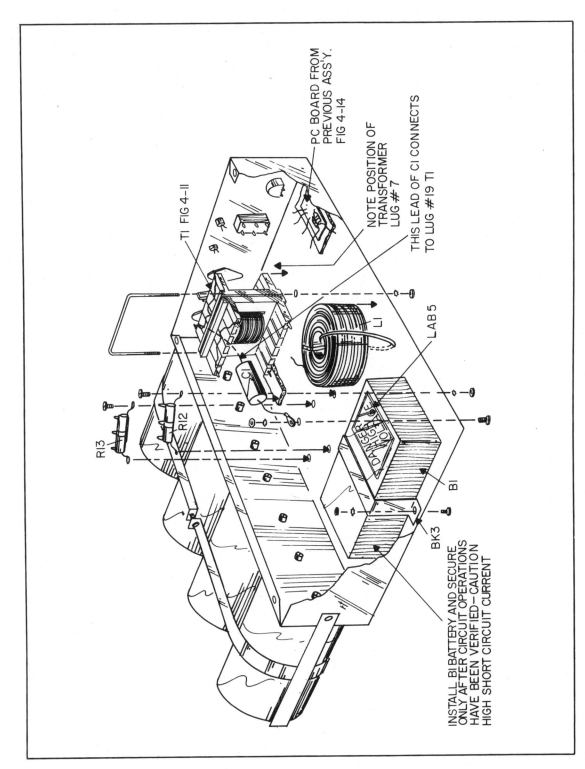

Fig. 4-13. Chassis case assembly.

PC BOARD FROM PREVIOUS ASS'Y. FIG 4-14

NOTE POSITION OF TRANSFORMER LUG #7

THIS LEAD OF CI CONNECTS TO LUG #19 TI

TI FIG 4-11

L1

LAB 5

R13

R12

CI

B1

DANGER

BK3

INSTALL B1 BATTERY AND SECURE ONLY AFTER CIRCUIT OPERATIONS HAVE BEEN VERIFIED—CAUTION HIGH SHORT CIRCUIT CURRENT

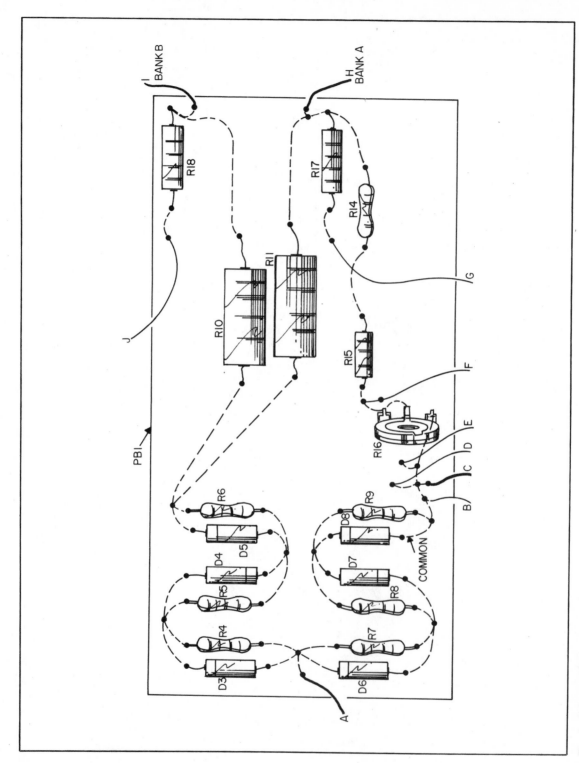

Fig. 4-14. Assembly board.

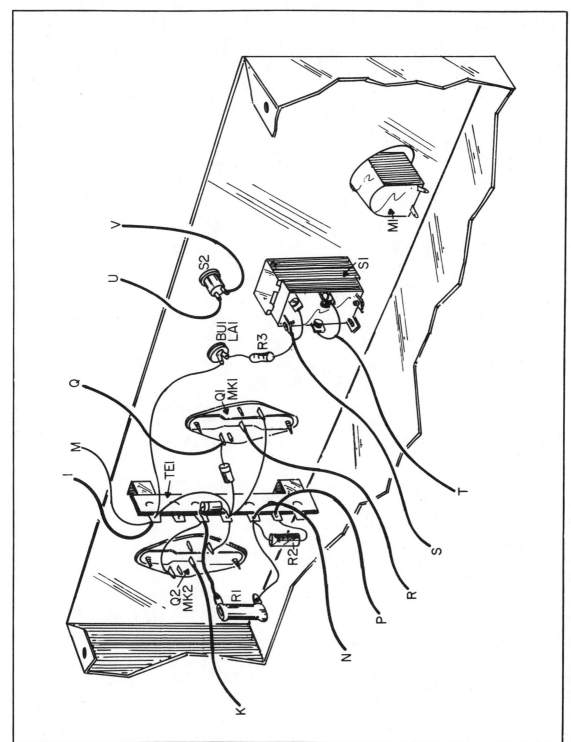

Fig. 4-15. Inverter and control panel.

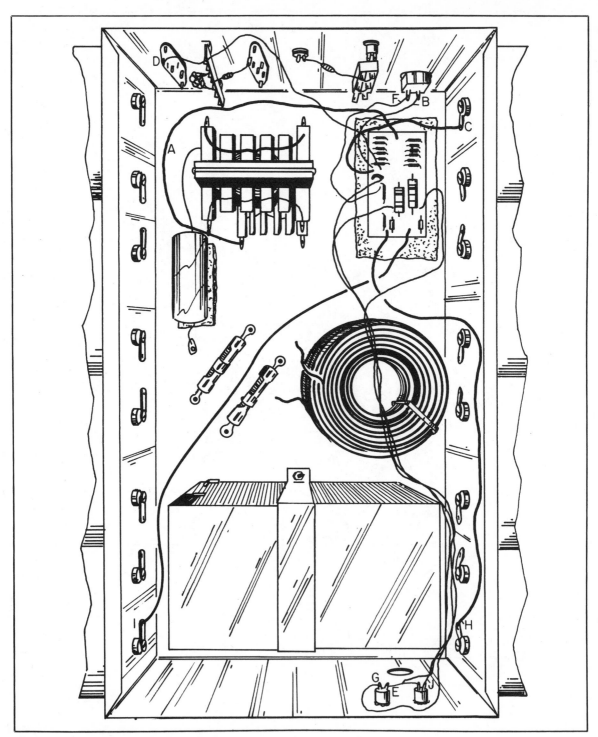

Fig. 4-16. Wiring aid for assembly board connections.

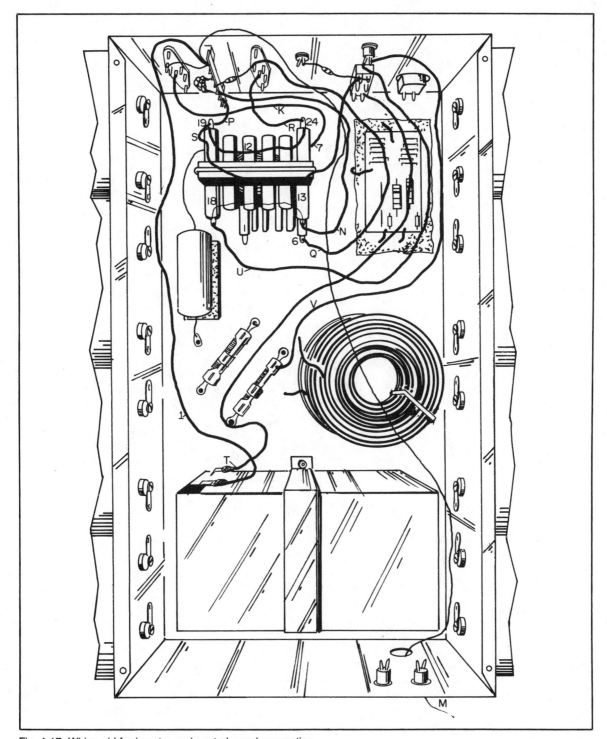

Fig. 4-17. Wiring aid for inverter and control panel connection.

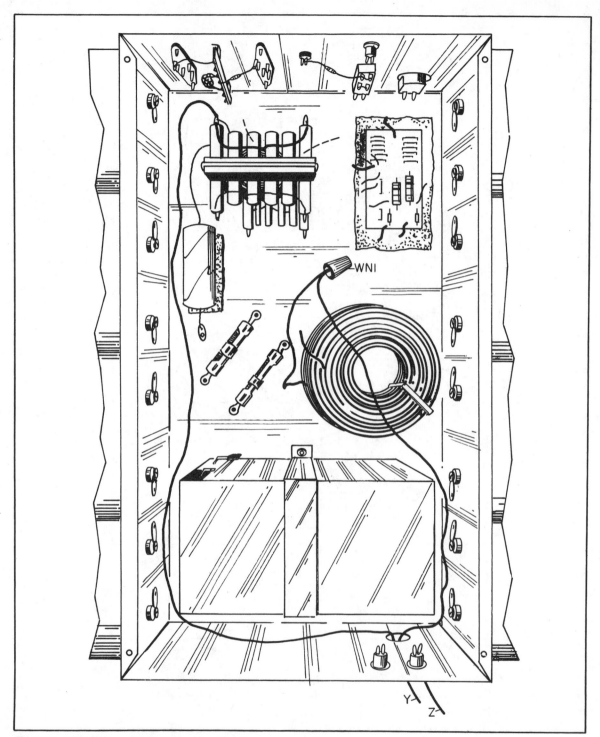

Fig. 4-18. Wiring aid common discharge connections.

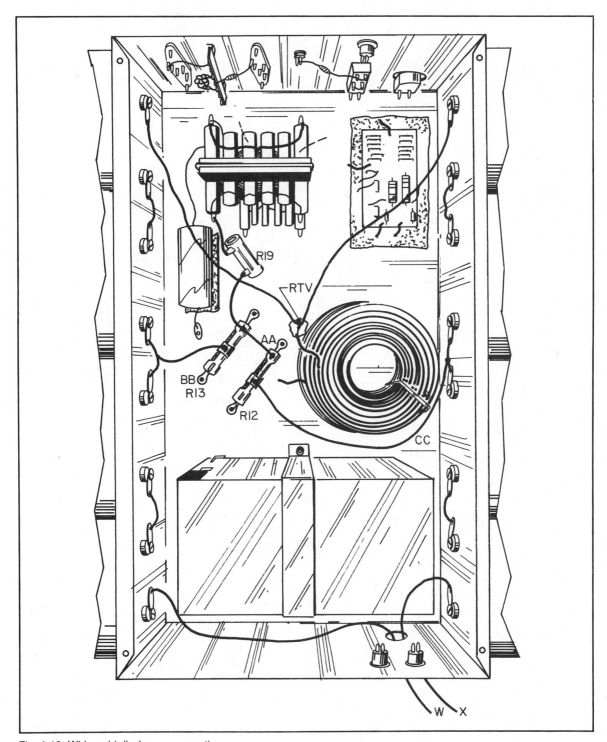

Fig. 4-19. Wiring aid discharge connections.

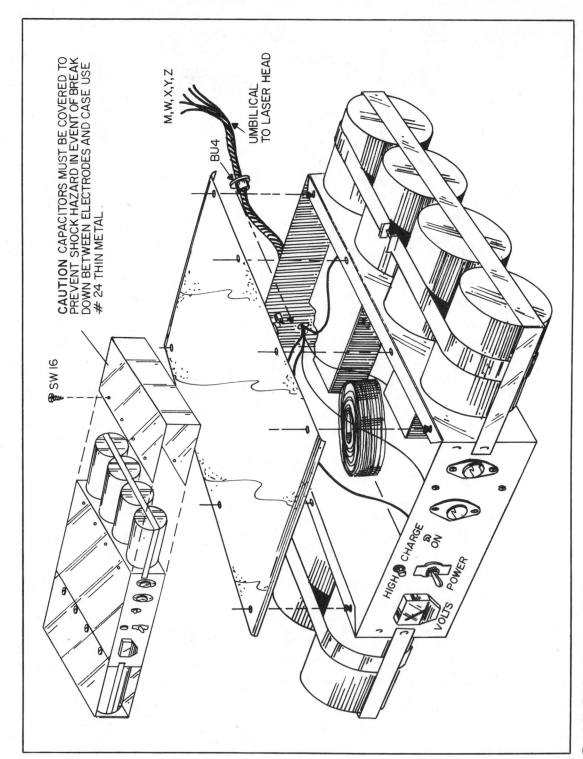

CAUTION CAPACITORS MUST BE COVERED TO PREVENT SHOCK HAZARD IN EVENT OF BREAK DOWN BETWEEN ELECTRODES AND CASE USE # 24 THIN METAL.

M,W,X,Y,Z

BU4

UMBILICAL TO LASER HEAD

SW 16

HIGH CHARGE ON

VOLTS POWER

Fig. 4-20. Final power pack assembly.

68

TESTING AND ADJUSTING YOUR
RUBY LASER POWER PACK SECTION

Again we must stress the deadly voltages present in this power supply and to use extreme caution in the following steps. Do not attempt unless experienced. Use the buddy system. Bodily contact may result in electrocution.

1. Make a thorough check of all circuitry for accuracy, loose connections, bad solder joints, correctly polarized components, short circuits, etc.

2. Open up the point shown at X by removing wire from R19 to pin 18 of T1 (Fig. 4-9).

3. Apply 6 volts to input from battery or other dc supply capable of 4 amps or more. Note operation of S1 and LA1 indicator lamp. Current draw should be less than 1 amp and power supply should produce an audible whine or pitch.

4. Connect an ac voltmeter across #1 and #18 of T1. Note a reading of about 700 to 1200 volts. This reading is not entirely accurate regarding absolute value due to shape of voltage waveform. (Most meters measure true rms only for sinusoidal waveforms.) If there is not an output, reverse connection #3 and #6 on T1 for proper phasing of feedback. If there is still a problem, recheck circuitry, transistor for heating, shorts, etc.

5. Turn input volts up to 12 and check waveshape at collector of Q1 and Q2. Note sketch of waveform in Fig. 4-9. Input current should be 1 amp.

6. *Reduce input to 6 volts* and reconnect point X. Monitor dc volts across Bank "A". Note input current now is over 2 amps and slowly decreases to about 1 amp as capacitor voltage builds up to 750 volts. Allow to charge for 2 minutes.

7. Discharge capacitor bank by connecting a 500 ohm 10 watt resistor across each of these several times. Finalize by shorting with a screwdriver. Always use the resistor *soft discharge* first or damage to the unit along with chunks of vaporized screwdriver will be flying around. Repeat for capacitor Bank "B".

8. Turn on S1 and observe meter to slowly rise indicating capacitors taking a charge. *Allow*

multimeter to just reach 200 volts and turn off S1. LA1 and LA3 should be glowing. Voltage should slowly drop to zero. Note how slowly these caps discharge and heed this as a warning. LA2 and LA3 will extinguish when capacitor bank voltage reaches about 120 V (still potentially dangerous). If it is desired to hard short the capacitors, do so through L1 to limit the current spike. Use a well insulated screwdriver with large clip leads, etc. Watch for magnetic reaction of ferrous material adjacent to field of L1.

9. Repeat Step 8. Quickly measure voltage across each storage capacitor (approximately 50 volts). These voltages must be all equal within 10% or resistor balancing will be necessary. However, resistor balancing robs power and slows charging rate. In some cases, shunting these capacitors with other capacitors of equal voltage may be feasible. Severe capacitor unbalance will produce a proportionally higher voltage across the smaller capacitor value, no doubt stressing it beyond the rating (BANG!).

10. Adjust calibration of M1 meter. Allow charge voltage to read 750 V as indicated by external meter. Adjust R16 so that M1 indicates ½ scale. Full scale should now indicate 1500 volts. Use care as 750 volts is present at the energy banks. R16 may be accessed through a hole allowing external adjusting for safety reasons.

11. Apply 12 volts and allow to charge for about 30 seconds. Push shorting switch S2 for about 15 seconds and note charging voltage speeding up. You will note here that one of the disadvantages of using the electrolytics is their property of retaining a charge memory. When using the system to maximum capacity it will be found that charging the banks in several steps steadily increasing the charge voltage after each discharge *will limit the charge time* necessary for maximum energy storage. Several successive charges and discharges may be necessary. This is only necessary as a warming up sequence. Also the point to speed up the charge by shorting S2 should be determined by experiment as different capacitors and other parameters may cause some variation here. We found in our lab unit that after three charge/discharge cycles that both

banks would reach the charge voltage of 1500 in about 60 seconds with shorting S2 at the 1000 volts point.

It is assumed that the power pack is operating correctly, transistor Q1, Q2 are not overheating and both capacitor banks equally charge to 1500 volts in a reasonable time (usually 1 to 2 minutes).

IGNITOR CIRCUIT MODULE (IGN1)

This circuit supplies a high voltage pulse of over 10 kV intended for preionizing the gas, initiating breakdown in a flash tube or between the discharge electrode in a spark gap switch. It can also be used to ignite gases and other flammable material (Figs. 4-10 and 4-21). The device is shown built as a separate module on a small 1⅝" piece of perfboard being placed in a plastic cap serving as a housing and potted with paraffin wax.

The circuit is nothing more than a simple capacitor C1 discharging into the primary inductance of pulse transformer T1. The HV output is the result of the high secondary to primary turns ration. The capacitor is switched via the silicon control rectifier SCR1 that in turn is triggered by the positive going pulses produced by UJT Q1. The pulse rate of Q1 is determined by the time constant of R7 and C2. Diode D3 recovers the negative portion of the discharge ringing pulses from the primary T1. The energy storage/dump capacitor C1 is charged through the resistor divider networks R1,R5 and isolation diodes D1 and D2. Note that only one charge line is required, however, when used in dual energy banks both lines are used for proper charge balance since energy is taken. One parameter that must be observed is that the time to charge up C1 is less than the time to charge up C2 in the UJT circuit. If this condition is reversed the SCR1 will switch C1 before it is fully charged reducing the output pulse.

Circuit layout is as shown and the only precaution is that T1 be physically located away from the other components to avoid breakdown as the output pulse can be in excess of 10 kV. The output lead of T1 is pin 4 and is removed from this base and brought out at the top of the assembly.

To test the unit involves positioning the HV output lead of T1 about ⅛" from the ground line and hooking up 12 volts to the UJT circuit and 200-400 volts to one of the input lines. Note a spark discharge occurring every second or so.

The entire tested perfboard assembly is placed inside of the plastic cap with the wire leads dressed and positioned for exiting as far away as possible from T1. Hot wax is now poured over the assembly with the output strain relieved in the hot wax. Lead should be several inches in length. Note pin 4 of T1 may be used for the output termination where the input voltage is 300 or less, therefore producing less potential stress.

LASER HEAD

The laser head section of your project is where most of the action takes place. It is where electrical energy is converted into spontaneous optical energy via the flash lamps and where this spontaneous optical energy is turned into stimulated emission via the optical properties of the laser rod. The laser system described in this project can generate relatively enormous amounts of power for a short period of time. Remembering from basic physics that power in (watts) = energy in (joules) × time in sec, therefore a joule = watt/sec. The power of our laser is in the thousands of watts, however, this power is in the form of a pulse where the duration time is in the order of a millisec or so. Solving for the energy per pulse, integrated over a period of one second now becomes ½ to 1 or 2 joules of energy.

Obviously when a laser of this type is quantitized in watts of power compared to joules of energy the figures are quite impressive. The pulse rate of this laser project should be about one shot per minute as there is no means for cooling the laser rod and hot rods will not work well and only get hotter causing total irreversible damage if this excessive heating is allowed to build up. Note that the flash lamps discharge energy is in the hundred of joules and will heat up to dangerous levels if not allowed to cool in-between pulses. For example, 500 joules of discharge energy results in nearly 500

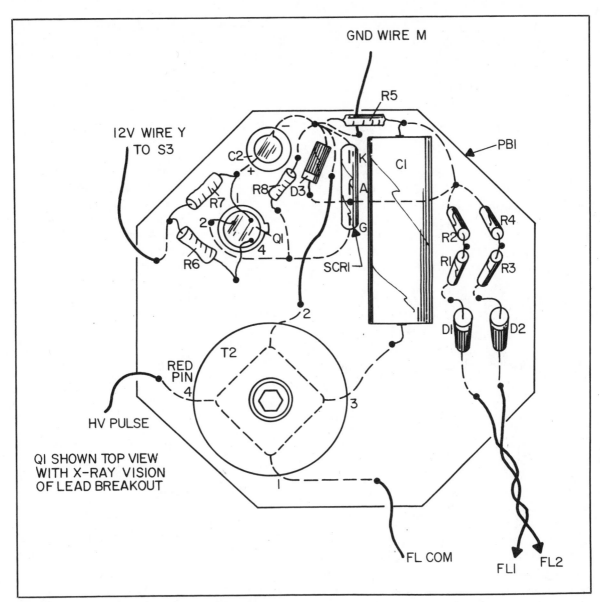

Fig. 4-21. IGN1 board layout. (Also see Fig. 4-10.)

watts being dissipated for one second inside of the cavity and this heat can cause temperature build up if not externally removed or sufficient time is allowed between these pulses.

The project is shown as a close optically coupled system which means simply that the laser rod is mounted as close to the flash lamps as possi- ble and the entire assembly carefully wrapped in a highly reflective thin metal foil that hopefully re- covers and returns most of the flash pulse to the rod. The metal foil also acts as a means of triggering the flash lamp via a spike of high voltage. This energy is electrically connected to the metal foil by a lead from the ignitor circuit described in this project and capacitively ionizes the flash lamps sufficient to sustain the flash voltage from the main energy storage capacitors.

CONSTRUCTION OF THE LASER HEAD (FIGS 4-22, 4-23, AND 4-24)

1. Verify proper operation of the *power pack section* (Fig. 4-20) and the *ignitor* (Figs. 4-10 and 4-21).

2. Fabricate TU1 from 8″ length of 3½″ OD Sked 40 PVC. Fabricate 1⅞″ hole at a slight angle for TU2 with center located 3″ from rear. Use a circle saw and firmly secure in drill press. See Fig. 4-24.

3. Fabricate TU2 handle from 6″ length of 1.9″ OD Sked 40 PVC. Note hole for trigger switch S3.

4. Fabricate plastic caps CA2, CA3 as shown. CA2 has a hole to accommodate BAR1 barrel (Fig. 4-22). CA3 is for retaining RP1 rear plate to TU1 and is fabricated by cutting out center with a sharp knife and using the walls of TU1 as a guide with the cap in place.

5. Fabricate HC1 from a 11″ × 2¾″ length of .035/20 gauge galvanized sheet or aluminum. Note dimension in Fig. 4-22. It must be rigid so that flash lamps will not distort and fracture therefore ¼″ folded lips along the length are necessary.

6. Fabricate RP1 rear plate from a 3¼ × 3¼ piece of .035/20 galvanized sheet or aluminum. Note hole for optional meter. Trim corners to make piece hexagonal as shown in Fig. 4-22.

7. Fabricate BAR1 barrel from a 3″ to 6″ length of copper tubing ½ to ¾ of inch in diameter. Position as shown so that tube is bore sighted and solder in place as shown in Fig. 4-22 as a final step.

8. Fabricate BLK1 and BLK2 from a 1.25″ × 1″ × ¾″ nylon, or preferably Teflon block as shown Fig. 4-23. Use exacto knife and file groove to nest 9/32 OD of PRC1,2. Mount as shown in Fig. 4-22 via screws SW12.

9. Position and drill holes for flash lamp clips FL1,2,3,4. These holes are for #4 sheet metal screws and must position flash lamps so that there is 9/32″ clearance between them. This allows the protective caps PRC1,2 just to clear or abut against

these tubes. There usually is enough clearance in the holes of the clips to allow some for this adjustment. The rod assembly also must nest into the grooves of BLK1,2.

10. Clip in ¼″ metal rods to simulate the flash lamps as sometimes they may not be properly aligned and this could cause lamp breakage. These rods also will aid in properly positioning these clips.

11. Connect wires W and X to proper clips and shape wires into position so that strain will be minimum when clips are snapped onto the flash tube ends. Wire Z is the common and is connected to the two clips as shown in Fig. 4-23. (This may be several looped pieces of buss wire.)

12. Connect the remaining circuit as shown Fig. 4-22. Position dress and strain relieve wires via a nylon clamp or similar retaining method.

13. Very carefully insert the laser rod into protective caps PRC1,2 as shown Fig. 4-23. Only insert until secure. Very lightly dab some RTV at several points. Be very careful not to scratch the laser rod. Remove any burrs from the caps. Note output end of laser rod when positioning in next step.

14. Cut a strip of silver (must be clean) or other highly reflective foil 2½″ width. Wrap around lamps and laser rod as shown in Fig. 4-23. Foil wraps also retain laser rod assembly. Be neat and wrap several layers with last layer on top being used for sandwiching and connecting the ignitor HV pulse output wire (Fig. 4-22). This lead must be short, direct, and properly insulated.

15. Complete the final assembly as shown Fig. 4-24.

CAUTION—CAUTION—CAUTION!

At this point it is necessary to wear your laser safety glasses as reflections both direct and indirect can cause eye damage. Obtain safety glasses from Glendale Optical Co., Woodbury L.I. N.Y. 11797. 1-516-921-5800. Make sure you specify the wavelength of your laser rod to obtain the right glasses.

16. Place a piece of carbon paper at output end of laser.

17. Turn on S1 and charge capacitor bank as in

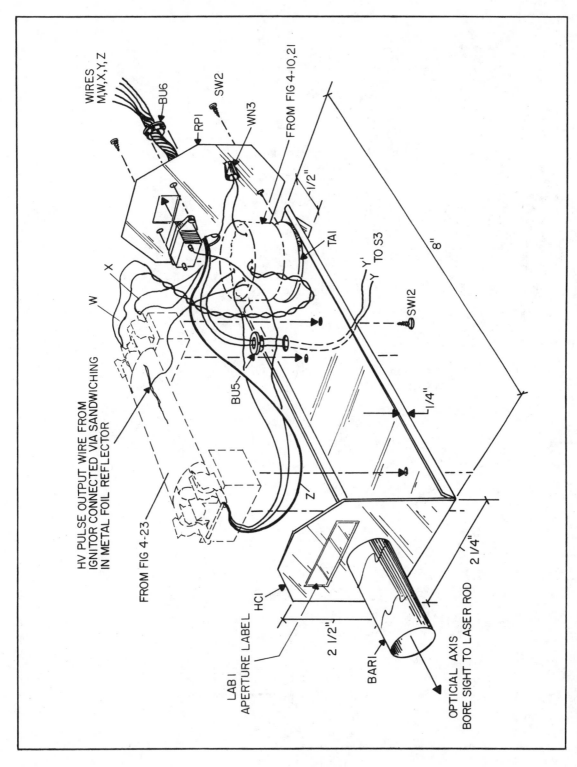

WIRES M,W,X,Y,Z

BU6

RP1

SW2

WN3

FROM FIG 4-10, 21

1/2"

TA1

Y'
Y TO S3

SW12

8"

X

W

1/4"

BU5

HV PULSE OUTPUT WIRE FROM IGNITOR CONNECTED VIA SANDWICHING IN METAL FOIL REFLECTOR

Z

FROM FIG 4-23

2 1/4"

HC1

LAB1 APERTURE LABEL

2 1/2"

BAR1

OPTICIAL AXIS BORE SIGHT TO LASER ROD

Fig. 4-22. Head subassembly.

73

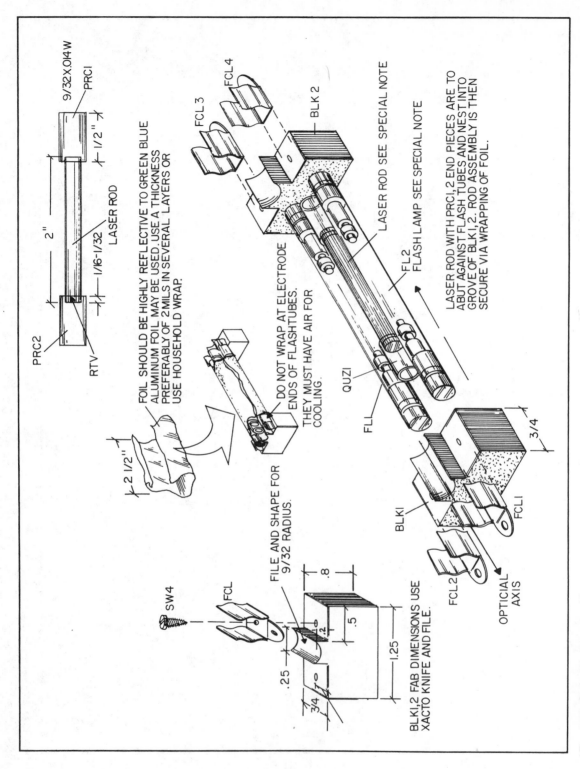

PRC1

9/32 × .014 W

PRC2

RTV

2"

1/2"

1/16-1/32

LASER ROD

FOIL SHOULD BE HIGHLY REFLECTIVE TO GREEN BLUE ALUMINUM FOIL MAY BE USED. USE A THICKNESS PREFERABLY OF 2 MILS IN SEVERAL LAYERS OR USE HOUSEHOLD WRAP.

DO NOT WRAP AT ELECTRODE ENDS OF FLASHTUBES. THEY MUST HAVE AIR FOR COOLING.

2 1/2"

FILE AND SHAPE FOR 9/32 RADIUS.

FCL 3

FCL 4

BLK 2

LASER ROD SEE SPECIAL NOTE

FL 2
FLASH LAMP SEE SPECIAL NOTE

LASER ROD WITH PRC1, 2 END PIECES ARE TO ABUT AGAINST FLASH TUBES AND NEST INTO GROVE OF BLK1,2. ROD ASSEMBLY IS THEN SECURE VIA WRAPPING OF FOIL.

QUZI

FLL

3/4

BLK1

FCL1

FCL2

OPTICIAL AXIS

SW4

FCL

.8

.5

.2

1.25

3/4

.25

BLK1,2 FAB DIMENSIONS USE XACTO KNIFE AND FILE.

Fig. 4-23. Optical cavity assembly.

74

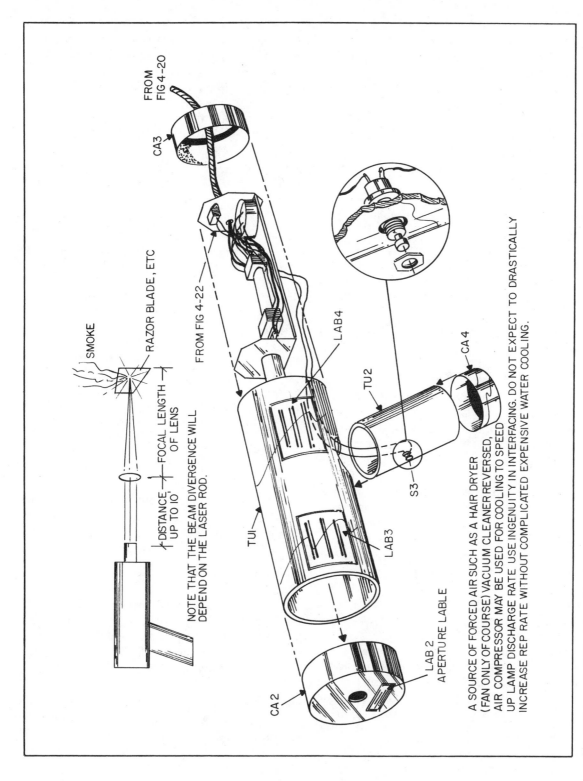

SMOKE

RAZOR BLADE, ETC

DISTANCE
UP TO 10'

FOCAL LENGTH
OF LENS

NOTE THAT THE BEAM DIVERGENCE WILL
DEPEND ON THE LASER ROD.

FROM
FIG 4-20

CA3

FROM FIG 4-22

LAB4

TU2

CA 4

S3

TUI

LAB3

CA 2

LAB 2
APERTURE LABLE

A SOURCE OF FORCED AIR SUCH AS A HAIR DRYER
(FAN ONLY OF COURSE) VACUUM CLEANER REVERSED,
AIR COMPRESSOR MAY BE USED FOR COOLING TO SPEED
UP LAMP DISCHARGE RATE USE INGENUITY IN INTERFACING. DO NOT EXPECT TO DRASTICALLY
INCREASE REP RATE WITHOUT COMPLICATED EXPENSIVE WATER COOLING.

Fig. 4-24. Laser head final assembly.

step 11 of the Test and Adjust section. Allow energy bank to charge up to 1000 volts and trigger the system. Note a loud bang and flash occurring. Check carbon paper for burns.

18. Allow several minutes and let energy bank charge to 1500 volts. Trigger and note laser burning a spot in carbon paper about the size of the rod. You may want to use a hair dryer or other source of cool air to speed up cooling the system between charges. Remember a hot laser rod doesn't like to lase.

19. Trigger discharge at various levels of lamp voltage and note lasing threshold point. This may vary with temperature.

20. In order to get your laser to burn holes, etc., a simple lens is required. This can be a conventional glass lens of a focal length 24″. See Fig. 4-24.

21. A good demonstration can be shown by placing a red balloon inside a clear one and puncturing it without apparent damage to the outer clear one.

SPECIAL NOTE ON FLASHLAMPS

The flash tubes in your laser provide the highly energetic optical pulse necessary to cause the lasing action in the rod to take place. It is important that the duration of the flash along with the spectral distribution of the output be taken into consideration for optimizing system output power. A ruby rod will favor an exciting pulse in the greenish/bluish end of the spectrum while a Yag rod will favor the reddish end. This wavelength distribution can be optimized by selecting certain operating parameters for the flash lamp regarding the combination of capacitance and discharge voltage relations.

The first consideration when adjusting parameters for optimizing output is that the explosion rating of the flashlamp is not exceeded. This parameter is a function of the discharge energy per inch of arc length and the pulse duration of this discharge. This relation depends on the particular tube used and is usually stated on the spec sheet. For the tubes we use here, (FX103C by ECG) we find this value to be 800 joules if the pulse length is 1 ms, dropping to 250 joules for a pulse length of 100 μsec.

Useful life should be no more than 30% of these ratings which allows somewhere between 10^3 to 10^4 flashes. The ruby rod as mentioned favors the flash energy to be in the greenish/bluish part of the spectrum. This means that the flash pulse should contain maximum current as the wavelength becomes shorter with increasing current density in the discharge. This means that for the given discharge energy, we use a relatively higher voltage on a lower value capacitor. The yag rod now favors the flash energy to be towards the reddish end of the spectrum. This means less current density or higher capacitance and lower voltage for the equivalent discharge energy pulse. Please note that we have experienced two extremes only to emphasize the advantage of proper operating parameters for your choice of applications. Ultimate selection should be determined with the help of the flash tube spec sheets and charts. The duration of the flash discharge requires an inductor used in series with the energy storage capacitors creating a discharge time equivalent to $2.5 \sqrt{LC}$ where L is the inductance in microhenrys and C is the capacity in microfarads. If we were to discharge the capacitor into the flash lamp without some current limiting impedance the current and duration would be determined by the low negative resistance of the flash lamp combined with any internal resistance in the capacitor. This would normally result in a very high current occurring in a short period of time no doubt exceeding the ratings of the flash lamp causing explosion. To select this inductor involves determining the proper voltage and capacity for the discharge circuit and simply using the formula

$$L = \frac{t^2}{6.25C}$$

It is suggested that the builder obtain ECG Data Sheet F1002C2 if he intends to deviate much because this is a simple comprehensive study on flash lamps with all the specification and charts that would be required.

LASER RODS

Your system is shown using the visible red ruby rod. This particular type was the laser that pioneered the field. The rod is made by first growing a crystal boule formed from molten aluminum oxide (Al_2O_3) and doped with the proper concentration of chromium (Cr). There are several methods of producing these boules and all processes may be used depending on cost and optical quality desired. The boule is machined into the required sized rod. The ends are either antireflective coated or may have integral dielectric mirrors eliminating the need for outside optics. The system shown in Figs. 4-22, 4-23, and 4-24 are intended for a laser rod with the integral mirrors coated on the ends.

Rods without the integral mirror require properly positioned external optics with a mechanically stable system for proper maintenance of their adjustment. An advantage to external mirrors is that damage is less frequent therefore they are considered more for higher energy systems. It has been reported that the mirrors used for helium-neon lassers may be substituted because the duty cycle of this unit is very low.

Performance of your system will depend on the type and quality of the rod used. Do not expect to obtain the beam quality of gas lasers. There is no way we can dictate the exact parameters that you may obtain from your system. The flashlamp energy of 1500 volts is 600 joules and using a conservative rating of flashlamp to output energy it is possible to obtain up to a joule of optical energy output. Note that this laser head may be lengthened to accommodate a three-inch rod rather than the two-inch rod shown. This will no doubt greatly increase output but will require the three-inch version of the flashlamps.

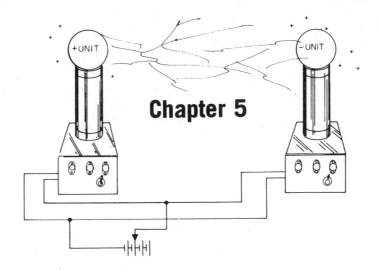

Chapter 5

High-Power
Continuous IR CO₂ Laser (LC5)

These plans describe a moderately powered CO_2 laser capable of producing a beam of heat at a wavelength of 10.6 microns with a beam energy of 30 or more continuous watts. This energy level is capable of burning, cutting, fabbing and welding thin sheet metal, plastics, wood, paper or just about any material that will absorb this wavelength.

The device is an axial flow device consisting of approximately 30″ active discharge length with a surrounding cooling jacket of water (see Fig. 5-1). Mirrors are internal cavity mounted on a flexible plate joined to the system via metal bellows. This method eliminates output windows and provides a convenient means of adjustment.

Two part construction consists of the laser head section connected to the support section via the necessary gas, vacuum, and high-voltage feed wires. Construction is straightforward with glassblowing and other specialized construction techniques minimized. The approach produces a stable reliable device with no frills or extras (Tables 5-1 through 5-4).

THEORY OF OPERATION

A carbon dioxide laser is by far one of the simplest lasing devices to assemble and operate. This asset is sometimes a disadvantage as this energy can be very dangerous in the hands of the inexperienced, both to the builder and his surroundings.

A CO_2 laser is a two level, vibrational device emitting in the 10.6 micron infrared region. Lasing is accomplished by electrically exciting nitrogen gas N_2 to an energy level close to that of the CO_2 molecule (see Fig. 5-2). The main dissipation of this energy is in the resonant transfer to the CO_2 molecules causing them to change from ground state to excited state 2. You will note that the lower laser level 3 is not prone to this transfer of energy, hence a population inversion now exists between levels 2 and 3. As the laser transition commences as a result of the stimulated energy transition from level 2 to 3, a means of returning level 3 energy to ground level is accomplished by quenching with a third gas, helium. Obviously, if level 3 was allowed

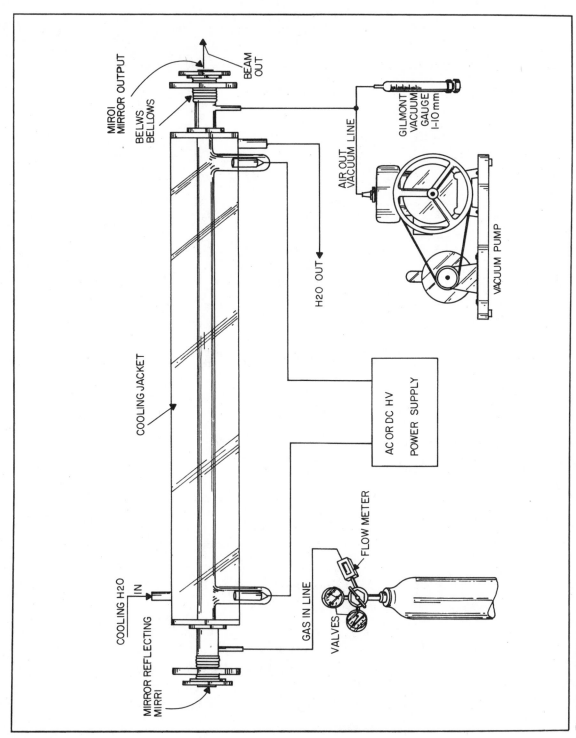

MIROI MIRROR OUTPUT

BELWS BELLOWS

BEAM OUT

MIRROR REFLECTING MIRRI

COOLING H2O IN

COOLING JACKET

H2O OUT

GILMONT VACUUM GAUGE 1-10 mm

AIR OUT VACUUM LINE

VACUUM PUMP

AC OR DC HV POWER SUPPLY

FLOW METER

GAS IN LINE

VALVES

Fig. 5-1. Laser diagram.

79

Table 5-1. CO$_2$ Laser Head Parts List.

DITU1	(1)	30″ length 3/4 OD × 1/16 wall pyrex tubing
ELTI,2	(1)	200 mA electrodes in pyrex 3 1/2 × 3/4 OD
LC5D	(1)	Assembled plasma discharge tube Fig. 5-6, includes above parts
LAWAS1	(2)	3 1/2 OD × 1 5/8 ID Large metal washer
ADR1	(2)	Fabbed above LAWAS per Fig. 5-7A Adj. rings
BRWAS	(4)	2 1/4 OD × 15/16 ID × 1/8 brass washer
MTGW1	(2)	Fabbed above BRWAS Fig. 5-7B mounting washer
MMT1	(2)	Fabbed above BRWAS per Fig. 5-7C mirror mount
ASW1	(6)	Screws 6 40 × 2″ fine thread, stainless grind to point
CPLS1	(2)	1″ long × 1 3/8 OD × 3/64 wall copper tubing fab per Fig. 5-7D
HNPS1	(2)	3/8 × 1 1/2″ copper tubing for hose nipples
BELWS	(2)	Bellows plated
ABP1	(2)	Abutting plates 4 × 4 × 1/2″ PVC with 1 3/4″ hole in center
ASMIM	(2)	Assembled mirror mounts as per Fig. 5-7E ready to use all soldered. Included above parts Line 4 to 13
ENP1	(2)	End plate 4 × 4 × 1/2″ PVC with 3/4″ hole in center
FM1	(4)	46″ length of 1 × 1 × 1 1/8″ A1 angle Fig. 5-10
SID1	(2)	29 × 3 1/2 × 1/4 Plexiglas/acrylic Fig. 5-11
TOBO1	(2)	29 × 4 × 1/4 Plexiglas Fig. 5-11
DN1	(2)	Hose to 3/8″ thread nipples brass
SCREWS	(72)	Screws 6 32 × 1/2″ flat head, Hardware Store
MIRO1	(1)	Output mirror ZnSe 80% plano/plano 25 mm
MIRR1	(1)	Reflector mirror Cu 98.5% plano/conv 10 meters 25 mm These are high powered CO$_2$ mirrors
PC1	(1)	Protective cover 36 × 4 × 4 1/4 #22 aluminum
BU1,2	(2)	Bushings strain relief
LAB1	(2)	HV Label
LAB2	(1)	Class IV Laser Invisible
LAB3	(1)	Aperture RTV Silicon #118, Hardware Store

Above items are available individually or as a complete parts package. Write Information Unlimited, Inc., P.O. Box 716, Amherst, N.H. 03031 or call 603-673-4730.

Table 5-2. Power Supply.

CO1	(1)	14/3 power cord
S1,2,3	(3)	dpdt 10 amp switch
LA1	(1)	115 V Ind Lite
VA1	(1)	7.5 A Variac
T1	(1)	9000 V 60 mA transformer
M1	(1)	10 amp ac meter
HVWIRE		7mm Ignition wire, Automotive Store
J1		Receptacle and plate. Hardware Store
CASE		Fab or chassis case
#14 HOOK-UP WIRE		#14 hook-up wire, Hardware or Electrical Store

Above items may be purchased from Information Unlimited.

Table 5-3. Required Support Equipment.

VGAUG	0-20mm (torr) Gilmont Vacuum Gauge - make a stand for (or use a 0-100 capsule gauge)
VACPMP	#1400 Rebuilt 2 stage vacuum pump
	#1399 Rebuilt 1 stage vacuum pump
VACHOSE	5/16ID Rubber vacuum hose
STOPCOK	Suitable for vacuum shutoff
REGUL	#TSA-751-320 two stage regulator
FLWVAL	#204-4421 flow meter and valve
GASCYL	CO_2 Laser mix gas cylinder
CLAMPS	Hose clamps, Hardware Store
"T"FITTING	3/8 Copper "T"

All the above items are available as individual parts or subassemblies, write Information Unlimited, P.O. Box 716, Amherst, N.H. 03031, or call 603-673-4730 for price and delivery.

to build up, the population inversion would soon be reduced between 2 and 3, hence terminating laser action.

It should be noted that the CO_2 gas molecule consists of many modes of vibration that contribute to the laser frequency. The main mode of vibration is around 10.6 microns and the system is often categorized in respect to wavelength as such. The earth's atmosphere offers a natural window for this wavelength, consequently a CO_2 laser is excellent in respect to application requiring propagation over long distances.

The laser described in these plans is referred to as a "low flow axial device". This is the simplest approach and is nothing more than an airtight glass discharge tube with gas being injected in one end

Table 5-4. Optional Items Available through Information Unlimited,Inc., Write or Call.

GBK3   Glassblowing kit—contains all the paraphernalia necessary to
              fabricate glass. A definite asset to the serious laser experimenter.

HNE1   Helium-neon laser—a definite asset for aligning mirror and other
              optical system.

VACPMP Vacuum pump—this vacuum system can be used for many laboratory
              functions. Pumps 1 CFM down to 20 microns.

Leak detector—electronic devices fingers out ultra small leaks. A must for those working into the fraction of microns.

Laser detection solution—detects leaks in oxygen and compressed gas systems.

Optical Systems for CO_2 laser are expensive and can be involved. We suggest the following sources of these specialized components for the particular operation you wish the laser to perform.

ADOLPH MELLER
P.O. Box 6001
Providence, RI 02904

HARSHAW CHEM CO
6801 Cochran Rd.
Solon, OH 44139

II-VI INC
Saxonburg Blvd.
Saxonburg, PA 16056

PTR OPTICS
145 Newton St.
Waltham, MA 02154 1-617-891-600

A GaAs Miniscus lens allows focusing the beam down for more precise and higher temperature work. These lenses come in a variety of diameter and focal lengths and are available through PTR OPTICS, 145 Newton St., Waltham, MA 02154. Ask for Reinhard Erdmann as he will assist you in determining the correct parameters for your application.

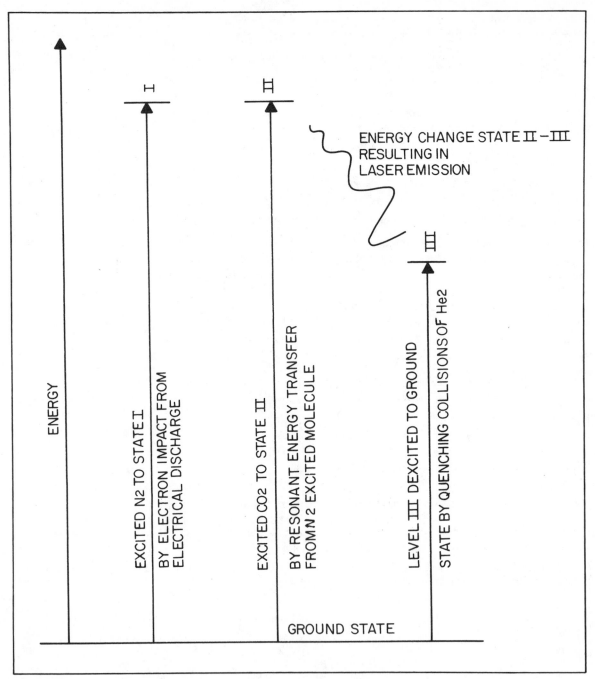

Fig. 5-2. Energy level diagram.

and exiting the other via a vacuum pump. Electrodes now placed into the flowing gas excite the nitrogen molecules commencing laser action.

CONSTRUCTION SUGGESTIONS

Note that we have used standard readily available parts and pieces wherever possible while still providing reasonably reliable operation. It is suggested the builder closely follow the plans and acquire through Information Unlimited, Inc. any parts he may not wish to fabricate or has problems locating.

Your CO_2 laser will be described in two sections: I. Laser head and II. Power supply and support equipment.

The laser head section is where the action takes place. It is where the beam exits, and the lasing phenomena occurs. Construction is started by fabricating the glass discharge tube with the builder having three choices regarding this step.

☐ He may attempt to master glassblowing and build it himself. A person interested in constructing lasers should seriously consider this craft along with acquiring the associated equipment necessary. See the parts list (Tables 5-1 through 5-4).

☐ He may purchase the necessary glass pieces and contract an experienced glassblower such as those that do work in neon advertising signs.

☐ Or he may purchase the ready made discharge tube from Information Unlimited, Inc. as described in the parts list of these plans.

If he should attempt the tube himself we offer the following basic glassblowing hints. It is also suggested that he acquire a manual or preferably a glassblowing kit available through Information Unlimited, Inc. This kit contains all the equipment and instructions necessary to do this project and of course *much much* more.

DISCHARGE TUBE
GLASSBLOWING CONSTRUCTION STEPS

1. Clean the glass with hot water with a softener added. Hands must be kept clean as fingerprints will easily burn in and dirty glass creates problems.

2. Refer to Fig. 5-3A. Cutting glass involves making a deep scratch using considerable pressure with a file where the break is to occur. Make the scratch with one pass. Wet the scratch and break tube with hands as shown in Fig. 5-3B. This method

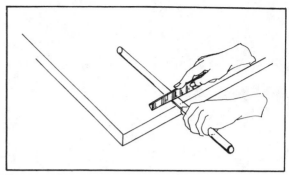
Fig. 5-3A. File nick.

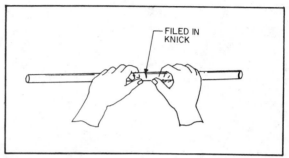
Fig. 5-3B. Break glass along nick.

is usually good for diameters up to 1″ and therefore suffices for these plans.

3. Learn to make a "T" seal. This is necessary for the joining of the premade glass electrodes to the main plasma tube. These seals are made by heating the desired location on the main tube using a sharp flame positioned so that the top edge of the tube is just beneath the flame. After heating gently blow a bulge that is slightly smaller in diameter than the tube which is to be sealed. See Fig. 5-4A, B, and C. Reheat the top of this bulge and blow it away as in Fig. 5-4C. Now, hold this tube in the left hand between thumb and just three fingers so that the hole is to the left of the flame. Take the electrode tubing in the right hand and heat it uniformly until the end is soft and just before joining the two, heat the other opening uniformly. See Fig. 5-4D.

Lift out of the fire and press the two together quickly, pulling slightly as soon as complete contact has been made and blow slightly to expand heated area. If both edges are soft enough they will flow together upon contact and the small amount of

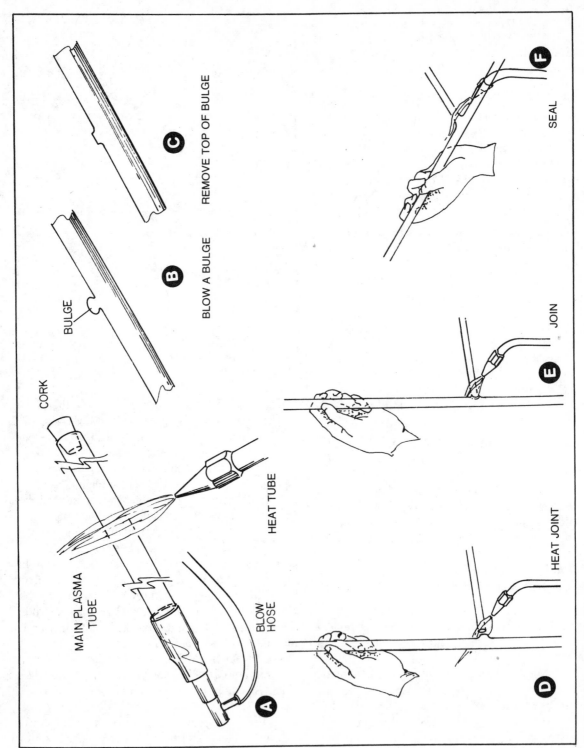

CORK

MAIN PLASMA TUBE

BULGE

HEAT TUBE

BLOW HOSE

A

B BLOW A BULGE

C REMOVE TOP OF BULGE

D HEAT JOINT

E JOIN

F SEAL

Fig. 5-4. Joining two glass tubes.

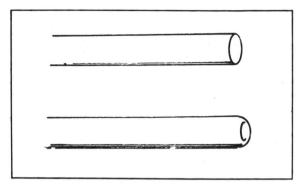

Fig. 5-5. Polished ends.

blowing and pulling will give you a nice smooth seal as in Fig. 5-4E. If, however, you find your seal heavy and uneven in wall thickness, heat one side at a time and blow gently only after the glass is sufficiently soft enough to work out the uneven area (Fig. 5-4F).

4. Fire polishing and end finishing is simply accomplished by rotating the end of the cut tube in the flame until it is smoothed out by the surface tension (Fig. 5-5).

This is a very, very brief section on glass handling and again it is strongly suggested that the laser builder seriously considers learning this useful art as most all types of lasers involve glass handling in one way or another. The builder may wish to attempt the

above steps several times until he feels proficient. Glass is not expensive and practice is advised.

CONSTRUCTION STEPS FOR THE LASER HEAD

Step 1. Construct or acquire the above *Plasma Discharge Tube* as shown in Fig. 5-6.

Step 2. Fabricate two *adjustment rings* ADR1 as shown Fig. 5-7A. These are made from large 3½ OD × 1⅝ ID steel washers LAWAS1. Tapped holes should be snug on adjustment screws ASW1.

Step 3. Fabricate two *mounting washers* MTGW1 as shown in Fig. 5-7B. These are fabbed from brass washers BRWAS.

Step 4. Fabricate two *mirror mounts* MMT1 as shown Fig. 5-7C. These are also fabbed from 2 brass washers BRWAS. It might be more convenient to have this step done by a machine shop for the 1/16″ ridge.

Step 5. Fabricate two *coupling sections* CPLS1 as shown Fig. 5-7D. These are made from 1″ length of 1⅛ OD × 3/64 wall copper tubing with a ⅜ hole for nipple sections. Clean pieces so they shine like a mirror.

Step 6. Fabricate two *hose nipple* sections HNPS1 as shown Fig. 5-7E. Clean them so they shine like a mirror.

Step 7. Insert HNPS1 nipples into coupling

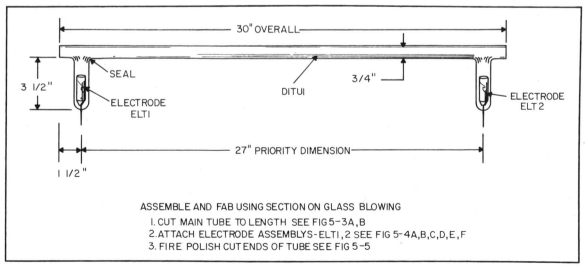

30" OVERALL

SEAL

ELECTRODE
ELTI

3 1/2"

DITUI

3/4"

ELECTRODE
ELT 2

27" PRIORITY DIMENSION

I 1/2"

ASSEMBLE AND FAB USING SECTION ON GLASS BLOWING
    I. CUT MAIN TUBE TO LENGTH SEE FIG 5-3A,B
    2.ATTACH ELECTRODE ASSEMBLYS-ELTI,2 SEE FIG 5-4A,B,C,D,E,F
    3. FIRE POLISH CUT ENDS OF TUBE SEE FIG 5-5

Fig. 5-6. Plasma discharge tube LC5D.

85

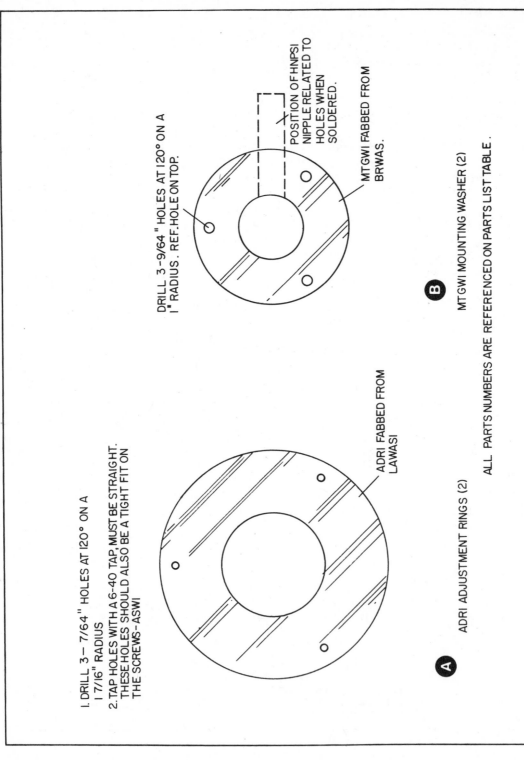

1. DRILL 3 – 7/64" HOLES AT 120° ON A
1 7/16" RADIUS

2. TAP HOLES WITH A 6-40 TAP, MUST BE STRAIGHT.
THESE HOLES SHOULD ALSO BE A TIGHT FIT ON
THE SCREWS-ASWI

DRILL 3 - 9/64" HOLES AT 120° ON A
1" RADIUS. REF. HOLE ON TOP.

POSITION OF HNPSI
NIPPLE RELATED TO
HOLES WHEN
SOLDERED.

MTGWI FABBED FROM
BRWAS.

ADRI FABBED FROM
LAWASI

MT GWI MOUNTING WASHER (2)

ADRI ADJUSTMENT RINGS (2)

ALL PARTS NUMBERS ARE REFERENCED ON PARTS LIST TABLE.

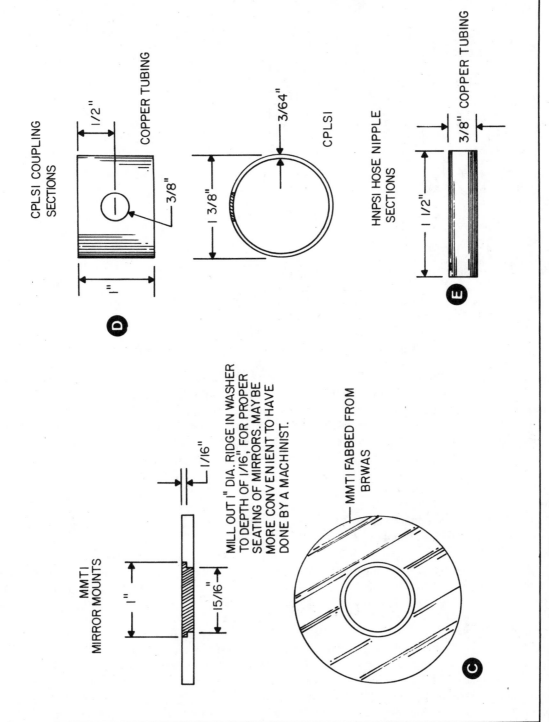

COPPER TUBING

1/2"

CPLSI COUPLING
SECTIONS

3/8"

1 3/8"

3/64"

CPLSI

D

HNPSI HOSE NIPPLE
SECTIONS

3/8" COPPER TUBING

1 1/2"

E

MMTI
MIRROR MOUNTS

1"

15/16"

1/16"

MILL OUT 1" DIA. RIDGE IN WASHER
TO DEPTH OF 1/16", FOR PROPER
SEATING OF MIRRORS. MAY BE
MORE CONVENIENT TO HAVE
DONE BY A MACHINIST.

MMTI FABBED FROM
BRWAS

C

Fig. 5-7. Adjustment rings and mounting washers.

87

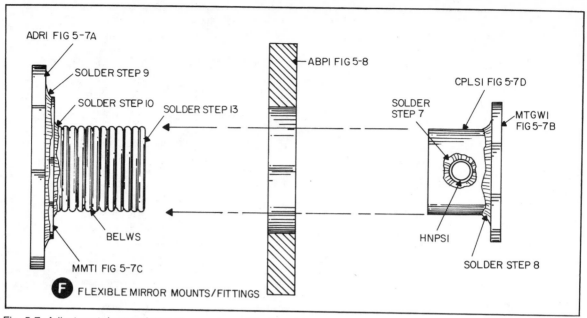

ADR1 FIG 5-7A

SOLDER STEP 9

SOLDER STEP 10   SOLDER STEP 13

ABP1 FIG 5-8

SOLDER
STEP 7

CPLS1 FIG 5-7D

MTGW1
FIG 5-7B

BELWS

HNPS1

MMT1 FIG 5-7C

SOLDER STEP 8

F FLEXIBLE MIRROR MOUNTS/FITTINGS

Fig. 5-7. Adjustment rings and mounting washers. (Continued from page 87.)

section CPLS1 and solder. Use plumbing techniques and make sure joint is *perfect*.

Step 8. Solder coupling sections CPLS1 to mounting washer MTGW1. Note Fig. 5-7E and 5-7B. These parts must be clean. Use plumbing techniques and make sure that the seal is *perfect*.

Step 9. Solder mirror mounts MMT1 to adjustment rings ADR1. Make sure that milled out mirror recess in MMT1 is on outward side and pieces are centered relative to one another.

Step 10. Solder bellows BELWS to above assembly as per Fig. 5-7F. *Joint must be centered and perfect. Do not* solder or connect adjustment ring and bellows sections to CPLS1 coupling section at this point.

Step 11. Fabricate two abutting plates ABP1 as shown Fig. 5-8. Dimensions are critical for proper alignment. Note locating holes for abutting against conical end of adjust screws ASW1.

Step 12. Fabricate two end plates ENP1 as shown Fig. 5-9. The correct dimensions are critical for proper fit.

Step 13. Place ABD1 abutting plate over bellows sections (Fig. 5-7E). Solder CPLS1 section to other end of bellows. Use a metal plate as a heat

shield to protect the PVC pieces from the flame of the torch when soldering these sections together. The flexible mirror mount and filter are now assembled and contain the butting plate ABP1 sandwiched between the adjusting ring ADR1 and the CPLS1 and HNPS1 coupler/nipple assembly. These solder joints must be *better than perfect* and all attempts to center and properly align the individual parts must result in near perfection. Remember solder is forgiving and can be redone should you foul up this assembly.

Step 14. Fabricate four frame members FM1 46″ long from 1 × 1 × ⅛ square aluminum angle Fig. 5-10.

Step 15. Fabricate Plexiglas sides SID1 29 × ½ × ¼″ (Figs. 5-11 and 5-12). Leave sticky paper on and be careful not to crack these pieces. They must be squared. Drill and tap holes for water drain nipples DN1, 2 in the side pieces (Figs. 5-13 and 5-14). Locate at bottom edge on one end and top edge on other (must be at opposite ends). Locate adjacent to electrodes in discharge tube and make sure there is clearance for frame members.

Step 16. Fabricate Plexiglas top and bottom TOP1, BOT1 29 × 4 × ¼″. Note holes for electrode

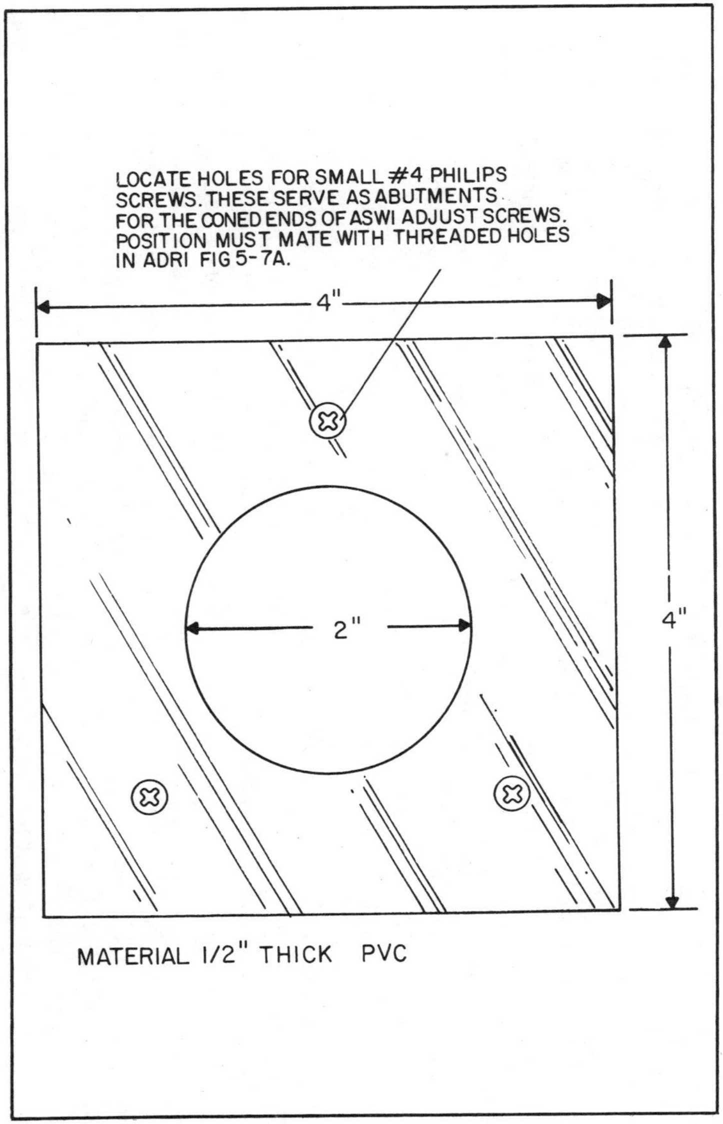

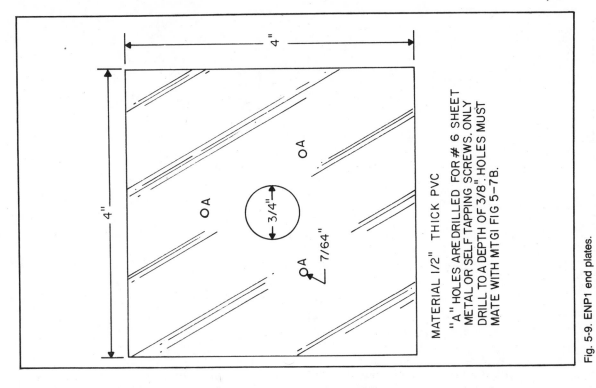

LOCATE HOLES FOR SMALL #4 PHILIPS SCREWS. THESE SERVE AS ABUTMENTS FOR THE CONED ENDS OF ASWI ADJUST SCREWS. POSITION MUST MATE WITH THREADED HOLES IN ADRI FIG 5-7A.
4"
4"
2"
MATERIAL 1/2" THICK PVC

Fig. 5-8. ABP1 abutting plates.

4"
4"
3/4"
A
A
A
7/64"
MATERIAL 1/2" THICK PVC
"A" HOLES ARE DRILLED FOR # 6 SHEET METAL OR SELF TAPPING SCREWS. ONLY DRILL TO A DEPTH OF 3/8". HOLES MUST MATE WITH MTGI FIG 5-7B.

Fig. 5-9. ENP1 end plates.

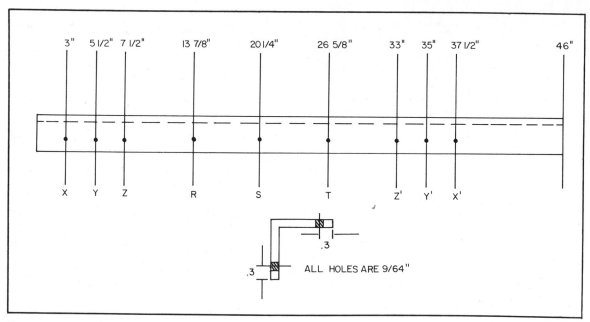

Fig. 5-10. FMI A, B, C, D, frame members (4).

EL1, 2 in top piece TOP1 (fabricate as per Fig. 5-11). It must be squared. Leave on sticky paper until larger holes are drilled and tapped. Surfaces or the Plexiglas and PVC should be abraded with sandpaper before applying sealant or adhesive.

Step 17. Lay out two bottom frame members FM1A and B onto a flat surface. Remove sticky paper and locate midpoint of BOT1 bottom piece of Plexiglas and align in FM1 with holes "S" (Fig. 5-10). Position and abut ENP1 and ENP2 end pieces with bottom piece onto the two frame members. Note holes Y and Y' on FM1 frame members aligning at middle of end pieces ENP1 and ENP2 when midpoint of bottom piece aligns with hole S. See Fig. 5-12. Drill and tap remaining holes for 6-32 screws. Make sure all pieces are fully in place. Temporarily screw together.

Step 18. Position two side pieces as shown Figs. 5-11, 5-12, and 5-13. Drill and tap remaining holes in ENP1/2, SID1/2, BOT1 pieces so they mate with holes in FM1A and B. Screw in place. Apply RTV to threads and "finger in" for water sealing. Do not install ENP2 permanently.

Step 19. Remove ENP2 end piece and proceed to seal all the abutting seams and screws of the system. Use clear RTV silicon and "finger in" for a good water tight seal. See Fig. 5-11 on abrading edges.

Step 20. Carefully position the plasma discharge tube DITU1 as you reinstall the removed end piece. Seal where the tube goes into the end pieces with RTV. These seals must be *perfect* or you will end up with water in the discharge tube. "Finger in" the RTV neatly and evenly.

Step 21. Carefully mark the top piece TOP1 for the electrode holes from the discharge tube. These should be slightly oversized. Drill carefully and do not crack the Plexiglas.

Step 22. Test fill with water and double check that all seams and screws are sealed against leakage or you will surely have problems. You may also want to seal all seams from the outside.

Step 23. Very carefully position the TOP1 top piece over the electrodes of the discharge tubes and make certain there is proper clearance when properly in place as shown in Fig. 5-14. Position frame members FM1C and D and screw to end pieces ENP1, 2 as in *Step 17*. Very carefully drill and tap remaining holes in TOP1 top piece. Screw together using RTV on threads. Please note at this point that

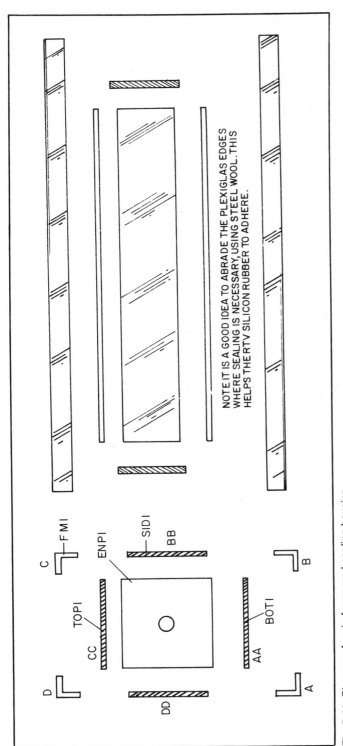

NOTE IT IS A GOOD IDEA TO ABRADE THE PLEXIGLAS EDGES WHERE SEALING IS NECESSARY, USING STEEL WOOL. THIS HELPS THE RTV SILICON RUBBER TO ADHERE.

Fig. 5-11. Blow-up of main frame and cooling housing.

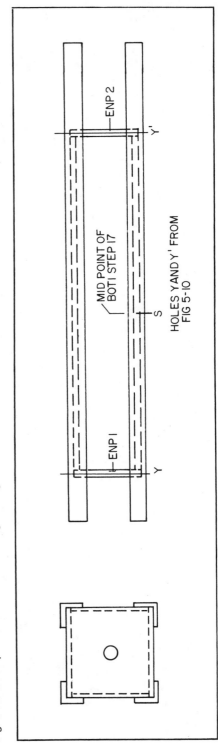

Fig. 5-12. Assembled main frame and cooling housing.

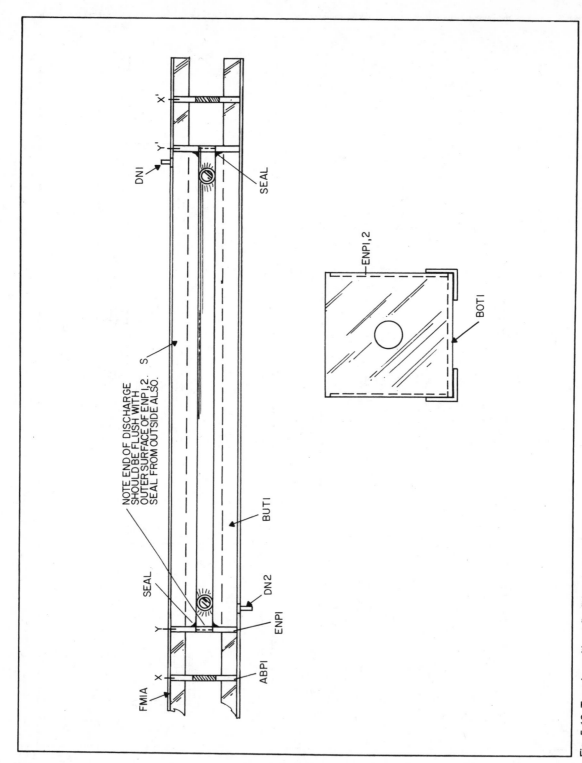

NOTE END OF DISCHARGE
SHOULD BE FLUSH WITH
OUTER SURFACE OF ENP1,2.
SEAL FROM OUTSIDE ALSO.

Fig. 5-13. Top view without flexible mirror mounts.

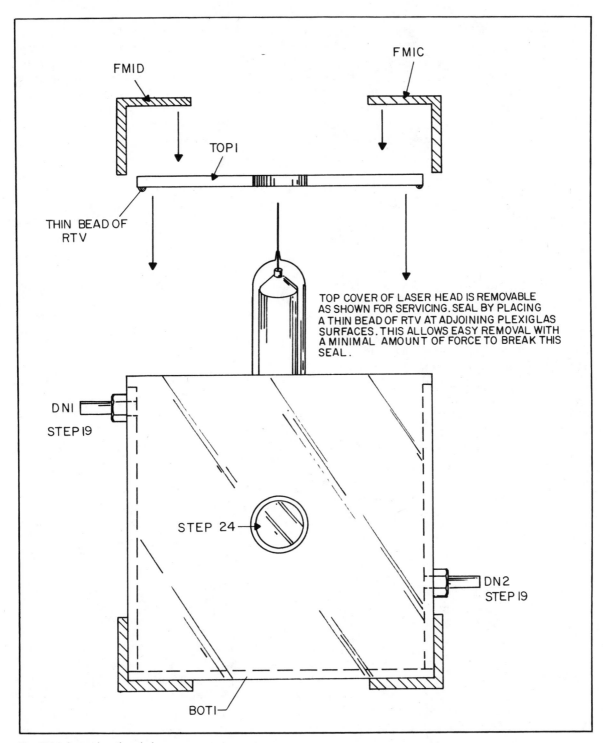

FMID

FMIC

TOP1

THIN BEAD OF
RTV

TOP COVER OF LASER HEAD IS REMOVABLE
AS SHOWN FOR SERVICING. SEAL BY PLACING
A THIN BEAD OF RTV AT ADJOINING PLEXIGLAS
SURFACES. THIS ALLOWS EASY REMOVAL WITH
A MINIMAL AMOUNT OF FORCE TO BREAK THIS
SEAL.

DN1
STEP 19

STEP 24

DN2
STEP 19

BOT1

Fig. 5-14. Laser head end view.

the top section consisting of FM1C, D and TOP1 are not water sealed to the main assembly as it may be necessary to remove the top section in the preliminary testing. The *top* screws in TOP1 may be sealed.

Step 24. Carefully seal the discharge tube to the end plates from the *outside* by literally "fingering in" a seal of RTV. Try not to get any inside of the tube. The edges of the hole in the end pieces should be abraded.

Step 25. Screw and seal assembly from Fig. 5-7E and Fig. 5-15 to end plates ENP1, 2. Make sure HNPS1 hose nipples are shown in their correct position Fig. 5-7B. Use RTV for sealing. Apply liberally but do not slop.

Step 26. Locate, drill, tap, and screw in abutting plates ABP1 to holes X and X' on FM1 members.

Step 27. Prepare total reflection end by cleaning MMT1 mirror mount flange in washer with a cotton swab and alcohol. Screw in the three adjust screws ASW1 and tighten just to abutting screws on ABP1 plate (Fig. 5-15). Check for proper seating. Turn each screw in three more full times for spring loading the bellows. If all has been built correctly this adjustment should place the mirror mount roughly in alignment with the tube axis.

Step 28. Position laser head in reference to a helium-neon laser as shown Fig. 5-16. The helium-laser should be mounted on a photographic tripod for ease in aligning. Bore sight laser to discharge tube as shown. This can be achieved by placing a piece of thin paper over the reflecting end along with a piece of paper with a small hole in the center over the exit end. Laser beam now enters center of exit end and aligns with center of reflecting end. This assures a true bore sight between laser head and the helium-neon laser. Firmly secure it in this position.

Step 29. Very carefully install copper reflecting mirror (MIRR1) and seal with RTV without slopping. "Finger in" the RTV.

Step 30. Note reflection from MIRR1 of He-Ne laser beam and adjust screws so that it reflects onto back of paper at "exit" end of laser head.

It is a good idea to "mike" the position of the mirror adjust plate ADR1 using verniers and noting distance between surfaces of ABP1 and ADR1. Make a record of these measurements adjacent to the adjust screws. If this is not done it may be difficult to obtain this bore sight when the mirrors are permanently in place.

Step 31. Install exit mirror MIRO1 using same procedure as reflection mirror. Note arrow indicating output direction. Adjust the mirror for back reflection into He-Ne laser output aperture. Make sure entering beam spot is still centered as this is the bore axis. Readjust screws for all back reflections to be incident at the He-Ne laser aperture. This adjustment should allow lasing to take place when system is powered up. Note setting being taken with verniers as done with reflection mirror in Step 30.

In the event that mirror adjustment becomes grossly misaligned, the following procedure is suggested without mirror removal. The reflection mirror adjustment *when exit mirror is in place* requires that the He-Ne laser be properly positioned so that it is as axial coincidental to the bore as possible. This requires a visual sighting using an external sight line or other means and may require trial and error *as the inner bore is not accessible at the reflecting end* to actually see the beam location. The reflecting mirror should now be adjusted by eye or with the above vernier measurement for approximate return reflection on the bore axis. The exit mirror may be in any reasonable position. Place a piece of white paper with a hole for the aperture around the He-Ne laser as shown in Fig. 15-17. There should be a reflection of the beam somewhere on the paper or thereabouts. Be careful not to confuse this reflection as due to the output exit mirror. Proper beam identification can be verified by turning one of the exit mirror adjust screws and noting no change in location. The reflection wanted is the one that the reflecting mirror adjust screw will change and hopefully this should be in rough alignment with the approximated eyeballed adjustments as described. If this reflection cannot be obtained it will be necessary to carefully move the He-Ne laser or the laser head (whichever is easier)

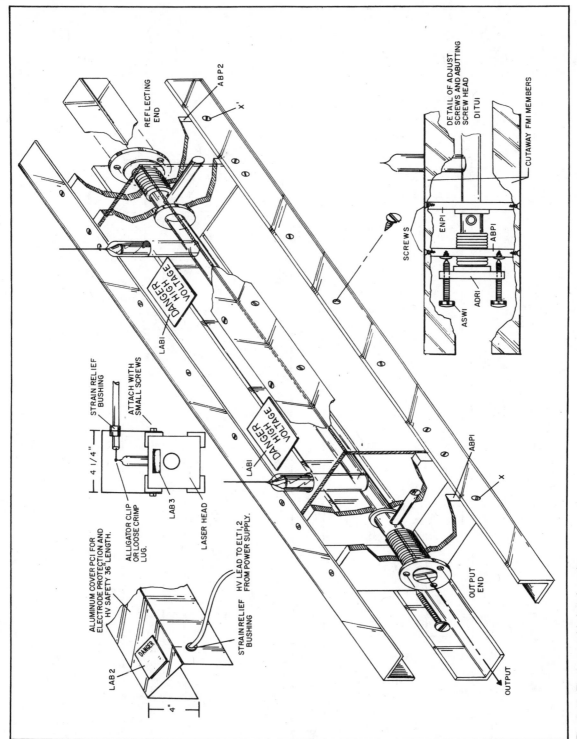

Fig. 5-15. Final assembly head section.

REFLECTING END

ABP2

DANGER HIGH VOLTAGE

LABI

STRAIN RELIEF BUSHING

ATTACH WITH SMALL SCREWS

4 1/4"

ALLIGATOR CLIP OR LOOSE CRIMP LUG.

LAB 3

LASER HEAD

ALUMINUM COVER PCI FOR ELECTRODE PROTECTION AND HV SAFETY 36" LENGTH.

HV LEAD TO ELT 1,2 FROM POWER SUPPLY.

STRAIN RELIEF BUSHING

DANGER

LAB 2

4"

DANGER HIGH VOLTAGE

LABI

ABPI

OUTPUT END

OUTPUT

DETAIL OF ADJUST SCREWS AND ABUTTING SCREW HEAD

DITUI

CUTAWAY FMI MEMBERS

SCREWS

ENPI

ABPI

ADRI

ASWI

95

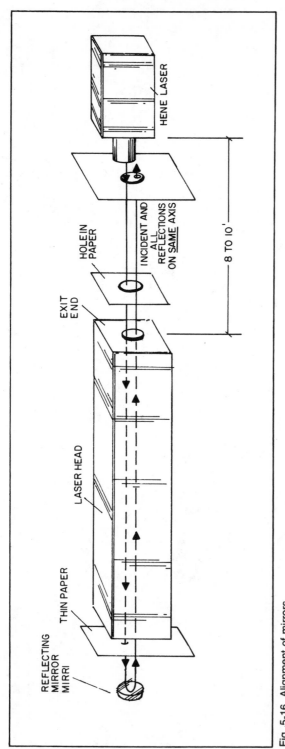

Fig. 5-16. Alignment of mirrors.

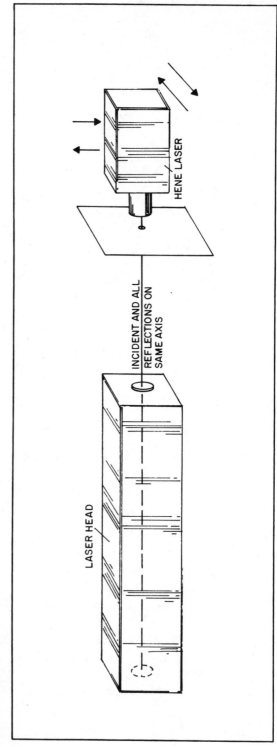

Fig. 5-17. Adjustment with mirrors in place.

very carefully search for the reflected spot attempting to get it as near bore sight as possible. Note that you may obtain reflections of diffracted light from the sides of the discharge tube when gross misalignment exists. When the reflection is located it should be sharp and clear and not contain erroneous reflections. Adjust reflection to center of He-Ne laser output aperture. This is the beam reflected back on itself by the reflecting mirror. The exit mirror is also now adjusted to reflect back to the aperture of the He-Ne laser. Alignment should be with all reflected beam spots centered in exit mirror and all reflection incident on the middle of the He-Ne laser aperture. This adjustment of mirrors should allow lasing to take place with final touch up being an indication of output power. Note that the reflecting mirror appears to require the most touch up when initially setting up the system.

Step 32. Fabricate cover PC1 as shown Fig. 5-15. Note that if cover is removed the discharge tube is automatically deenergized by the action of the alligator clip or loose crimp lugs disconnecting from EL1, 2. This by no means is to be considered as a safe method of cover removal as a high voltage hazard will exist. Further safety may be built in by electrically connecting PC1 to grounded frame via a flexible piece of wire or braid.

Step 33. Attach appropriate warning labels before using device. Please note this laser is a radiation hazard and has the capability of burning and cutting. Obviously it can cause severe burns to human tissue and is an optical hazard.

Proper posting of the area the device is used in and a warning to personnel is required. A special room with warning lights should be allocated for use in experimenting and using this laser with all personnel present wearing protective goggles.

CONSTRUCTION OF POWER SUPPLY AND SUPPORT EQUIPMENT

The power supply shown produces a variable 0 to 9000 Vac at 60 mA controlled by variac VA1. It is this electrical output that supplies energy to the system by exciting the nitrogen gas in the plasma discharge tube. This ability to control system power will be an obvious advantage when the operator realizes the many combinations of gas pressure and plasma tube current to be experimented with for obtaining either maximum or most efficient output. *These states do not occur simultaneously.*

You will note from Fig. 5-18 that the output to the discharge tube is pure ac, however, rectifiers are available as an option for those who wish dc power for their system. Dc may provide electrical heating of the positive electrode acting as a plate or collector similar to cold cathode or vacuum tube. At the time of this writing we have many hours on the prototype laser and have not detected any adverse effects of using ac.

Metering of the supply is done via meter (M1) that reads transformer (T1) primary current. This approach eliminates the high voltage insulation and mounting requirements involved in placing a current meter in the HV circuit and measuring discharge tube current directly. However, the direct method may be required for certain applications.

Power supply assembly uses standard conventional wiring procedures. Primary wiring should be #14 with the vacuum system hard wired or connected via a convenient plug and mating receptacle. A separate switch (S1) controls the pump allowing "pump down" before power is applied to the laser discharge tube. An optional receptacle for powering Pirani gauges or other electrical equipment may be conveniently placed on the front or rear panel.

The HV circuit is energized by switch (S2) that also powers indicator *danger lamp* (LA1). This warns the operator of the potential optical and HV hazards. The power leads to the laser head can be automotive ignition cable. This wire is not that stiff and offers a good voltage safety margin as it usually is rated at 30 kV plus.

The power supply is built into a metal box that can either be completely fabricated or may be built in a standard chassis with front panel and cover. In any event pay attention to HV points for safety and clearance reasons as this current is dangerous and can cause fires if allowed to break down in an arc. Input requirements are standard 120 Vac single-phase at 15 amps.

MI AMMETER PROVIDES A RELATIVE MEASUREMENT OF POWER INTO THE SYSTEM. ACTUAL CURRENT TO TUBE MAY BE MEASURED BY PLACING METER IN SERIES WITH ONE OF THE HV LEADS. THIS IS CUMBERSOME AND REQUIRES METER INSULATION.

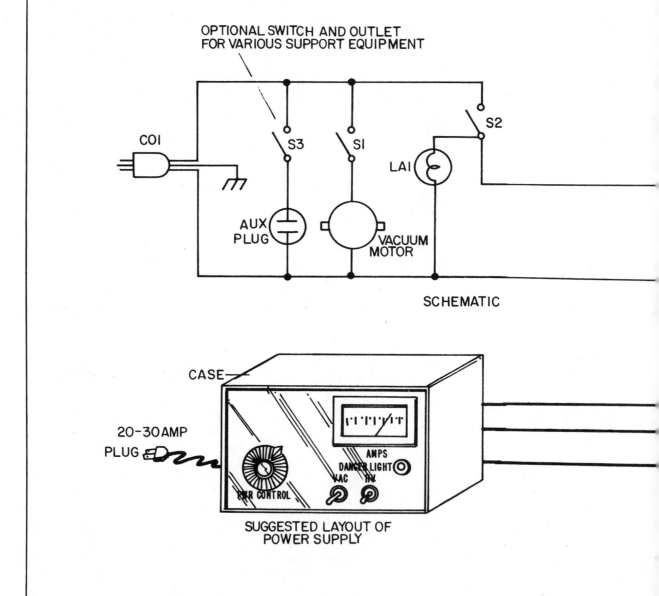

Fig. 5-18. Suggested power supply for CO_2 laser.

CAUTION MOST LUMINOUS TUBE TRANS-
FORMERS ARE MID POINT GROUNDED WHICH
MEANS A DANGEROUS SHOCK CONDITION
EXIST IF CONTACT IS MADE WITH EITHER
OF THE LASER ELECTRODES TO GROUND.

USE #14 WIRE FOR HOOK UP

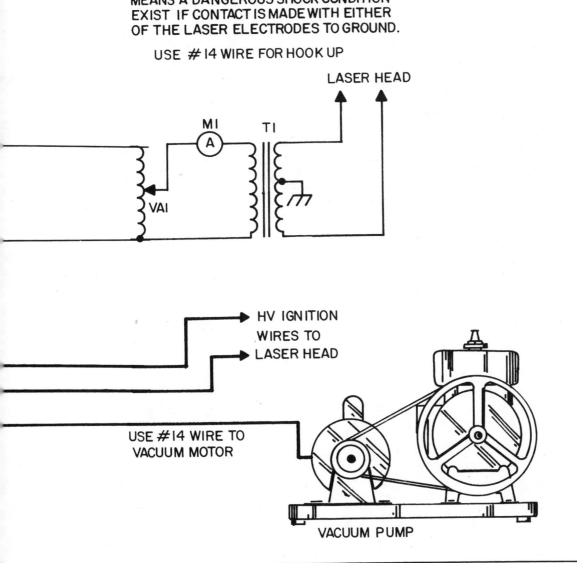

LASER HEAD

M1

T1

VA1

HV IGNITION
WIRES TO
LASER HEAD

USE #14 WIRE TO
VACUUM MOTOR

VACUUM PUMP

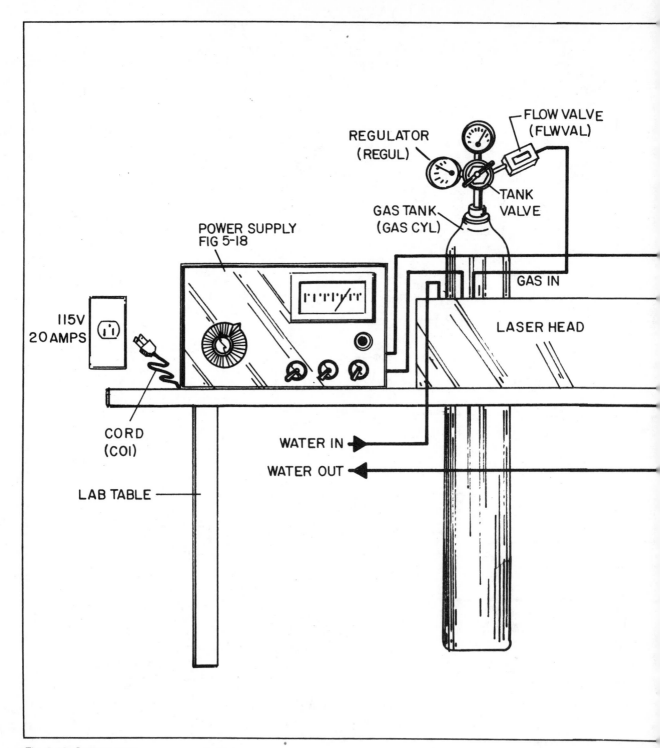

Fig. 5-19. System setup.

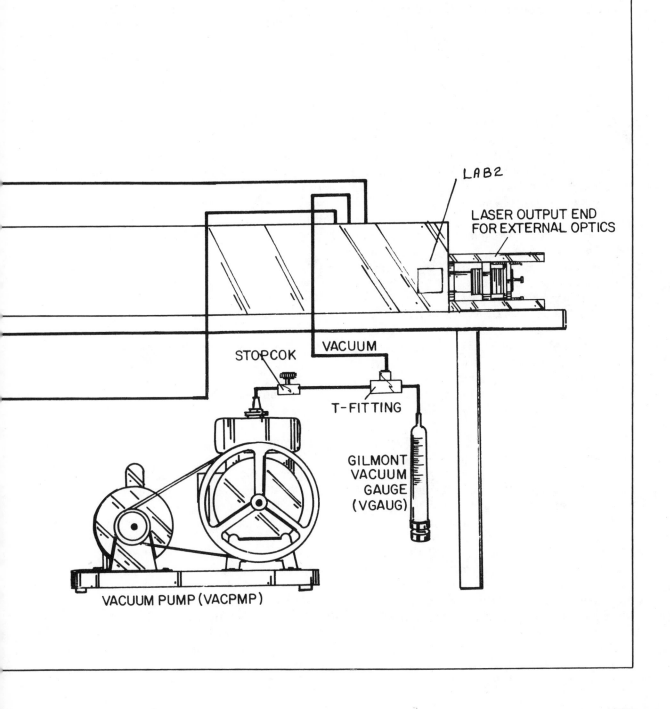

LAB2

LASER OUTPUT END
FOR EXTERNAL OPTICS

VACUUM

STOPCOK

T-FITTING

GILMONT
VACUUM
GAUGE
(VGAUG)

VACUUM PUMP (VACPMP)

SYSTEM SETUP

The system is shown (Fig. 5-19) with the laser head and power supply placed on a table. The vacuum pump and gas cylinders are placed on the floor to avoid vibration from the pump motor that could cause beam jitter or modulation. The setup shown is only a suggestion and the user/builder may have his own ideas.

Step 1. Hook up water supply and allow to fill. (Note for limited use of the laser, a running water supply may not be needed.) Fill water jacket so that discharge tube is totally immersed. Plug input and outlet hole with a cork stopper to prevent leakage. If running water is required use tygon tubing and limit water pressure as cooling jacket is not made to take pressure but just to provide adequate flowing for cooling.

Step 2. Connect vacuum lines as shown in Fig. 5-19. Use vacuum grease and hose clamps on all fittings. Attach regulator (REGUL) and flow meter (FLWVAL) to gas tank (GASCYL) and check as shown on instructions that are included with regulator valve. *Make sure regulator is set to less than 5 pounds or gas may blow up this system from overpressure. Turn "off" flow meter.* This must be verified before connecting to laser.

Step 3. Set up vacuum gauge as shown on included instructions with the Gilmont vacuum gauge. You may want to make a wooden stand with hose strain relieved.

Step 4. Turn on vacuum pump and allow to pump down for 15 minutes. Note gauge reading less than 2 torr (1 mm). Turn off valve to pump and note vacuum holding. If not, check for faults using conventional leak detection methods. Pinching "off" hoses at various stages sometimes helps to locate the leaking section. Our system pumps down easily to 50 millitorrs in 10 minutes. In the event that there is any leaks in the system they must be fixed for proper operation. There is no quick and dirty method for correcting leaks in a vacuum system. Careful workmanship in constructing and hookup is the best approach.

Step 5. With system pumped down slowly open valve on gas tank. *Caution—Caution*—again regulator must be set at less than 5 pounds and the

flow meter set to "off". Remember if you allow gas pressure to build up faster than the pump can take it down, a positive pressure will eventually build up to that which the regulator is set. This could blow out the mirrors from their mounts if allowed to exceed 5 pounds.

Step 6. Carefully admit gas into the system via the flowmeter needle valve. Observe the Gilmont vacuum gauge and adjust 8-20 toors. Allow it to stabilize. Note that the pump will take on a different sound as gas is admitted to the system.

Step 7. *At this point it is necessary to obtain safety glasses.* We use the plastic protection face masks used in shops, etc. Place a piece of wood several inches from the exit mirror. Turn on power and slowly adjust variac (VA1) until tube glows a soft pinkish, purplish. If you are lucky and did all your homework (mirror alignment in head section) the piece of wood should immediately start smoldering with a spot size of about ¼" or 5 to 7 mm. Adjust beam current for 5 amps input to T1 or if reading "direct tube current" to about 50 mA.

If output is not detected by the charring of wood obtain some carbon paper and check for effect. Very carefully tweak the top adjust screw of the reflecting mirror and note carbon paper discoloring or burning. The trick at this point is to carefully tweak the mirrors for maximum output as indicated by the burning effect of the beam. Experience will show that the reflecting mirror is usually the one requiring more frequent touching up as the system is used.

Once mirror settings are optimized it is suggested that various gas pressures and combinations of tube current be experimented with to determine the best parameters for output power. Keep a chart for further reference as each system will be a little bit different. Results will vary but parameters should be close to these specified. (Our test unit worked well at 20 torrs.)

This laser system is capable of projecting a beam of energy across a good size room and immediately burn a hole in whatever it touches. It is very hazardous and should always be terminated into a block of wood as a misdirected beam can start fires and seriously burn flesh. As an example, I was burned on my upper leg accidentally when attempting to reflect the beam onto a target. While I was standing behind the device, part of the beam missed the mirror and hit me. Before I knew what happened my pants were burnt and a painful burn that eventually blistered was received providing an excellent lesson in safety.

APPLICATIONS

In order for your laser to do useful work such as drilling, cutting, etc., it is usually necessary to focus or shape the beam depending on the work required. Lenses and optics for the wavelength output of these lasers are expensive and relatively

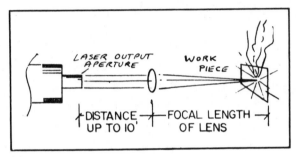

Fig. 5-20. Simple lens setup.

specialized and we leave the decision to the user when selecting the optics for the particular laser function. We have listed several suppliers of optical components for CO_2 laser systems in Table 5-4. A means for mounting a basic lens system is built into the laser head output end consisting of the extended frame members and a PVC mounting block that easily fits into place and is adjusted by a sliding action.

The laser is capable of cutting plastic, cloth fabric, cutting styrofoam, etching wood and many other applications where moderate power is required. Beam direction and position can also be controlled by using a second reflecting mirror and moving it for beam positioning. This sometimes is feasible as the laser head and work piece may now be stationary with the mirror moved for beam placement on the work piece. A simple suggested lens system is the use of a galium arsenide miniscus lens. This lens is available through PTR Optics (see Table 5-4) and comes in several different focal lengths and diameters. Spot size will not be as small as that obtained by first expanding the beam and then focusing down, however, this method is very expensive.

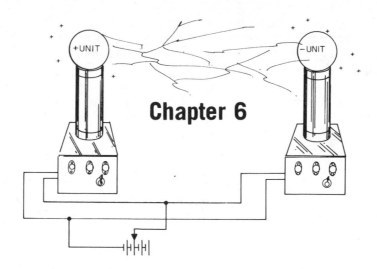

Chapter 6

Laser Light Detector (LLD1)

This project shows how to construct an electro-optical receiver capable of detecting light pulses from a considerable distance and activating and controlling external devices via this light beam. When used with the low-powered Simulated Beginners Laser, optical control can be achieved up to several hundred meters. When used with the IR Laser Rifle, control is possible over 4 to 6 km. The device lends itself to becoming the target system for simulated weapons practice. It demonstrates long range optical control, long range intrusion alarms, (for property protection, perimeter protection, announcer, etc.). It also provides a good example of an optical control link for classroom demonstration purposes.

CIRCUIT DESCRIPTION

Your laser light detector (Fig. 6-1 and Table 6-1) utilizes a sensitive photo transistor (Q5) placed at the focal point of a lens (LE2). The output of Q5 is fed to a sensitive amplifier consisting of array (A1) and is biased via the voltage divider consisting of R14 and R1. The base is not used. Q5 is capacitively

coupled to a Darlington pair for impedance transforming and is further fed to a capacitively coupled cascaded pair of common-emitter amplifiers for further signal amplification. Sensitivity control (R7) controls base drive to the final transistor of the array and hence controls overall system sensitivity. Output of the amplifier array is capacitively coupled to a one-shot consisting of Q1 and Q2 in turn integrating the output pulses of Q2 onto capacitor C8 through D1. This dc level now drives relay drivers Q3 and Q4 activating K1 along with energizing indicator D3, consequently controlling the desired external circuitry. The contacts of K1 are in series with low ohm resistor R13 to prevent failure when switching capacitive loads.

J2 allows "listening" to the intercepted light beam via headsets. This is especially useful when working with pulsed light sources such as GaAs lasers or any other varying periodic light source.

CONSTRUCTION STEPS

1. Identify all components. Note that indi-

Table 6-1. Laser Light Detector Parts List.

R1,4,8	(3)	390 k 1/4 watt resistor
R2	(1)	5.6 M 1/4 watt resistor
R3,5,6	(3)	6.8 k 1/4 watt resistor
R10,16	(2)	5.6 k 1/4 watt resistor
R11,15,17,14	(4)	100 k 1/4 watt resistor
R13, A & B	(2)	10 ohm 1/4 watt resistor
R18	(1)	220 ohm 1/4 watt resistor
R19	(1)	100 ohm 1/4 watt resistor
R20	(1)	1 k 1/4 watt resistor
R7/S1	(1)	5 k pot & switch combination
C1	(1)	.047/μF/25 V disc cap
C2,3,5,6,7,9	(6)	1 μF/25 V elect
C4	(1)	.01 μF/25 V disc cap
C8	(1)	4.7 μF/25 V elect
Q1,2,3,4	(4)	PN2222 npn silicon transistor
A1	(1)	CA3018 amp array
Q5	(1)	Photo transistor L14G3 (A PIN photodiode is better suited when detecting the fast laser pulses. Response time now becomes an important factor.)
D1	(1)	1N914 sig diode
D2	(1)	1N4007 1000 V diode
D3	(1)	LED indicator FLV106
K1	(1)	Mini dip relay 6 V spst
J1,2,3	(3)	RCA phono jacks
P1,2	(2)	RCA phono plug
CL1,2	(2)	Battery snap clips for 9 V rect.
CA1	(1)	Case 4 × 2 1/8 × 1 5/8 al mini box
PB1	(1)	Perfboard 1 1/4 × 2 5/8
KN1	(1)	Small plastic knob for R7/S1
BU1	(1)	3/8" plastic bushing for D3
BU2	(1)	Cord clamp bushing for wire from K1
LE2	(1)	Lens 54 × 89 mm
DO1	(1)	Mtg dowel 2 × 2 soft wood
EN2	(1)	Enclosure 6 1/2 × 2 3/8 sked 40 PVC (fabbed as shown)
CA3,4	(2)	2 3/8 plastic caps
WR1	(12")	Shielded mike cable
WR4	(24")	24" plastic hook-up wire
	Optional Items	
B1,2	(2)	9-volt transistor batteries
PC1	(1)	Printed circuit board or use perforated board
Headset	(1)	8 ohm monophonic or equivalent
T1	(1)	Matching 8 ohm/1 k transformer
FTR1	(1)	3" IR filter

Complete kit with PC board available through Information Unlimited, Inc., P.O. Box 716, Amherst, N.H. 03031. Write or call 1-603-673-4730 for price and delivery.

cated layout must be followed for proper performance (Figs. 6-2 and 6-3).

2. Fabricate an aluminum minibox as shown for J1, J2, BU1, BU2 and R7/S1. Assemble these parts in place as shown. Note ground lug under J1. Note mating holes for securing Q5 housing (Fig. 6-4).

3. Reference corner of assembly board and mark as shown in Fig. 6-2.

4. Insert CA3018 as shown and carefully position as shown. Bend over several leads to keep from falling out. This insertion of the CA3018 may require several attempts before aligning the 12 leads with the appropriate holes in perfboard or PC

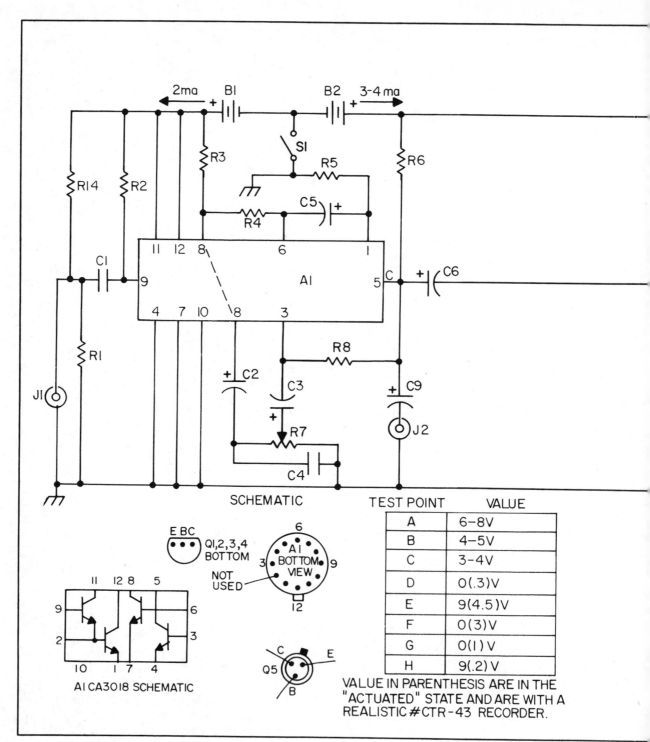

Fig. 6-1. Circuit schematic.

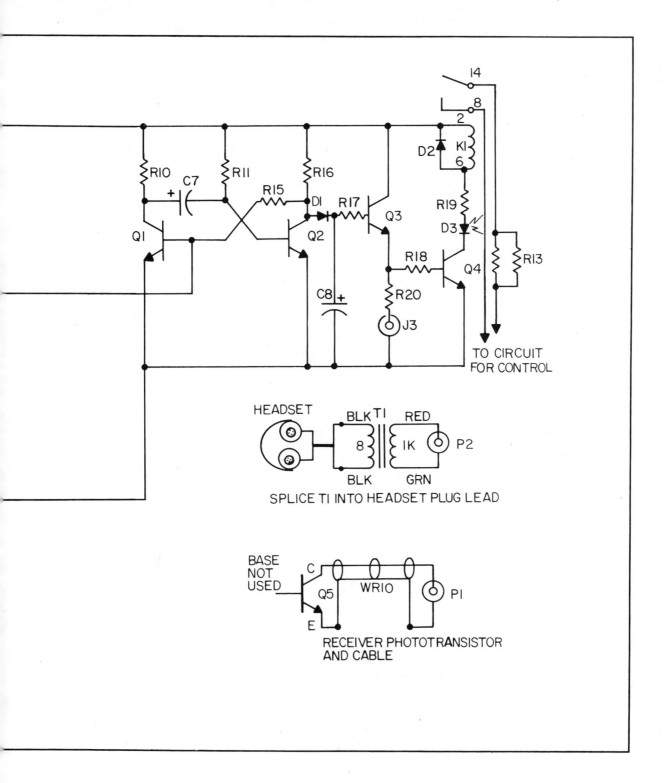

14

8

D2 K1
2
6

R10 C7 R11 R15 R16 D1 R17 Q3 R19

Q1 Q2 D3

R18 Q4

R13

C8

R20

J3

TO CIRCUIT
FOR CONTROL

HEADSET BLK T1 RED

8 1K P2

BLK GRN

SPLICE T1 INTO HEADSET PLUG LEAD

BASE
NOT
USED C

Q5 WR10 P1

E

RECEIVER PHOTOTRANSISTOR
AND CABLE

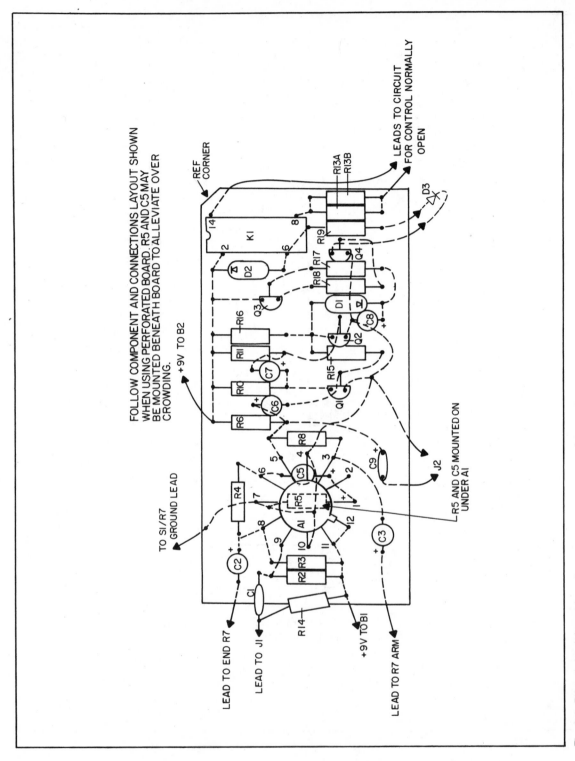

Fig. 6-2. Printed circuit board layout.

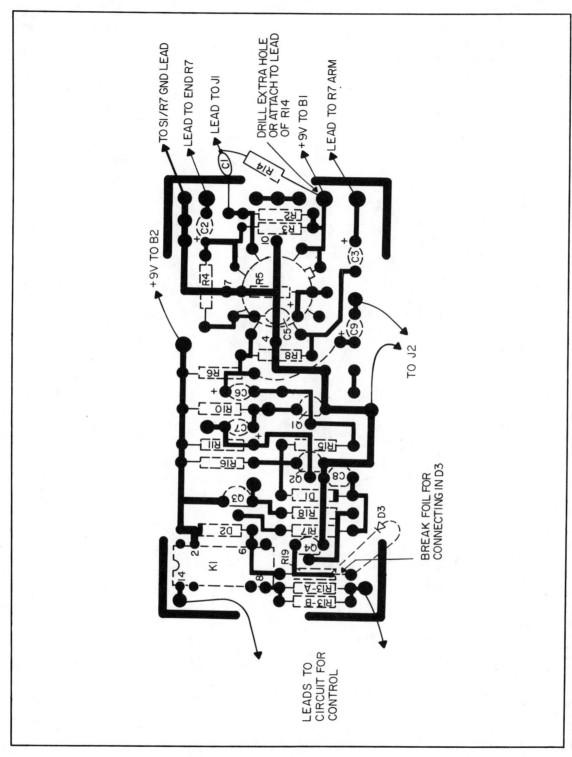

TO SI/R7 GND LEAD

LEAD TO END R7

LEAD TO JI

DRILL EXTRA HOLE
OR ATTACH TO LEAD
OF RI4

+9V TO BI

LEAD TO R7 ARM

CI

RI4

+9V TO B2

C2

RI2
RI4

10

C3

R4

7

R5

+

4

C5

C9

+

R8

TO J2

R6

C6

+

RIO

C7

+

9I

RI

RI5

RI6

Q2

C8

Q3

DI

RI8

D2

RI7

D3

K1

2

9

RI9

Q4

BREAK FOIL FOR
CONNECTING IN D3

I4

8

RI3-A

RI3-B

LEADS TO
CIRCUIT FOR
CONTROL

Fig. 6-3. Printed circuit board foil side.

109

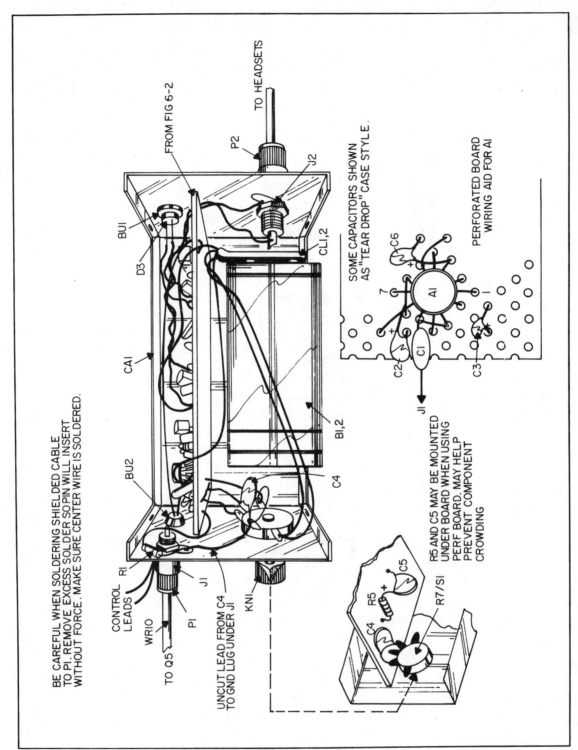

BE CAREFUL WHEN SOLDERING SHIELDED CABLE TO P1. REMOVE EXCESS SOLDER SO PIN WILL INSERT WITHOUT FORCE. MAKE SURE CENTER WIRE IS SOLDERED.

FROM FIG 6-2

TO HEADSETS

P2

J2

BU1

D3

CL1,2

CA1

SOME CAPACITORS SHOWN AS "TEAR DROP" CASE STYLE.

C6

7

A1

C2

C1

C3

PERFORATED BOARD WIRING AID FOR A1

J1

BU2

B1,2

C4

CONTROL LEADS

R1

J1

WR10

P1

KN1

TO Q5

UNCUT LEAD FROM C4 TO GND LUG UNDER J1

R5

C5

C4

R7/S1

R5 AND C5 MAY BE MOUNTED UNDER BOARD WHEN USING PERF BOARD. MAY HELP PREVENT COMPONENT CROWDING

Fig. 6-4. Electronic assembly.

110

board as shown. Note C5 and R5 shown mounted under the CA3018. Use Figs. 6-2 and 6-3 when using printed circuit board.

5. Using standard audio frequency wiring techniques, proceed to wire and solder starting with C1 inserting the designated components and soldering point by point. Carefully check for accuracy and quality of solder joints. Remember mistakes can ruin the CA3018. Observe correct polarity of electrolytic capacitors and position of relay.

6. Attach battery clips by inserting leads through holes in perfboard adjacent to their termination points as shown. This method strain-relieves these wires.

7. Connect C4 across R7 as shown. Connect ground end of R7 to ground lug J1 via uncut lead from C4. Connect buss jump from this point to one of the lugs of S1 as shown. Connect R1 between J1 and ground lug of J1 with above and solder this point. Note these leads must be as short and as direct as possible to prevent noise and hum.

8. Note that the assembled board should have the following leads for connection to the components in the aluminum minibox.

☐ Input lead of C1 along with R14 for connection to (short as possible) use component lead.

☐ Ground lead from pins 4, 7 and 10 of CA3018 for connecting to ground of case at S1/R7.

☐ Two buss leads to R7 end and R7 arm (use uncut leads of components if possible). These leads must be as short as possible.

☐ Leads to D3. Note foil must be broken on PC board as shown for inserting D3 in series.

☐ Leads to J2. Use C9 lead if possible.

☐ Control leads from K1.

9. Visual check for solder, wiring errors, and shorts.

10. Fabricate EN2 from a 6½″ piece of 2⅜″ OD schedule 40 PVC tubing. Drill ¾″ hole as shown approximately 3″ from rear end. This hole is for optical alignment. Note two mating holes for securing to CA1. These holes are also in top of this piece for access with a screwdriver. The bore axis of this tube should be parallel with CA1. It may be more convenient to slot the rear hole to allow slight side movement for final alignment. The large ¾″

hole can be covered with plug, tape, etc., when not needed.

11. Fabricate DO1 centering dowel from a 2″ length of 1½″ OD or thereabouts for smooth sliding fit into EN2. Q5 is mounted for optical centering, via small pin holes in wood for securing via its leads. Cable WR1 is fed to Q5 via a slight off-center feed hole in dowel. Connection is made by soldering to exposed leads of Q5 (watch for overheating) and then securing with RTV or equivalent. Leads to Q5 should be left long enough to allow touch-up repositioning to true optical axis for final alignment.

12. Fabricate CA3 and CA4 from a 2⅜″ plastic cap. Remove end with exception of ⅜″ lip to retain lens (LE2) and optional filter (FTR1) against end of EN2. CA4 is also a 2¾″ plastic cap. Place small hole for cable (WR1). Hole should create friction hole to prevent DO1 from sliding once set.

Secure with RTV when complete and finally aligned. It is assumed that the assembled unit to this point has been wired correctly with no shorts and good solder connections.

13. Install P1 into J1.

TESTING

1. Turn R7/S1 full ccw (off position).

2. Connect one terminal of a fresh 9-volt battery to CL1 and connect a millimeter between the unused contact of the battery and the clip. Turn on R7/S1 and note current reading of approximately 2 mA. Fully connect battery and designate B1.

3. Repeat above using second battery designated B2. Note current reading of 3-4 mA. Turn up gain R7 and note B2 current increasing to 12 mA and D3 lighting when light is detected. This is the relay current and indicates an "on" state. Current should drop back to 3 mA when relay turns "off" in several seconds. Note that R7 must not be set at too high a gain or the unit will not turn off. R7 will have to be set way down in normal background light if unit is used without FTR1.

4. Adjust Q5 to focal length of LE2. This is accomplished by pointing unit at a distant source of light and placing a piece of paper over Q5. Adjust DO1 position for a sharp image of light source over Q5 lens. This is easily accomplished through access hole in enclosure.

111

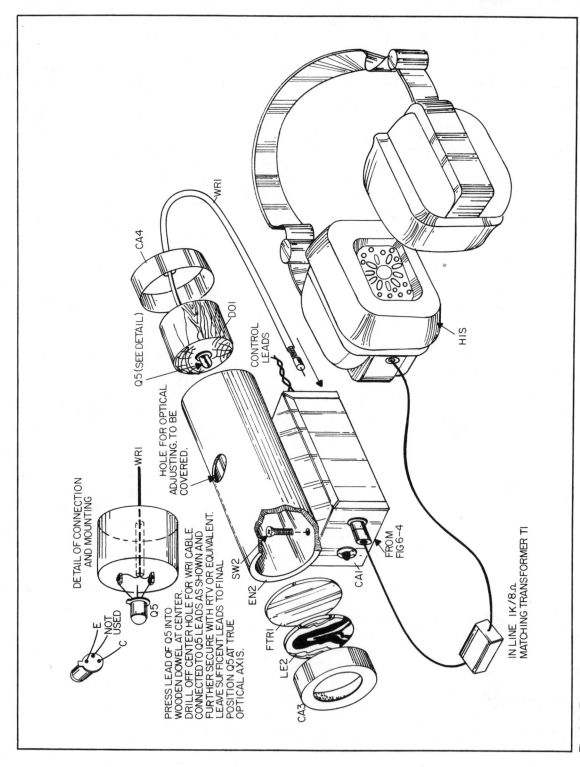

DETAIL OF CONNECTION AND MOUNTING

WRI

Q5

E

NOT USED

C

PRESS LEAD OF Q5 INTO WOODEN DOWEL AT CENTER. DRILL OFF CENTER HOLE FOR WRI CABLE CONNECTED TO Q5 LEADS AS SHOWN AND FURTHER SECURE WITH RTV OR EQUIVALENT. LEAVE SUFFICENT LEADS TO FINAL POSITION Q5 AT TRUE OPTICAL AXIS.

WRI

CA4

Q5 (SEE DETAIL)

DOI

CONTROL LEADS

HOLE FOR OPTICAL ADJUSTING. TO BE COVERED.

SW2

EN2

FTRI

LE2

CA3

CAI

FROM FIG 6-4

HIS

IN LINE IK/8Ω MATCHING TRANSFORMER TI

Fig. 6-5. Final assembly.

THE LLDI LIGHT DETECTOR IS INSTALLED AS THE TARGET WITH ITS LENS BEING APERTURED BY SUCCESSIVELY SMALLER COVERS. THESE COVERS ARE REPLACED AS THE USER BECOMES MORE ACCURATE. RANGE MAY BE UP TO 20 METERS WITH THE LHP2 SIMULATED LASER, OR UP TO SEVERAL HUNDRED METERS USING THE GENUINE Ga As LASER SUCH AS OUR LRG3, LP3, OR SSLI. USE INGENUITY IN SIMULATING THE DEVICE INTO A GUN CONFIGURATION. TRIGGER IS OBVIOUSLY THE SWITCH.

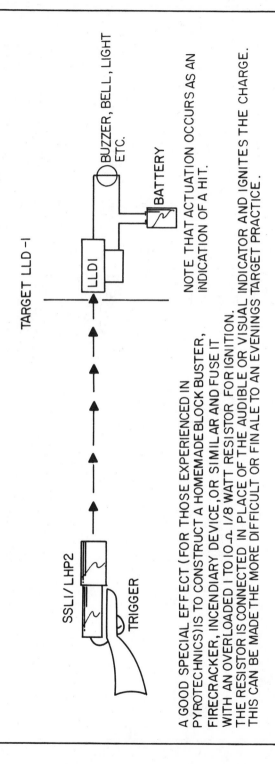

TARGET LLD -I

BUZZER, BELL, LIGHT ETC.

BATTERY

LLDI

NOTE THAT ACTUATION OCCURS AS AN INDICATION OF A HIT.

SSLI/LHP2

TRIGGER

A GOOD SPECIAL EFFECT (FOR THOSE EXPERIENCED IN PYROTECHNICS) IS TO CONSTRUCT A HOMEMADE BLOCK BUSTER, FIRECRACKER, INCENDIARY DEVICE, OR SIMILAR AND FUSE IT WITH AN OVERLOADED 1 TO 10 Ω 1/8 WATT RESISTOR FOR IGNITION. THE RESISTOR IS CONNECTED IN PLACE OF THE AUDIBLE OR VISUAL INDICATOR AND IGNITES THE CHARGE. THIS CAN BE MADE THE MORE DIFFICULT OR FINALE TO AN EVENINGS TARGET PRACTICE.

Fig. 6-6. Laser shooting gallery.

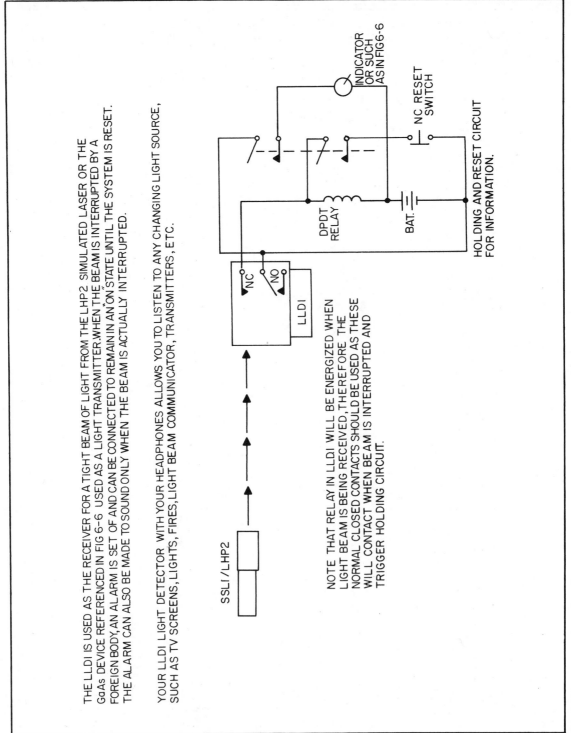

THE LLDI IS USED AS THE RECEIVER FOR A TIGHT BEAM OF LIGHT FROM THE LHP2 SIMULATED LASER OR THE GaAs DEVICE REFERENCED IN FIG 6-6 USED AS A LIGHT TRANSMITTER. WHEN THE BEAM IS INTERRUPTED BY A FOREIGN BODY, AN ALARM IS SET OF AND CAN BE CONNECTED TO REMAIN IN AN ON STATE UNTIL THE SYSTEM IS RESET. THE ALARM CAN ALSO BE MADE TO SOUND ONLY WHEN THE BEAM IS ACTUALLY INTERRUPTED.

YOUR LLDI LIGHT DETECTOR WITH YOUR HEADPHONES ALLOWS YOU TO LISTEN TO ANY CHANGING LIGHT SOURCE, SUCH AS TV SCREENS, LIGHTS, FIRES, LIGHT BEAM COMMUNICATOR, TRANSMITTERS, ETC.

INDICATOR OR SUCH AS IN FIG 6-6

NC RESET SWITCH

HOLDING AND RESET CIRCUIT FOR INFORMATION.

DPDT RELAY

BAT.

NC

NO

LLDI

SSLI / LHP2

NOTE THAT RELAY IN LLDI WILL BE ENERGIZED WHEN LIGHT BEAM IS BEING RECEIVED, THEREFORE THE NORMAL CLOSED CONTACTS SHOULD BE USED AS THESE WILL CONTACT WHEN BEAM IS INTERRUPTED AND TRIGGER HOLDING CIRCUIT.

Fig. 6-7. Long-range intrusion alarm.

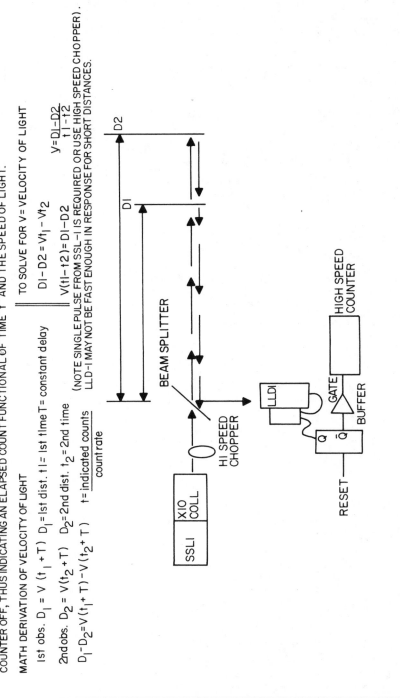

THIS APPLICATION IS TO DEMONSTRATE A METHOD OF MEASURING THE SPEED OF LIGHT. IN ORDER TO OBTAIN ACCURATE RESULTS SPECIALIZED EQUIPMENT WILL BE REQUIRED. FRONT END OF LLDI WOULD REQUIRE A HIGH SPEED LOW CAPACITANCE PIN PHOTO DIODE TO OBTAIN THE RESPONSE NECESSARY FOR THE SHORT DURATION LASER PULSE.

$\overline{Q}$ GATES COUNTER "ON" WHEN COLLIMATED LIGHT PULSE FROM LASER IS DETECTED BY LLD-I VIA DIRECT FROM BEAM SPLITTER. PART OF LIGHT TRAVERSES TO MIRROR LOCATED AT PRESENT DISTANCE AND RETURNS TO LLD-I AT SOME TIME LATER, INDEXING FLIP-FLOP TO GATE COUNTER OFF, THUS INDICATING AN ELAPSED COUNT FUNCTIONAL OF TIME "t" AND THE SPEED OF LIGHT.

MATH DERIVATION OF VELOCITY OF LIGHT

1st obs. $D_1 = V(t_1 + T)$ D_1 = 1st dist. t_1 = 1st time T = constant delay

2nd obs. $D_2 = V(t_2 + T)$ D_2 = 2nd dist. t_2 = 2nd time

$D_1 - D_2 = V(t_1 + T) - V(t_2 + T)$ t = $\dfrac{\text{indicated counts}}{\text{count rate}}$

TO SOLVE FOR V = VELOCITY OF LIGHT

$D_1 - D_2 = Vt_1 - Vt_2$

$V(t_1 - t_2) = D_1 - D_2$

$V = \dfrac{D_1 - D_2}{t_1 - t_2}$

(NOTE SINGLE PULSE FROM SSL-I IS REQUIRED OR USE HIGH SPEED CHOPPER). LLD-I MAY NOT BE FAST ENOUGH IN RESPONSE FOR SHORT DISTANCES.

SSLI X10 COLL

HI SPEED CHOPPER

BEAM SPLITTER

D1

D2

LLDI

Q $\overline{Q}$

RESET

GATE BUFFER

HIGH SPEED COUNTER

Fig. 6-8. Light speed indicator setup.

115

5. Plug in high impedance headphone into J2. (Note Fig. 6-1 showing standard 8-ohm headsets with spliced-in matching transformer for stepping up to 1000 ohms. High impedance headsets are scarce and usually uncomfortable to wear.)

6. Turn on R7/S1 and slowly turn up gain until a loud 60 cycle hum is heard. This is the normal lighting frequency being picked up by Q5 and at normal ambient light conditions will completely block the amplifier. Reduce the gain and attempt to point Q5 at various objects indicating different levels of signal depending on reflection characteristics of surfaces, etc. Point at TV screen, scope or any periodically changing source of light. You will note that the circuit is relatively prone to power line hum pick-up. This is because of the plastic enclosure EN2 and the floating base lead of Q5. This may be biased with a resistor to the emitter for use with high signal levels. It is assumed that testing will be done in normal electrical lighting for this step. If not, you may not obtain the 60-Hz hum from the varying light. If you troubleshoot a faulty circuit, it may be convenient to use the test points shown in Fig. 6-1 and thoroughly familiarize yourself with the circuit description given in the beginning of the plans. For final assembly see Fig. 6-5. Also see Figs. 6-6, 6-7, and 6-8.

Section II
A Little Information on Ultrasonics

Acoustical ultrasonics is a gray area with practically little or no information available on the subject. Ultrasonics using liquids and solids as a medium are not new and are being used for cleaning purposes, heat-treating, welding, aiding in chemical reactions, etc., with many companies manufacturing various products utilizing this form of high-frequency mechanical energy. Unlike acoustical ultrasonics where air is the transmission medium, energy transfer from the transducer to the target object is easily accomplished with high efficiency.

The reason is that all forms of waves or cyclic energy including electromagnetic radiation must be properly terminated to a load to accept or obtain maximum energy transfer. A vibrating transducer of a given area will transfer many times as much energy to a highly dense medium such as a liquid than it would to a much less dense medium such as air. To those who are familiar with what happens with the improper termination of a radio frequency transmission line to an antenna knows that a standing wave is produced that is functional of this

energy mismatch. Using this analogy to a mechanical vibrating surface transferring energy to the air medium instantly makes one realize the severe energy mismatch that exists.

Efficient transducers for acoustical ultrasonic energy transfer must displace a relatively large volume of air and yet not inherit a large mass or inertia, regarding the vibrating diaphragm or reed. Unfortunately ultrasonic acoustical transducers are very inefficient both in the conversion of electric energy into mechanical and the mechanical into acoustical.

The power merit of a transducer is best related to its resultant production of a referenced sound pressure level not its power handling capabilities. As an example, electromagnetic transducers such as tweeters rated at 50 watts seldom produce the equivalent sound pressure level that their more efficient piezoelectric counterparts do using only a fraction of this input driving power. Economics also favor the piezoelectrics.

Sound pressure levels produced are usually

measured in decibels which is a logarithmic function. The human ear has such a large range of hearing relative to pressure intensities that use of logarithmic measurement is justified.

There are two ways of assigning a decibel level to a given sound pressure level and we shall use the 0 dB reference as the threshold of hearing. This corresponds to a sound pressure level of .0002 microbars. A bar is atmospheric pressure of about 14.7 lbs/in². Another accepted reference point is 0 dB being at a sound pressure level of 1 microbar or one dyne. This corresponds to a 74 dB difference between the two reference levels. We shall use the 0 dB threshold of hearing reference as a sound pressure level of .0002 microbars when mentioning sound pressure levels.

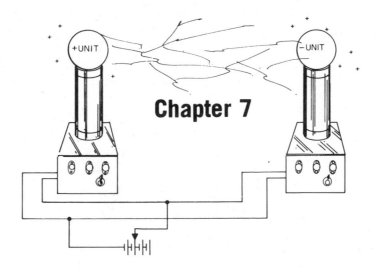

Chapter 7

Pocket Pain Field Generator (IPG5)

This project shows how to construct a hand-held pocket-sized electronic device producing waves of ultrasonic sound capable of controlling horses, dogs, rodents and many other animals. It also produces an extreme discomfort to women and children, therefore, please bear this in mind when using the device. Some people cannot tolerate this ultrasonic sound and therefore, the device *sometimes* may be used to discourage an attack. However, we do not recommend this particular use of the device since it is not that foolproof and may provoke rather than discourage.

CIRCUIT THEORY

A 555 timer (I1) is connected as an astable oscillator set somewhere between 16-21 kHz (Fig. 7-1 and Table 7-1). R1 along with R2, R3, and C2 determines the frequency and symmetry of this waveform. Adjustment of this frequency is by trimmer (R1). Output is taken via pin 3 and is resistively coupled to the base of Q1, that is operated in a class-C mode. Positive pulses occurring at the col-

lector of Q1 drives TR1 through inductor (L1). L1 forms a resonant circuit with the internal capacity of TR1. It should be noted that TR1 has an inherent capacitance of about .15 μF and this must be tuned out via the series inductance of L1 for efficient power transfer. Note that L1 may be tunable for maximizing results at a set frequency. T1 is the 8-ohm section of a transformer and serves as an audio choke while offering only its dc resistance for feeding the collector of Q1. Battery (B1) is a standard 9-volt transistor radio battery or can be a rechargeable ni-cad.

CONSTRUCTION STEPS

1. Layout perfboard as shown in Fig. 7-2. Identify all parts and pieces.

2. Carefully locate holes for the odd pins of IC1, transformer T1 mounting tabs, and drill perfboard.

3. Assemble components to board using component leads whenever possible and insulating tubing whenever a lead bridges one another. Ob-

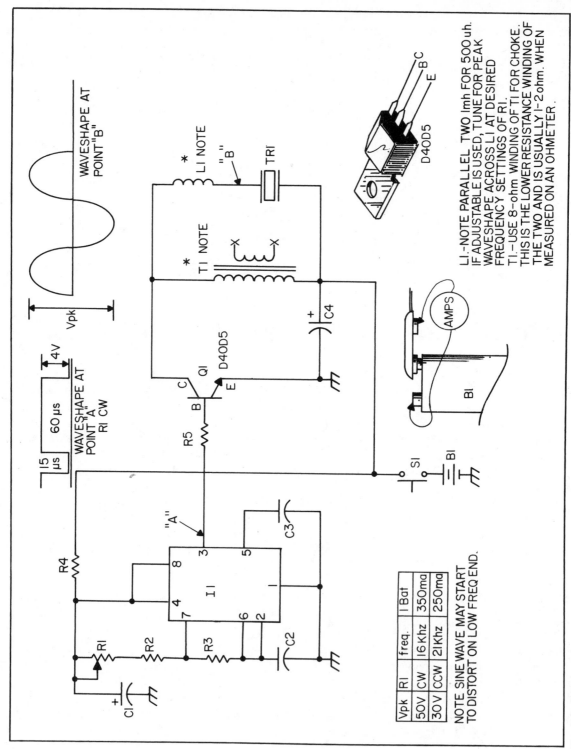

WAVESHAPE AT POINT "B"

WAVESHAPE AT POINT "A" RI CW

L1. - NOTE PARALLEL TWO 1mh FOR 500 uh. IF ADJUSTABLE IS USED, TUNE FOR PEAK WAVESHAPE ACROSS L1 AT DESIRED FREQUENCY SETTINGS OF R1.

T1. - USE 8-ohm WINDING OF T1 FOR CHOKE. THIS IS THE LOWER RESISTANCE WINDING OF THE TWO AND IS USUALLY 1-2ohm. WHEN MEASURED ON AN OHMETER.

Vpk	RI	freq.	I Bat
50V	CW	16Khz	350ma
30V	CCW	21Khz	250ma

NOTE SINE WAVE MAY START TO DISTORT ON LOW FREQ END.

Fig. 7-1. Schematic.

Table 7-1. Pocket Pain Field Generator Parts List (IPG5).

R1	(1)	2 k vert. trim pot
R2,3	(2)	2.2 k ¼ watt resistor
R4	(1)	10 ohm ¼ watt resistor
R5	(1)	1 k ¼ watt resistor
C1	(1)	100μF/25 volt electrolytic capacitor
C2	(1)	.01 μF 50 volt poly capacitor
C3	(1)	.01 μF 50 volt disc capacitor
C4	(1)	1 μF 50 volt electrolytic capacitor
I1	(1)	555 timer dip pack
Q1	(1)	D4OD5 npn power tab transistor
T1	(1)	1 k 8 ohm transformer
L1	(2)	1 mH inductor
TR1	(1)	Piezo driver #1023
S1	(1)	Push-button switch
CL1	(1)	6″ snap clips
CA1	(1)	4.4″ × 2.44″ × 1.06 plastic box
PB1	(1)	2″ × 1.3″ × .1 grid perfboard
SCR1	(1)	2″ × 2″ fine screen mesh
WR3	(10″)	#24 wire, plastic hook-up
B1	(1)	9 volt transistor battery (not included in parts kit)

Complete kit with printed circuit board available through Information Unlimited, Inc., P.O. Box 716, Amherst, N.H. 03031. Write or call 1-603-673-4730 for price and delivery.

served polarity of C1, C4 and position of I1. Wire and solder as in Figs. 7-1 and 7-2 showing connections as dashed lines. Connect TR1, CL1 and S1 using wire leads.

4. Remove the high impedance (1000 ohm leads) of T1. These can be identified by measuring with an ohmmeter and are the higher resistance windings (usually about 30 ohms).

5. Carefully check wiring for accuracy and shorts especially around I1.

TESTING

Obtain a Simpson Multimeter and check for absence of a short circuit across battery connection when S1 is depressed. This test verifies any gross errors such as shorts, etc., and helps prevent damage that could be done by the ni-cad battery. Rotate R1 full ccw (lowest frequency end). Insert a meter set on amps across the contacts of S1 and note a current reading of about 250 mA. Rotate R1 to full cw and note current increasing to approximately 350 mA. You will also note a piercing, uncomfortable sound in your ears or back of your neck usually depending where R1 is set. It may be necessary to

wear ear protection as this can be extremely painful to most people. A further check of circuit operation can be accomplished by connecting an oscilloscope across the terminals of TR1 and note the waveforms as shown in Fig. 7-1.

FINAL ASSEMBLY

1. Fabricate plastic case CA1 as shown in Fig. 7-3. Carefully fabricate a 1¼″ to 1½″ hole for transducers. Drill hole for S1 and small access hole for adjustment of R1.

2. Place screen and glue in place using RTV. Attach TR1 transducer also using RTV or similar sealer.

3. Install assembly board and battery. Attach rear cover and note position and dress of wires.

Please take note that this device is intended for intermittent use. There is inadequate heatsinking of the components in this compact layout for continuous use.

OPERATION AND APPLICATIONS

Your IPG5 Ultrasonic Control Device is an unique, patented electronic device that when used

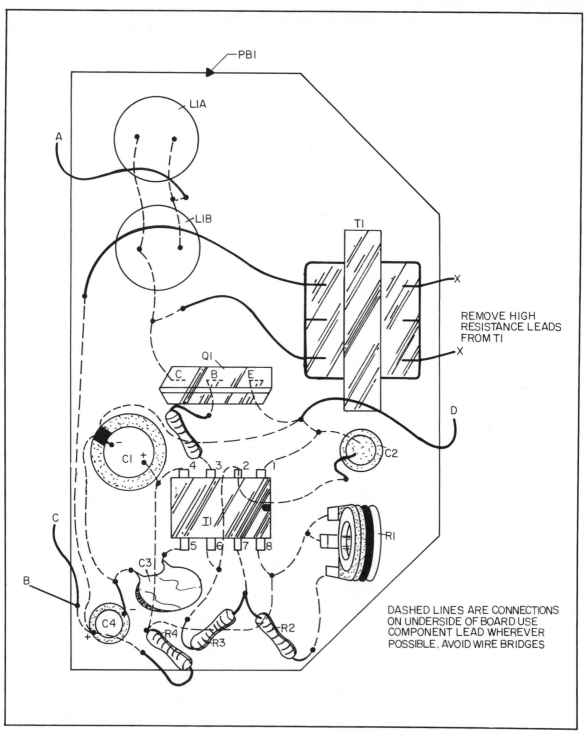

Fig. 7-2. Board assembly.

122

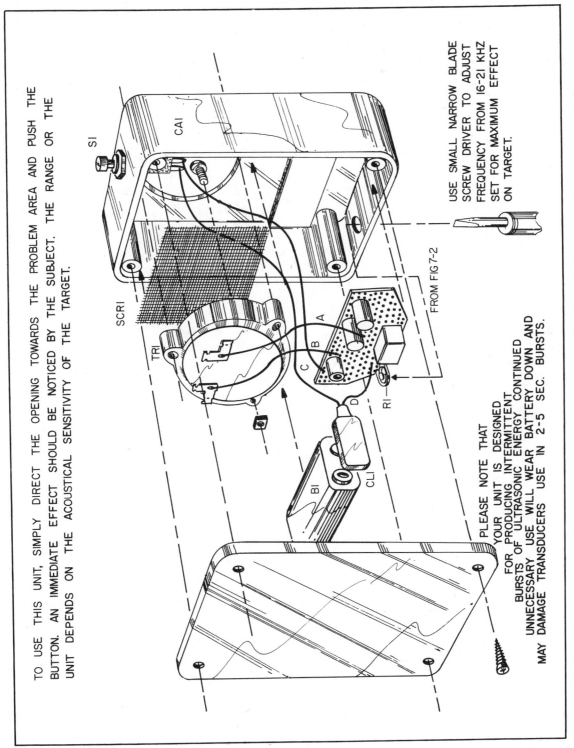

TO USE THIS UNIT, SIMPLY DIRECT THE OPENING TOWARDS THE PROBLEM AREA AND PUSH THE BUTTON. AN IMMEDIATE EFFECT SHOULD BE NOTICED BY THE SUBJECT. THE RANGE OR THE UNIT DEPENDS ON THE ACOUSTICAL SENSITIVITY OF THE TARGET.

USE SMALL NARROW BLADE SCREW DRIVER TO ADJUST FREQUENCY FROM 16-21 KHZ SET FOR MAXIMUM EFFECT ON TARGET.

PLEASE NOTE THAT YOUR UNIT IS DESIGNED FOR PRODUCING INTERMITTENT BURSTS OF ULTRASONIC ENERGY CONTINUED UNNECESSARY USE WILL WEAR BATTERY DOWN AND MAY DAMAGE TRANSDUCERS USE IN 2-5 SEC. BURSTS.

FROM FIG 7-2

CAI

SI

SCRI

TRI

A

B

C

D

RI

BI

CLI

Fig. 7-3. Final assembly.

123

properly will prevent harassment from unruly, uncontrolled dogs. The device emits an acoustical beam of energy that dogs find intolerable, but most humans cannot hear but will feel (Fig. 7-3). This energy causes discomfort to these animals and obviously becomes more severe the closer the dog approaches. Most dogs may *only* be affected by this device when in a hyperactive state such as chasing, fighting, attacking, etc.

Children and young people are more prone to hearing this energy than adults and consequently it should be used bearing this in mind. *The unit could be used to discourage certain unwanted personal encounters, however, we do not recommend it as personal defense device in a hostile situation.*

You may adjust the tone from 16-21 kHz using a small screwdriver. *Adjust to maximum effect on particular target subject.* To use your unit, you must remember and observe the following: Your unit emits this energy from the wire mesh cut out in the front. It should be continually pushed on and off repeatedly until the animal retreats and should not be left on constantly in any single situation. An aggressive dog may approach several times at which you give him a shot each time until he retreats.

The unit should only be used when the need arises. Do not press the button or use around house pets as these animals *usually will not appear to be affected* in the same way as an unruly or strange dog, but will accept it as a reprimand, etc., because the sound will cause discomfort to these house pets for no reason at all.

Use common sense, do not approach and use on a known vicious or malicious dog. Indicate by thrusting the unit towards the menacing dog letting him realize where the discomfort is coming from. Do not provoke a dog and then use this device. This is unnecessary and cruel and if brought to the attention of the authorities could result in a stiff penalty. Do not expect to control dogs at great distances because the energy drops off increasingly with distance.

Caution—This is not a toy—keep away from children. Do not place near your ear. Do not use on deaf or extremely menacing dogs. Do not use on command attack dogs. Use common sense at all times.

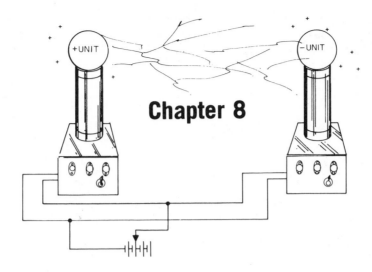

Chapter 8

Phaser Shock-wave Pistol (PSP3)

This project shows how to construct a moderately-powered directional source of continuous adjustable high frequency acoustical shock waves. The output energy of the device is conservatively rated at 115 dB. Frequency is variable from an audible 10 kHz to 20 kHz. (More range is possible by several minor component value changes.) The unit is intended for use as a research tool in the study of animal behavior, acoustical experimentation, or as a source of intense directional acoustical high frequency sound for other scientific and laboratory applications.

The unit is to be used with discretion and not treated as a toy, caution must be used as exposure to most people causes pain, headache, nausea and extreme irritability. (Younger women are especially affected.) Do not under any circumstances point the unit at a persons ears or head at close range or severe discomfort and possible ear damage may result. Usage on dogs and other animals must be done with discretion.

CIRCUIT THEORY

A 555 timer (I1) is connected as an astable oscillator set somewhere between 10-20 kHz (Fig. 8-1 and Table 8-1). R1, along with R2, R3, and C2 determines the frequency and symmetry of this waveform. Adjustment of this frequency is by control (R1). Output is taken via pin 3 and is resistively coupled to the base of Q1, operated in a class-C mode. Positive pulses occurring at the collector of Q1 drives TR1 through inductor (L1). L1 forms a resonant circuit with the internal capacity of TR1. It should be noted that TR1 has an inherent capacitance of about .15 μF and this must be tuned-out via the series inductance of L1 for efficient power transfer. Note that L1 may be tunable for maximizing results at a set frequency. T1 is the 8-ohm primary section of a matching transformer and serves as an audio choke while offering only its dc resistance for feeding the collector of Q1. Battery (B1) consists of eight standard 1.5 volt AA cells or they can be rechargeable ni-cads.

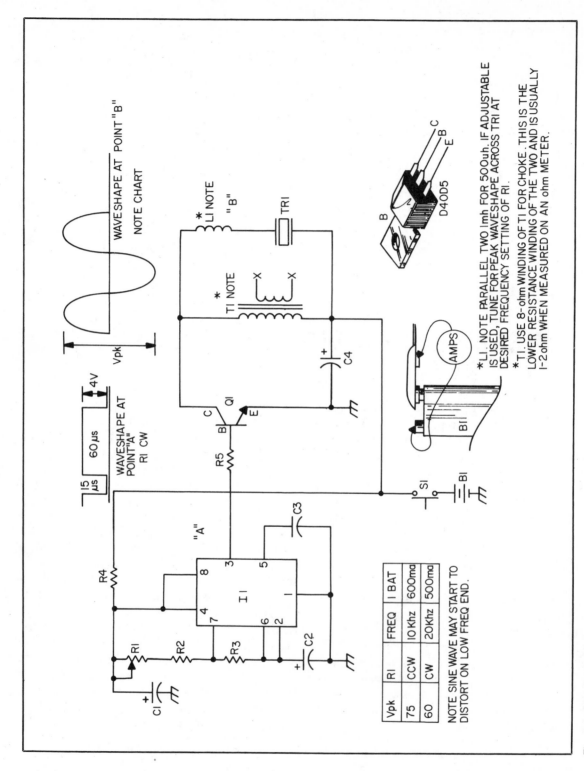

Fig. 8-1. Circuit schematic.

Table 8-1. Phaser Shock-wave Pistol Parts List.

R1	(1)	5 k pot-linear
R2	(1)	2.2 k ¼ watt carbon resistor
R3	(1)	3.9 k ¼ watt carbon resistor
R4	(1)	10 ohm ¼ watt carbon resistor
R5	(1)	1 k ¼ watt carbon resistor
C1	(1)	100 μF 25 volt electrolytic capacitor
C2	(1)	.01 μF 50 volt poly capacitor
C3	(1)	.01 μF 50 volt disc capacitor
C4	(1)	1 μF 50 volt electrolytic capacitor
I1	(1)	555 timer dip configuration
Q1	(1)	D40D5 npn power tab transistor
T1	(1)	1 k 8 ohm transformer (rework)
L1	(2)	1mH inductor
TR1	(1)	Piezo transducer #KSN1001 selected
S1	(1)	Push-button switch
CL1	(1)	10″ snap clips
WR3	(15″)	#24 plastic hook-up wire
PB1	(1)	2″ × 1.3″ × .1 grid perf board
CA1,2	(2)	3½″ plastic cap
CA3	(1)	2″ plastic cap
EC1	(1)	4″ × 3½″ sked 40 PVC tube
HA1	(1)	6″ × 2″ sked 40 PVC tube
KN1	(1)	Small knob ¼″
TA1	(1)	1″ × 3″ × ⅛″ two sided tape
BH1	(1)	8 "AA" cell holder
B1	(8)	1.5 "AA" cells (not included in kit)

Complete kit with printed circuit board available through Information Unlimited, Inc., P.O. Box 716, Amherst, N.H. 03031. Write or call 1-603-673-4730 for price and delivery.

CONSTRUCTION STEPS

1. Layout perfboard as shown in Fig. 8-2. Dimensions are approximately 2″ × 1½″. Trim corners as shown. Identify all parts.

2. Carefully locate and drill holes for the odd pins of transformer (T1) mounting tabs.

3. Assemble components to board using their leads whenever possible and insulating tubing wherever a lead bridges one another. Observe polarity of C1, C4, and position of I1. Wire and solder as shown using Figs. 8-1 and 8-2 showing connections as dashed lines. Connect leads to S1 and CL1.

4. Remove the high impedance (1000 ohm leads) of T1. These can be identified by measuring with an ohmmeter and are the higher resistance windings (usually about 30 ohms). Carefully check wiring for accuracy and shorts, especially around I1.

5. Connect 8-ohm wires from T1 as shown.

These are the lower resistance readings, usually only several ohms.

TESTING

Obtain a Simpson Multimeter and check for absence of a short circuit across battery connection when S1 is depressed. This test verifies any gross errors such as shorts, etc., and helps prevent damage done by the ni-cad battery should a gross mis-wire exist. Insert eight fresh AA cells into holder and clip in TR1 using test clips, rotate R1 to full ccw (low frequency end). Insert meter set on amps across the contacts of S1 and note a current reading of about 600 mA. Rotate R1 full cw and note current changing to approximately 500 mA. You will also note a piercing, uncomfortable sound in your ears or back of your neck depending upon where R1 is set. It may be necessary to wear ear protection since

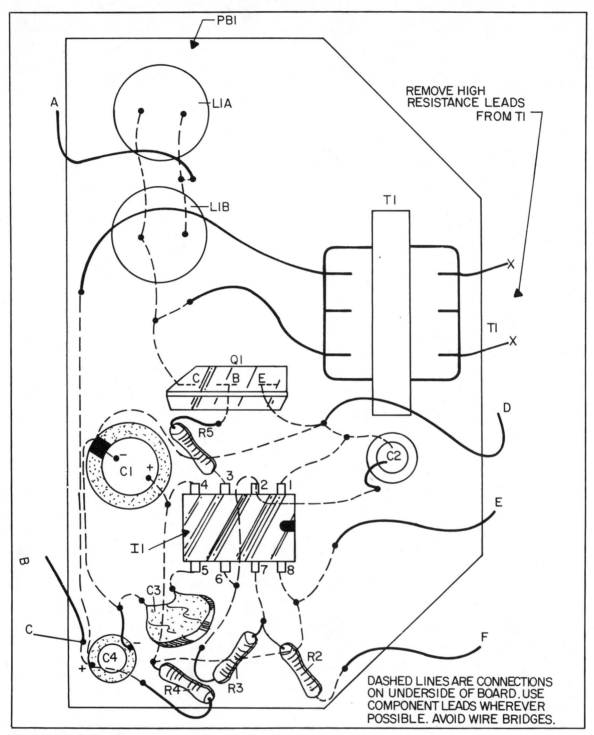

Fig. 8-2. Board assembly.

128

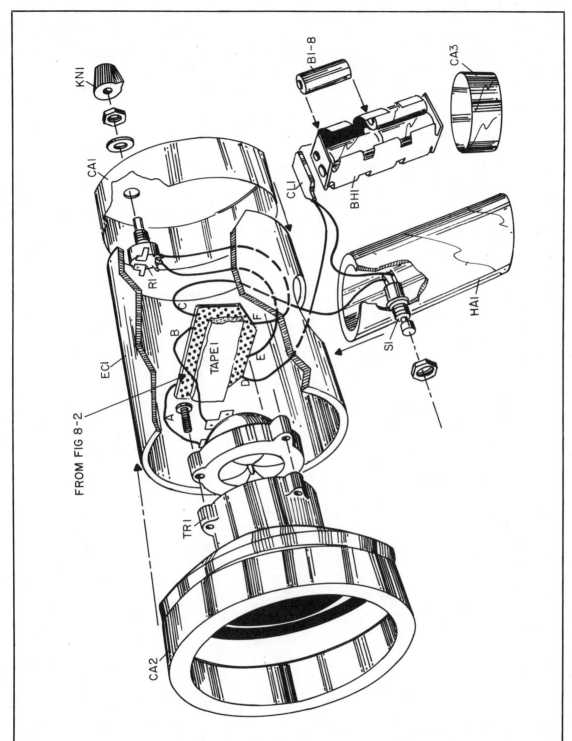

FROM FIG 8-2

Fig. 8-3. Final assembly.

129

this can be extremely painful to most people. A further check of circuit operation can be accomplished by connecting an oscilloscope across the terminals of TR1 and noting the waveforms as shown in Fig. 8-1.

FINAL ASSEMBLY

1. Attach circuit board to TR1 via two-sided tape as shown in Fig. 8-3. Use short leads for wiring.

2. Fabricate a hole in EC1 for handle HA1. This may be difficult and should be done in a drill press using a circle cutter saw. (Do not attempt with fly cutter.) Use a 1⅞" circle cutter set at a slight angle to achieve a pistol grip effect. Deburr and steel wool to a clean finish. Place in middle of EC1.

3. Fabricate HA1 handle with a ⅜" hole for S1 and screw retainer SW1. HA1 is secured via epoxy, RTV, etc., to EC1.

4. Fabricate a hole in CA1 for attachment to R1.

5. Fabricate CA2 by securing to enclosure and cutting out the center with an exacto knife, using inner edge of EC1 for a guide. This method allows a nice even flange or lip for retaining TR1 transducer. Remove flange of TR1 transducer for proper fit to EC1.

6. Final assembly is as shown using own ingenuity. Leads should allow servicing without cutting or unsoldering. Battery holder BH1 is placed in handle HA1 and abutted against S1. Cap CA3 now holds BH1 in place and is secured via a small screw. This method allows easy replacement of the batteries.

Note that this unit is intended for intermittent use only. There is inadequate heatsinking of the components in this compact layout for continuous use.

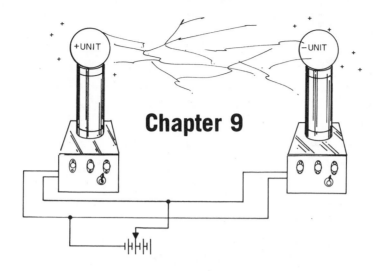

Chapter 9

High-Power Shockwave Gun (PSP5)

This project plans describe the construction of a relatively high-powered acoustical ultrasonic device with adjustable frequency limits of 10 to 20 kHz. Sound pressure levels are 120 to 128 dB based on the .0002 microbar O db threshold of hearing reference.

Assembly consist of two sections, the output head that houses five piezoelectric high-efficiency transducers arranged in a circular array. This housing is approximately 6″ to 7″ in diameter and need only be about 1″ in depth. The power and driving section is shown enclosed into a 5″ cylindrical 3½″ OD tubular enclosure. A handle consisting of a 6″ length of 2″ diameter tubing is now attached to the above via a large hole and also houses the battery pack for the unit. The assembly resembles a handgun with a 6″ flat output head placed at the end of the main enclosure.

The device is intended for laboratory studies or wherever a source of high acoustical ultrasonic energy is required. It can be used to flush out certain species of animals (especially rodents). It can also be used to discourage human encounters and in *some instances* may discourage a potential attack by inducing a combination of paranoia and pain.

CIRCUIT DESCRIPTION

Q1 and Q2 form a symmetrical free-running astable multivibrator whose frequency is controlled by R2 and R5 simultaneously charging both RC circuits of Q1 and Q2 respectively (Figs. 9-1 and 9-2). R4 and R7 are trimpots, that allow symmetry adjustment as well as the high-frequency limit. This limit is necessary since too high a frequency may destroy the transducers at the high drive voltage being used. R3 and R6 are limit resistors preventing accidental shorting of the bases to the collector. C2 and C3 are poly capacitors and determine the C component of the oscillator period. R1 and R8 are the collector load resistors. The outputs of Q1 and Q2 are fed to the bases of Q3 and Q4 respectively through current limiting resistors (R9 and R10). These transistors are the buffer drivers for (Q5 and

ADJUST R2,5 FOR FULL CW
ADJUST R7 FOR 25 μsec OF FIRST HALF OF WAVE
ADJUST R4 FOR 50 μsec OF TOTAL PERIOD
MEASURE AT COLLECTORS OF Q5 AND Q6

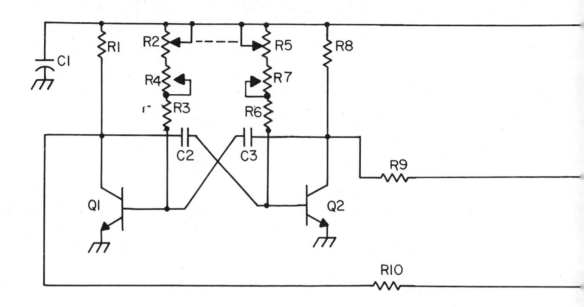

PN2222

D40)5

Fig. 9-1. Circuit schematic.

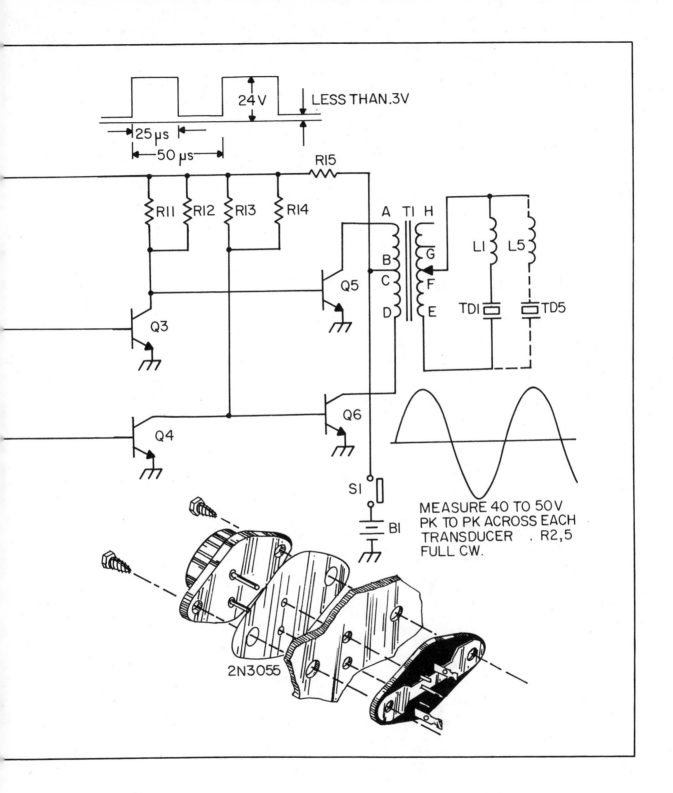

24 V

LESS THAN .3V

25 μs

50 μs

R15

R11 R12 R13 R14

Q3

Q5

A TI H

B

C

D

G

F

E

L1 L5

TD1 TD5

Q4

Q6

S1

B1

2N3055

MEASURE 40 TO 50V
PK TO PK ACROSS EACH
TRANSDUCER   R2,5
FULL CW.

133

Table 9-1. High-Powered Ultrasonic Shock-wave Sun Parts List.

R1,8	(2)	470 ohm ¼ watt resistor
R2,5	(1)	10 k dual linear pot
R3,6	(2)	4.7 k ¼ watt resistor
R4,7	(2)	2 k trimmer
R9,10	(2)	100 ohm ¼ watt resistor
R11,12, 13,14	(4)	220 ohm 1 watt resistor
R15	(1)	1 ohm ½ watt resistor
C1	(1)	100 µF/25 V electrolytic capacitor
C2,3	(2)	.02 µF/50 V poly caps or use (2) .01 in parallel
Q1	(2)	Npn transistor PN2222
Q3,4	(2)	Npn pwr tab transistor D40D5
Q5,6	(2)	Pwr transistor TO3 2N3055
T1	(1)	Transformer Fig. 9-2
E Core	(2)	"E" Cores for T1
MEDBOB	(1)	Nylon bobbin for T1
L1,2,3,4,5,	(5)	1 mH inductors (use two in parellel for favoring high frequency operator)
TD1,2,3,4	(5)	Selected transducers #1020
S1	(1)	Push-button switch
CL1	(1)	Battery chip #20 wire
BM1	(1)	8AA cell battery holder
B1,2,3,4,5, 6,7,8	(8)	1.5 V AA cells or rechargeable ni-cads (not incl'd.)
WR7	(18")	#16 or #18 buss wire
WR20	(8')	#20 plastic hook-up wire
MK1,2	(2)	TO3 mounting kits for Q5,6
RP1	(1)	3¼ × 3¼ #22 fab plate Fig. 9-4
PB1	(1)	5¼ × 2½" perfboard Fig. 9-3
BK1	(1)	2" × ½ × ½ bracket Fig. 9-4
FP1	(1)	8⅝" fab #22 round plate for transducer mounting (OR FAB A RECTANGULAR PLATE)
SGR1	(1)	6" round window screen (Fab into a rectangular houson)
FEN1	(1)	Front enclosure use pie plate or equivalent
CA1	(1)	Rear retaining cover 3½ plastic cap fab
BU1	(1)	½" plastic bushing
HA1	(1)	6" × 1.9" OD PVC sked 40 Fig. 9-6
EN1	(1)	5½ × 3½" OD PVC sked 40 Fig. 9-6
KN1	(1)	Knob
CA2	(1)	1⅞ plastic cap
NU1	(15)	#6 32 kep nuts
SW12	(3)	#6 32 self tapping
SW2	(10)	#6 32 × ½" screw
SW1	(3)	#6 32 × ¼ screw
SW8	(6)	#6 × ¼ sheet metal screw
L1A	(5)	Optional variable inductors for tuning and selection of a particular frequency. 360-580 microhenrys

Complete kit available through Information Unlimited, Inc., P.O. Box 716, Amherst, N.H. 03031, write or call 1-603-673-4730 for pricing and delivery.

Q6) final output power transistors and supply the necessary base current for full collector to emitter saturation while providing a stage of isolation to the free-running multivibrator from the final amplifiers. The outputs of Q5 and Q6 are fed into (T1) transformer as a push-push square wave. The output of T1 is now fed to the individual transducers TD1-TD5 through individual inductors. The inductors

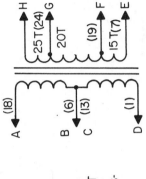

T1 SCHEMATIC

NUMBERS IN PARENTHESIS
ARE BOBBIN PIN DESIGNATIONS

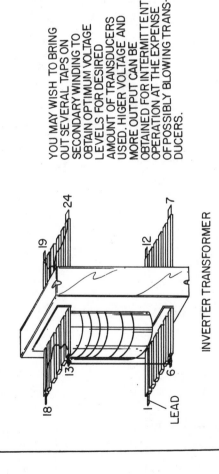

INVERTER TRANSFORMER

YOU MAY WISH TO BRING OUT SEVERAL TAPS ON SECONDARY WINDING TO OBTAIN OPTIMUM VOLTAGE LEVELS FOR DESIRED AMOUNT OF TRANSDUCERS USED. HIGER VOLTAGE AND MORE OUTPUT CAN BE OBTAINED FOR INTERMITTENT OPERATION AT THE EXPENSE OF POSSIBLY BLOWING TRANSDUCERS.

STEPS FOR WINDING T1 BOBBIN

1. PARALLEL OR BIFILAR WIND 10 TURN PAIRS OF #20 ENAMEL COVERED MAGNET (OR VINYL) COVERED WIRE AS SHOWN EVENLY SPACED ALONG ENTIRE BOBBIN LENGTH. THIS FULLY UTILIZES ALL OF THE CORE.
2. SECURE AND SOLDER LEADS, A, B, C, & D TO LUGS AS SHOWN. MAKE SURE THAT THE ENAMEL COVERING IS ENTIRELY REMOVED AND LEADS ARE TINNED BEFORE SOLDERING. IDENTIFY BY MARKING.
3. WIND A SINGLE LAYER OF TAPE OVER THIS WINDING.
4. WIND 25 TURNS OF #20 PLASTIC COVERED WIRE NEATLY OVER ABOVE AND CONNECT LEADS TO LUGS ON BOBBIN. BRING OUT TAPS, AT 15, & 20 TURNS FOR SELECTION OF VOLTAGE OUTPUT. IDENTIFY BY MARKING. WHEN THE DESIRED TURNS ARE DETERMINED,
5. CAREFULLY PLACE "E" CORES TO BOBBINS AND TAPE TOGETHER. NOTE IT IS MORE CONVENIENT TO USE YOU MAY WISH TO PLACE THE SINGLE WINDING E & H FIRST ON THE BOBBIN. PLASTIC JACKETED WIRE AS THE INSULATION IS EASIER TO REMOVE, WHEN SELECTING THE OPTIMUM TAP ON THE OUTPUT WINDING.

Fig. 9-2. T1 assembly.

135

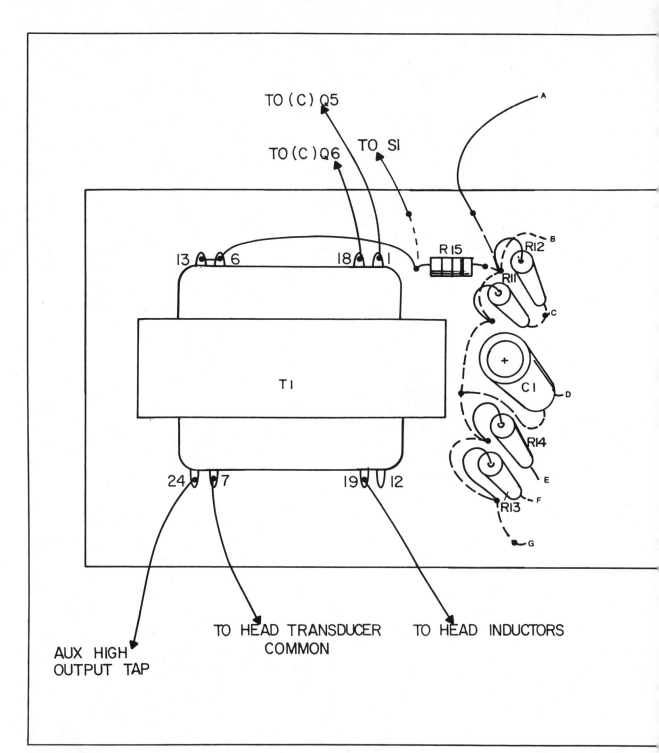

Fig. 9-3. Assembly board.

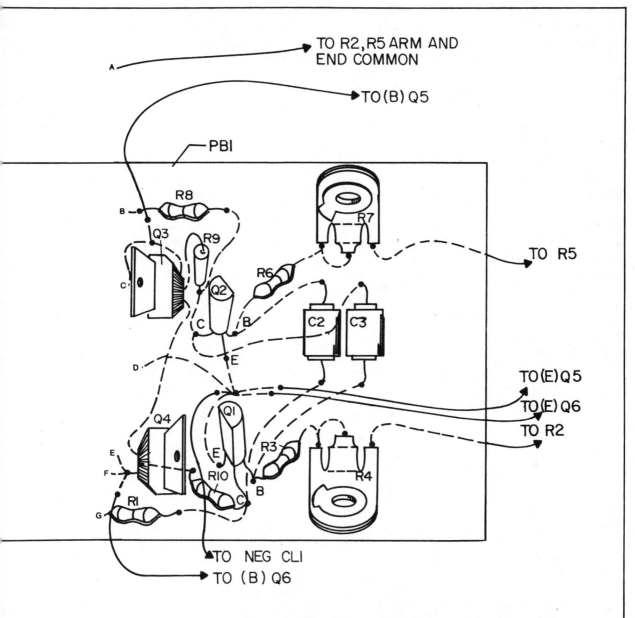

TO R2, R5 ARM AND END COMMON

TO (B) Q5

PBI

A

R8

B

Q3

R9

R7

TO R5

C

R6

Q2

C B

C2 C3

E

D

TO (E) Q5

TO (E) Q6

Q1

TO R2

Q4

R3

E

F

E

R10

B

R4

C

R1

G

TO NEG CLI

TO (B) Q6

NOTE: IT MAY BE NECESSARY TO PARALLEL CONNECT (2) .01 MFD POLY STYRENE CAPS TO OBTAIN THE .02 MFD REQUIRED. THIS CONNECTION IS NOT SHOWN ON THIS FIGURE FOR CLARITY.

combine with the inherent capacity of the transducers and form a resonant circuit at approximately 13 to 15 kHz. The waveshape is now more sinusoidal across the transducer rather than square. The sine shape is more desirable as it allows a higher voltage waveform consequently more output per the rms value than does the original square wave.

CONSTRUCTION STEPS

1. Assemble T1 transformer (Fig. 9-2).

2. Layout and assemble perfboard as shown (Fig. 9-3). Observe correct polarity of components. Use component leads wherever possible for making connections beneath the board. Avoid wire bridges.

3. Fabricate rear plate (RP1) as shown (Fig. 9-4). Use MK1 mounting kit for template to locate holes for Q5 and Q6. Make sure these holes are free from burrs as they will cut through the mica insulation washer and short out to the chassis.

4. Fabricate bracket (BK1) as shown in Fig. 9-4. Bracket secures the assembly board (Fig. 9-3) to the rear plate.

5. Mount components on rear plate and connect all wires from assembly board.

6. Wire CL1 and C1 to assembly board. These wires should be 20 gauge as they are the power input leads.

7. Fabricate (FEN1) rear enclosure using a pie plate or equivalent shaped enclosure.

8. Fabricate (FP1) transducer mounting plate from a diameter circle that will fit FEN1 in Step 7. Locate and punch out (5) 1⅞" holes located on a 1¾" radius. Evenly space and use #22 gauge aluminum.

9. Mount transducers (TD1-5) as shown in Fig. 9-5 with protective screen (SCR1) as shown.

10. Wire and solder in inductors (L1-5) as shown. Form some buss wire into two circular configurations and complete wiring using these wire circles as connection points. Note the polarity markings on transducers.

Do not connect transducers to T1 at this point.

11. Fabricate (EN1) main enclosure as shown in Fig. 9-6 from a 5½" length of 3½ OD PVC Schedule 40. Use either black or white. Locate and drill a 1⅞" hole 2¼" from end. Use a circle saw and drill at a slight angle to obtain a pistol grip effect.

12. Fabricate (HA1) handle from a 6" length of 1.9" OD PVC Schedule 40 as shown in Fig. 9-6.

13. Fabricate (CA1) rear retaining cover by placing plastic cap over EN1 and carefully remove center of cap using the wall of EN1 as a guide. Use a sharp exacto knife or equivalent.

TEST AND CALIBRATING

1. Connect 12 Vdc between transformer buss jump and ground. Rotate R2/R5 full cw and note a current draw of between 250 to 500 milliamperes.

2. Connect scope to output of T1 and adjust R4 so that first *half* of wave is 25 microseconds. Adjust R7 for *total* period of 50 microseconds. Repeat until waveform is exactly symmetrical in time and has a total period of 50 microseconds or 20 kHz. Peak-to-peak volts should be 40 and both transistors Q5 and Q6 should be cool running. Input current should be 500 milliamperes or less.

3. Rotate R2/R5 ccw and note period increasing to approximately 150 microseconds or 6 to 7 kHz. Range of control should be smooth and stable. Symmetry may change slightly but should always favor the higher frequency end of the frequency adjustment.

4. Remove power and connect transducer head as shown. Apply *6 volts* and measure waveshape across transducer with R2/R5 full cw. It should be sinusoidal and be between 30 and 40 volts peak-to-peak. Check all transducers. Repeat with R2/R5 full ccw. Wave may be distorted but still sinusoidal. You will note a definite resonance occurring when this frequency control is varied. This is due to the fixed inductors resonating with the transducers capacity. *These inductors may be variable and adjusted to favor a certain frequency for optimizing output.* This is usually indicated by the waveshape across the particular transducer peaking and becoming a more perfect sinusoidal shape when the adjustable inductor is tuned. Note that these tuned inductors will only favor the frequency that they are tuned for.

5. At this point the builder has a choice of increasing output at the cost of blitzing the trans-

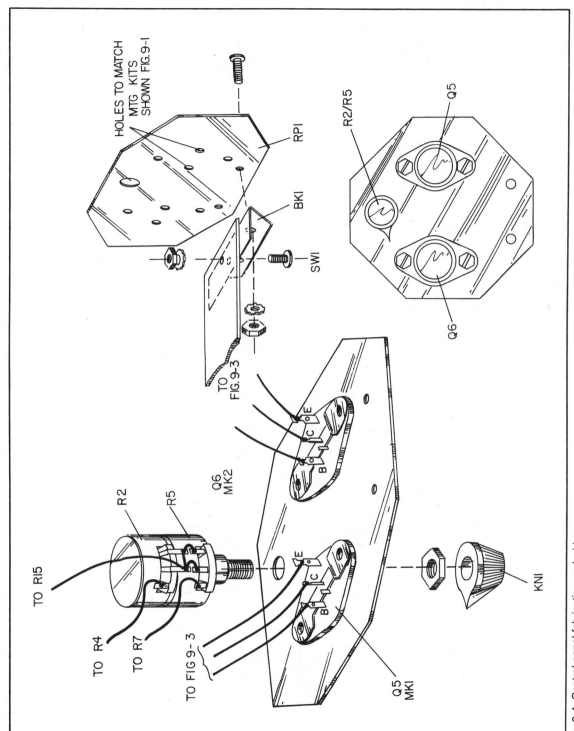

HOLES TO MATCH
MTG KITS
SHOWN FIG. 9-1

RPI

BKI

SWI

TO FIG. 9-3

R2/R5

Q5

Q6

Q6
MK2

R2

R5

TO RI5

TO R4

TO R7

TO FIG 9-3

E
C
B

E
C
B

KNI

Q5
MKI

Fig. 9-4. Control panel fabrication and wiring.

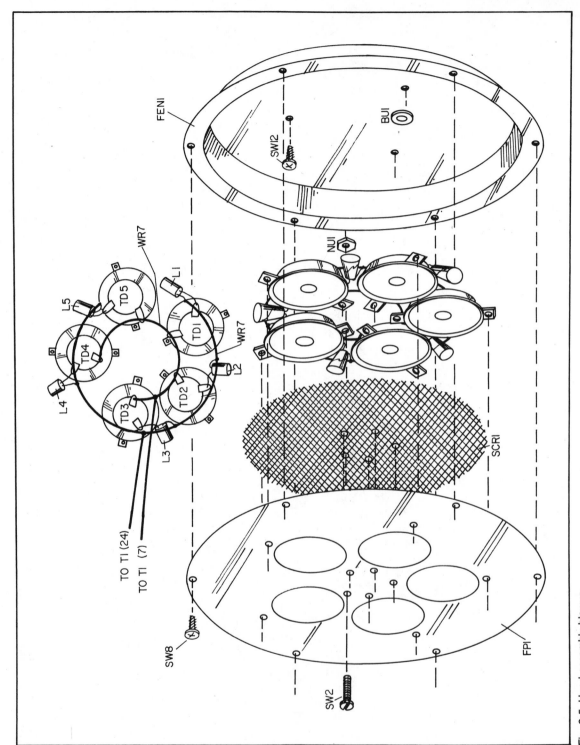

Fig. 9-5. Head assembly blowup.

140

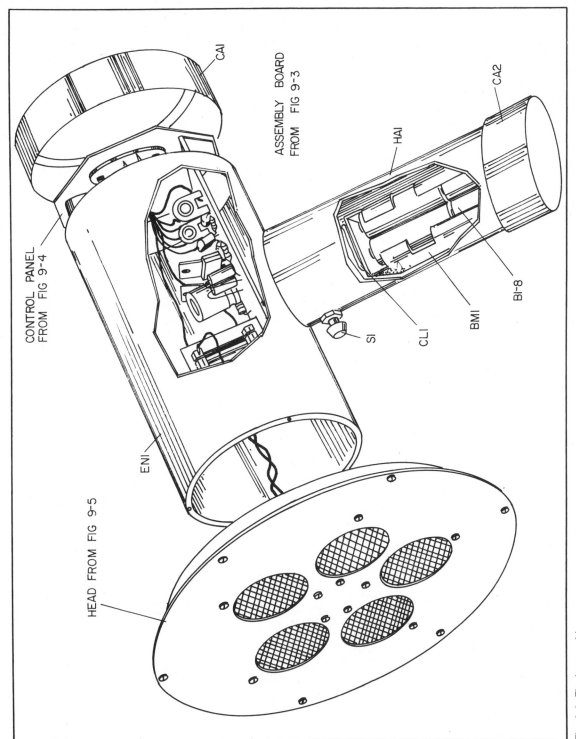

CAI

ASSEMBLY BOARD
FROM FIG 9-3

CA2

HAI

CONTROL PANEL
FROM FIG 9-4

BI-8

BMI

CLI

SI

ENI

HEAD FROM FIG 9-5

Fig. 9-6. Final assembly.

141

ducers. If this frequency is going to favor the low end, higher output voltage may be applied to the transducers without failure. However, if this frequency is adjusted on the high side, overheating will definitely occur and may cause failure. The power charts on the transducer claim a maximum allow-able (rms) voltage of approximately 50 volts. If the frequency of operation is well below 20 kHz, more voltage can be applied up to where the output starts to divide as indicated with a receiving transducer observing the waveshape on a scope.

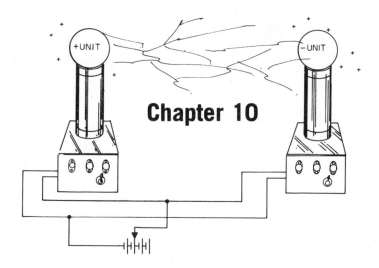

Programmable High-Power Ultrasonic Generator (PPG1)

This project can be used to prevent unauthorized intrusion by creating an invisible field of highly irritating and uncomfortable ultrasonic energy at moderate sound pressure levels. The basic unit consists of a centrally located power-controlled oscillator whose output is continually changing in tone, feeding four remotely located ultrasonic transducers positioned for maximum protection of the area in question. This system is not to be confused with the motion detectors where low-power ultrasonics are only used as a means of sonic radar. The system can be made to automatically turn on via detection circuitry so that an intruder entering an area would immediately encounter a high-pitched squealing noise. This noise would steadily increase in frequency and cause an uncomfortable feeling in the back of the head. This pitch can be automatically changing which adds to the discomfort. The system can also be adjusted to produce a high volume audible alarm at the low end for further effect producing the usual paranoia and intense nervousness necessary to cause an immediate vacating of the premises. Frequency limits and sweep

times are easily controlled via the front panel knobs along with an auto/manual selector switch that enables either a steady or constantly varying tone. System contains on/off indicator lamp, fuse, auto/manual selector switch, tone adjust control, sweep rate control. Remote control and transducer connection strips are conveniently located on the rear of the unit. Power requirements are 120 Vac at less than 60 watts.

This project shows how to construct a moderately high-powered, variable-sweep frequency, ultrasonic-acoustical generator capable of handling the equivalent of 400 watts of resultant power obtainable from an equivalent conventional dynamic transducer system. This is possible due to a recently developed piezoelectric tweeter speaker developed by Motorola. These devices produce six times the acoustical output obtainable from conventional dynamic methods for the same required driving power. This feature allows the use of relatively lightweight, low-powered equipment to obtain these high acoustical energies. Effective animal or rodent control with a device such as this is the result of automatically varying the frequency so these animals cannot develop a tolerance to it (such as deafness, immunity or slot hearing). Many fixed frequency ultrasonic devices lose their effectiveness after several weeks because of this immunity problem.

It should be understood that certain people are affected more than others, some to a point where they will vomit, experience severe headaches, and cranial pains. Some people will experience severe pain in the ear, teeth, lower head, etc. Statistically, women and younger children are many times more sensitive to this device than are average male adults. With this in mind, the builder must exercise consideration when testing and using the device as many people will not be aware of the source of this pain or uncomfortable feeling and will attribute it to a headache or other physical ailment. Also certain people are affected mentally to a point of actually losing their tempers completely or becoming extremely quick tempered. Some people will experience a state of extreme anxiety when overly exposed. Therefore, consideration must be used at all times when testing or using this or similar devices.

Construction using the tweeters in an array configuration is more hazardous to people than the individual placement of these tweeters when used to control a large area. The array approach is intended when the unit can be directed towards known intrusion points.

BRIEF DESCRIPTION OF OPERATION

We shall now proceed to give a brief description of the electronic circuitry referencing the "block diagram" layout (Fig. 10-1 and Table 10-1). The device is essentially a high-powered source of ultrasonic acoustical energy of a frequency constantly varying between preset low and high limits at an operationally adjustable sweep rate when operated in the "auto" mode. "Manual" mode allows presetting a continuous frequency to a fixed value and also the lower limit at the swept frequency when in the "auto" mode. Please note the abbreviation of the stages in the block diagram. They will be used from here to the end of the book. An array or individually located tweeter transducers (TR) of from 4 to 16 are driven by a transformer coupled final amplifier (FA) consisting of two class-"B" operated power transistors mounted on a heatsink. These transistors are driven by a pulse driver and inverter (PID). This stage is driven by a voltage control oscillator (VCO) with an adjustment for the lower frequency limit. The VCO is driven by either a varying dc level generator by the staircase generator (SCG) in the "auto" mode or an adjustable dc level termed "manual" frequency control (MFC).

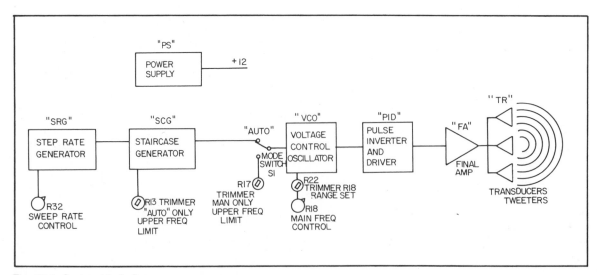

Fig. 10-1. System block diagram.

144

Table 10-1. Programmable High-Powered Phaser Property Guard Parts List.

Ref	Qty	Description
R1,4,27 28,29	(5)	1 k ¼ watt resistor
R2,3,15	(3)	470 ohm ¼ watt resistor
R5	(1)	5.6 M ¼ watt resistor
R6	(1)	1.2 k ¼ watt resistor
R7,16	(2)	470 k ¼ watt resistor
R8	(1)	560 k ¼ watt resistor
R9	(1)	1.2 M ¼ watt resistor
R10	(1)	1.5 M ¼ watt resistor
R11	(1)	1 M ¼ watt resistor
R12	(1)	1.8 M ¼ watt resistor
R22,13	(2)	25 k trimpot hor.
R14	(1)	68 k ¼ watt resistor
R17	(1)	100 k trimpot hor.
R18,S2	(1)	10 k pot/switch
R19	(1)	39 k ¼ watt resistor
R20	(1)	10 ohm ¼ watt resistor
R24	(1)	5.6 k ¼ watt resistor
R25	(1)	220 ohm ¼ watt resistor
R21	(1)	2.2 k ¼ watt resistor
R30,R31 A & B	(4)	220 ohm 1 watt resistor
R32 A & B	(1)	50 k dual pot
R33,34	(6)	1 ohm ½ watt resistor (connect 3 in parallel)
R35 A & B	(2)	40 ohm 10 watt power
R26,23	(2)	4.7 k ¼ watt resistor
C1,2	(2)	10 μF@25V electrolytic cap
C6	(1)	1000 μF@25 V electrolytic cap
C8	(1)	.001 μ@50 V poly cap
C11	(1)	4700-8000 μF@16 V electrolytic cap
C5	(1)	470 pF/50 V disc cap
C9	(1)	.01 μF/50 V disc cap
C4,7	(2)	.1 μF/50 V disc cap
C3	(1)	1000 pF/50 V disc cap
I2	(1)	LM566 Vco dip
I1	(1)	LM3900 dip quad amp
Q1,2,3,4,5	(5)	PN2222 npn transistor
Q6	(1)	PN2907 pnp transistor
Q7,8	(2)	D40D5 npn pwr tab transistor
Q9,10	(2)	2N3055 npn TO3 transistor
D2	(1)	50 V 1 amp rect. 1N4002
D1,3	(2)	1N914 signal diode
CR1,2,3,4	(4)	50 V 3 amp rect. 3AQ5
T2	(1)	Power transformer 12 V/3 amps
T1BOB	(1)	Medium bobbin for T1
T1CORE	(1)	Medium core "E" for T1
HDWRE	(1)	U bolt bracket for T1
LUGS	(4)	Solder lugs
T1	(1)	Assembled per Fig. 10-18
S1	(1)	Small slider switch dpdt
HS1	(1)	Dual TO3 heatsink
FT1	(4)	Rubber feet stick-on
WN1	(1)	Small wire nuts
TE1	(1)	4 screw terminal
MK1,2	(2)	TO3 mounting kits
SW2	(12)	6/32-½" long screw
KN1	(2)	¼" knob
BU1	(1)	Line cord bushing
NU1	(8)	6/32 locknuts
TD1	(4)	Piezo transducers, see Fig. 10-26
NE1	(1)	Neon lamp short body long leads
LB1	(1)	Lamp bushing red
CA1	(1)	7 × 21" 22 ga metal case fab per Fig. 10-15
PB1	(1)	2.5 × 5" perfboard
FH1	(1)	Lamp fuse 3 ga panel hold and 1 amp sloblo fuse
CO1	(1)	Molded 3 wire line cord
WR3	(6')	#24 hook-up wire red
WR4	(6')	#24 hook-up wire black
WR1	(2')	#18 hook-up wire red
WR2	(2')	#18 hook-up wire black

Use low voltage wire such as that suitable for speakers, etc., for connecting up remotely located transducers. Use ingenuity in mounting and positioning these stations and protect against rain if used outside. *Connect transducers stations in parallel.*

Complete kit with optional PC board available from Information Unlimited, Inc., P.O. Box 716, Amherst, N.H. 03031. Write or call 1-603-673-4730 for pricing and delivery.

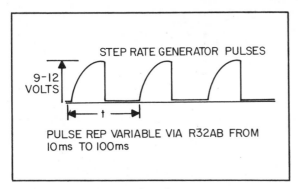

Fig. 10-2. Waveshape at Q1 collector.

These two functions are selected by the "mode" switch. The SCG also contains the adjustment for the upper frequency limits of the VCO. The SCG is driven by the step rate generator (SRG) that determines the rate of frequency sweep between the present "auto" limits. The SRG is controlled by the sweep control rate (R32). Please note that the test points and readings were obtained with the device operating from a 12 volt power supply. These readings will correspondingly change when using the internal ac power supply as shown. Full-bridge method produces a operating voltage of 15 while the fullwave center-tap version produced only 8 volts.

CONSTRUCTION STEPS

Observe the waveforms in Figs. 10-2 through 10-12 and refer to Fig. 10-13 (schematic) and Fig. 10-14 (assembly board). The construction steps to this device will be divided into sections, each containing assembly steps, circuit theory and functions. Note: use component leads for wiring whenever possible.

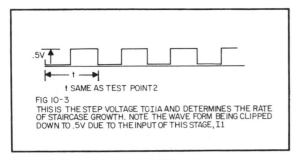

FIG 10-3
THIS IS THE STEP VOLTAGE TO I1A AND DETERMINES THE RATE OF STAIRCASE GROWTH. NOTE THE WAVE FORM BEING CLIPPED DOWN TO .5V DUE TO THE INPUT OF THIS STAGE, I1

Fig. 10-3. Waveshape at pin 2 I1A.

Step-Rate Generator (SRG)

Lay out the 2.5″ × 5″ piece of perfboard as shown Fig. 10-14. Use for component interconnection sketch aid. Insert and wire R1, R2, R3, R4, C1, C2, Q1, Q2, and C3 as shown. Observe polarity of C1 and C2 and position of Q1 and Q2.

Insert the three leads (J,K,M) each about 12″ as shown to R32A and R32B (sweep control rate). Use holes in perfboard for strain-relieving of these wires. Tape together and wire to R32A and B as shown in schematic (Fig. 10-13). Check wiring and soldering for errors and quality.

Apply 12 Vdc to respective points and measure. Observe test points with scope as shown Fig. 10-2. Vary R32B and note frequency varying from 10 to 100 pps.

This circuit is nothing more than a free-running multivibrator consisting of an astable switch with Q1 and Q2 switching from an "off" to "saturating" mode producing a square wave output of voltage equal to approximately V_{CC} (+12 V).

This circuit performs as follows: Q1 (for reference sake) starts to conduct causing its collector voltage to decrease consequently producing a negative voltage via C1 at the base of Q2 completely "off." C1 now must discharge through R32A and R4 to a point where the base of Q2 will cause conduction repeating this sequence of events through C2 and the base of Q1. You will note that the waveform may be deteriorated at the high repetition rate end. This is due to the charging time of C1 and C2 through R2 and R3 respectively. The dual pot R32A and B determines the discharge time of C1 and C2 consequently the pulse repetition rate.

Staircase Generator (SCG)

Drill extra holes for pins of I1 and arm of R13 perfboard. Insert and wire I1 and other related components, R5, R6, R7, R8, R9, R10, R11, R12, R13, D1, and C4. Observe polarity and position of D1 and I1. Wire and solder as shown. Mount C5 under board as shown.

Insert and wire D2 and C6 and connect +12 and ground buss lines together with SRG section. Note that D2 provides protection of I1 and I2 from reversed polarity during testing. Check wiring and

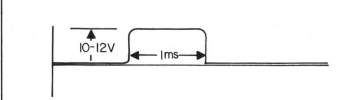

SCHMIDT OUTPUT PULSE IS THE RESET FOR THE STAIRCASE GENERATOR AND OCCURS WHEN THE STAIRCASE GROWS TO ABOUT 75 % OF THE 12 VOLTS VC OR ONCE EVERY SWEEP. THIS IS THE OUTPUT OF THE SCHMIDT DISCRIMINATOR IIB. IT IS TRIGGERED BY IIC CONNECTED AS A COMPARATOR SAMPLING THE OUTPUT LEVEL AT PIN 4, IIA. IT MAYBE NECESSARY TO EXTERNALLY TRIGGER THE SCOPE FOR A CLEAR PICTURE OF THIS WAVEFORM.

Fig. 10-4. Waveshape pin 9 I1B.

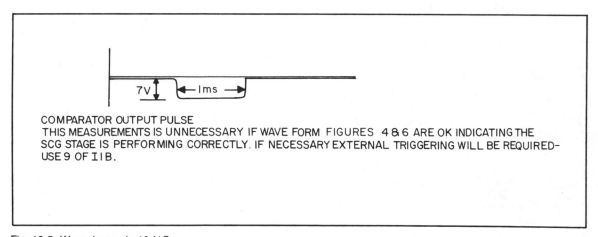

COMPARATOR OUTPUT PULSE
THIS MEASUREMENTS IS UNNECESSARY IF WAVE FORM FIGURES 4 & 6 ARE OK INDICATING THE SCG STAGE IS PERFORMING CORRECTLY. IF NECESSARY EXTERNAL TRIGGERING WILL BE REQUIRED- USE 9 OF IIB.

Fig. 10-5. Waveshape pin 10 I1C.

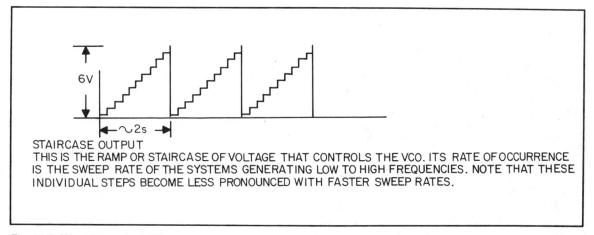

STAIRCASE OUTPUT
THIS IS THE RAMP OR STAIRCASE OF VOLTAGE THAT CONTROLS THE VCO. ITS RATE OF OCCURRENCE IS THE SWEEP RATE OF THE SYSTEMS GENERATING LOW TO HIGH FREQUENCIES. NOTE THAT THESE INDIVIDUAL STEPS BECOME LESS PRONOUNCED WITH FASTER SWEEP RATES.

Fig. 10-6. Waveshape pin 4 I1A.

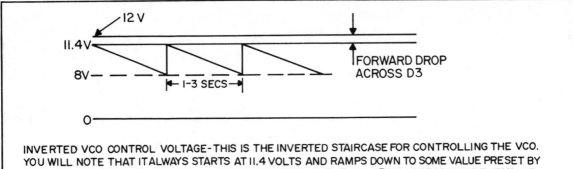

INVERTED VCO CONTROL VOLTAGE-THIS IS THE INVERTED STAIRCASE FOR CONTROLLING THE VCO. YOU WILL NOTE THAT IT ALWAYS STARTS AT 11.4 VOLTS AND RAMPS DOWN TO SOME VALUE PRESET BY R13. DIODE D3 SUPPLIES OFFSET AND PREVENTS PIN 5 OF THE VCO FROM APPROACHING THE VALUE VC 12V AND CONSEQUENTLY LOCKING UP. THE RAMP OUTPUT OF Q3 IS FAIRLY LINEAR OVER THE REQUIRED RANGE OF SWEEP VOLTAGE.

Fig. 10-7. Waveshape at pin 5 I2 auto position.

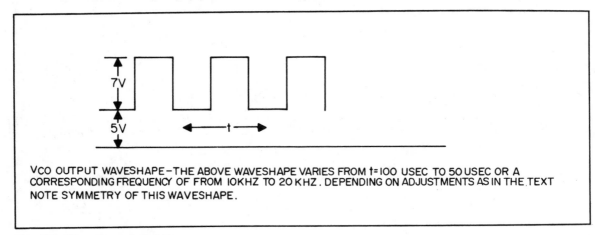

VCO OUTPUT WAVESHAPE-THE ABOVE WAVESHAPE VARIES FROM t=100 USEC TO 50 USEC OR A CORRESPONDING FREQUENCY OF FROM 10 KHZ TO 20 KHZ. DEPENDING ON ADJUSTMENTS AS IN THE TEXT NOTE SYMMETRY OF THIS WAVESHAPE.

Fig. 10-8. Waveshape at pin 3 I2.

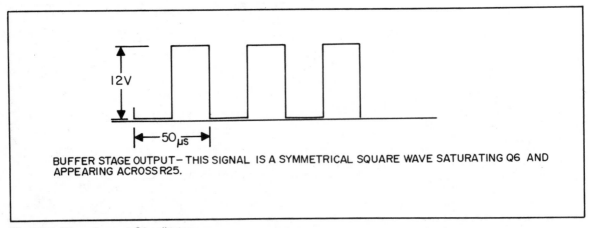

BUFFER STAGE OUTPUT- THIS SIGNAL IS A SYMMETRICAL SQUARE WAVE SATURATING Q6 AND APPEARING ACROSS R25.

Fig. 10-9. Waveshape at Q6 collector.

148

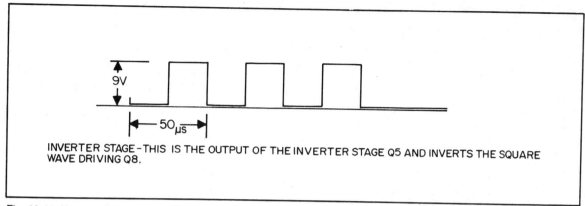

INVERTER STAGE – THIS IS THE OUTPUT OF THE INVERTER STAGE Q5 AND INVERTS THE SQUARE WAVE DRIVING Q8.

Fig. 10-10. Waveshape at Q5 collector.

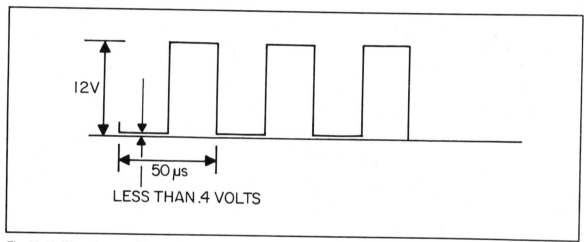

LESS THAN .4 VOLTS

Fig. 10-11. Waveshape at Q7 collector final amp not connected.

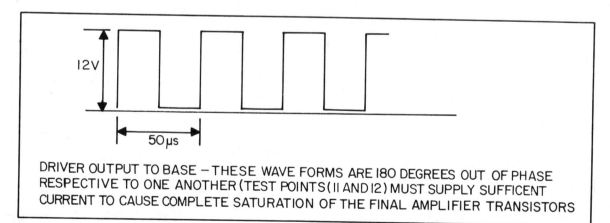

DRIVER OUTPUT TO BASE – THESE WAVE FORMS ARE 180 DEGREES OUT OF PHASE RESPECTIVE TO ONE ANOTHER (TEST POINTS (II AND I2) MUST SUPPLY SUFFICENT CURRENT TO CAUSE COMPLETE SATURATION OF THE FINAL AMPLIFIER TRANSISTORS

Fig. 10-12. Waveshape at Q8 collectors final amp not connected.

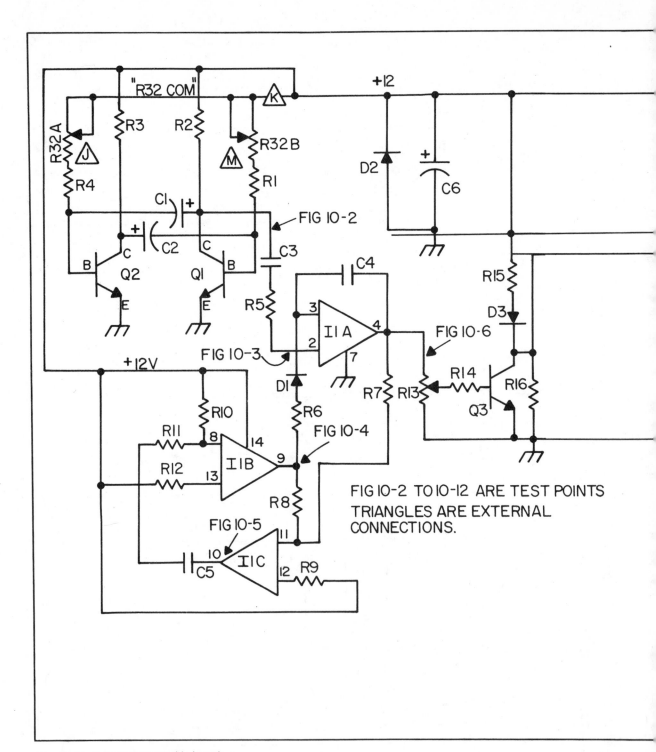

Fig. 10-13. Schematic assembly board.

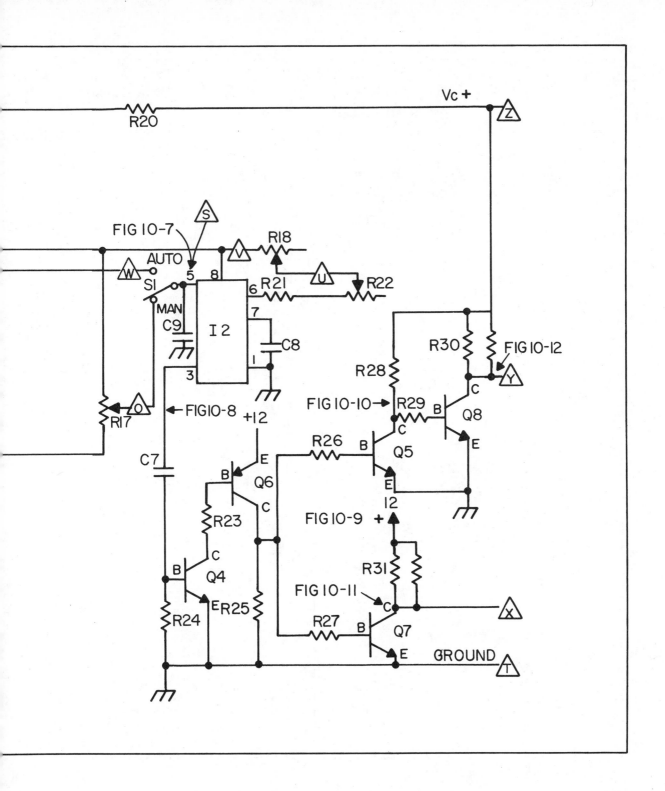

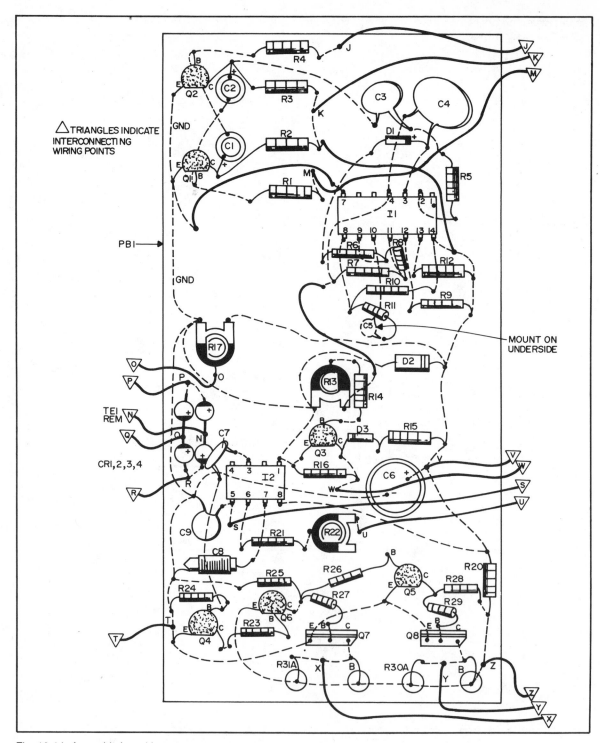

Fig. 10-14. Assembly board layout.

soldering for errors remembering that mistakes can cost you the I1 integrated circuit.

Connect 12 volts as before and observe test points 3, 4, 5, and 6 shown. Set R32AB at midrange. Observe respective waveforms shown in Figs. 10-3, 10-4, 10-5 and 10-6. This circuit utilizes a quad operational amplifier. Three of these circuits are used. The other remains as a spare.

The first amplifier I"1A" functions as an integrator where the current pulses from the step-rate generator are integrated and held on C4. It is this voltage that builds a step at a time determined by the step-rate generator (SRG). Discharging C4 is necessary to reset the circuit and again start from the bottom of the staircase and resetting at this upper limit. This is accomplished by amplifier I"1C" functioning as a comparator, sampling the staircase level when it reaches approximately 75% of 12 volts and triggering a Schmidt-discriminator consisting of I"1B" that resets I"1A" through blocking diode D1 and resistor R6 commencing the sequence once again. The staircase voltage variation produced at pin 4 of I"1A" is used as the control voltage for a "voltage-controlled oscillator" (VCO).

Voltage-Controlled Oscillator (VCO)

Drill extra holes in perfboard for I2 and the arms of R17 and R22. Insert R14, R15, R16, R21, D3, C7, C8, and C9. Insert R17 and R22 and bend over tabs to secure in place. Insert Q3 and I2 and note proper position. Wire and solder as shown.

Insert "12" wire leads to points indicated on board and connect to external controls R18 and S1. Strain-relieve and twist these leads via inserting through holes in perfboard. Connect 12 V and ground buss lines as shown. Carefully check all wiring and soldering for shorts, etc. Note errors can be costly to the integrated circuits.

Connect 12 volts as before. Place S1 in "auto" position and R18 at midrange. Observe test point Fig. 10-7 and note dc level starting from 11.4 and stepping down to about 8 V. (inverse of waveform at the test point in Fig. 10-6).

Preset R22 to midrange and connect scope to the test point in Fig. 10-8. This point is the pulse output of the VCO and should be constantly varying

along with the staircase ramp voltage noted at test point in Fig. 10-7. It is this varying frequency that is amplified and used to drive the transducers of this system. When in the "auto" mode of S1 you will note that this frequency value is constantly varying between certain limits at a sweep rate determined by R32B. When in the "manual" mode of S1, the VCO frequency output is adjusted by R18, "Manual Frequency Control" and does not vary or sweep back and forth.

To adjust the upper and lower frequency limit, perform the following: Adjust R32 for slowest sweep rate. Determine limits in this case 10 kHz to 25 kHz while observing the test point in Fig. 10-8. Suggested frequency limits are as follows:

☐ *Anti-intrusion*—Set R18 to 5 kHz sweeping to 10 kHz.

☐ *Crowd control*—Set R18 to 10 kHz sweeping to 15 kHz.

☐ *Rodent control*—Set R18 to 15 kHz sweeping to 25 kHz.

If inhabited by people set R18 at lower 20 kHz and sweep to 25 kHz.

A. Preset all trimpots midrange, R18 full cw (highest frequency).

B. Sweep switch to "auto".

C. Set R22 to 20 kHz—50 μsec (low end).

D. Set R13 to 25 kHz—40 μsec (high end) sets frequency window.

E. Set sweep switch to manual.

F. Set R17 to 20 kHz—50 μsec.

G. Check range of R18 low end to less than 10 kHz both auto and manual. You will note that when in the "auto" sweep position that a frequency window of 5 kHz exists with the low point being set by R18.

This circuit is the heart of the device. It determines the operating frequency as a function of the level dc voltage at pin 5 of I2. This voltage is the result of the staircase generator. You will note that the frequency excursions are controlled via R13 (upper limit) and R22 (lower limit).

The VCO is the integrated circuit I2 and contains an internal current source that charges external capacitor C8. When C8 reaches a certain voltage a Schmidt triggers and produces a square wave output at pin 3 (Fig. 10-8). An external resistor connected at pin 6 along with the external capacitor C8 determines the center frequency of the device.

Pulse Inverter and Driver (PID)

Drill extra holes for the bases of Q7 and Q8.

Insert R23, 24, 25, 26, 27, 28, 29, 30, 31, Q4, Q5, Q6, Q7 and Q8 as shown. Note Q6 is a 2N2907 pnp. Note position and polarity of transistor. Wire and solder as shown connecting V_c+, grounds and R20 as shown. Check for accuracy soldering and shorts. Connect 12 Vdc as before. Observe test point Fig. 10-8. Set S1 to "manual" and adjust R18 for 20 kHz.

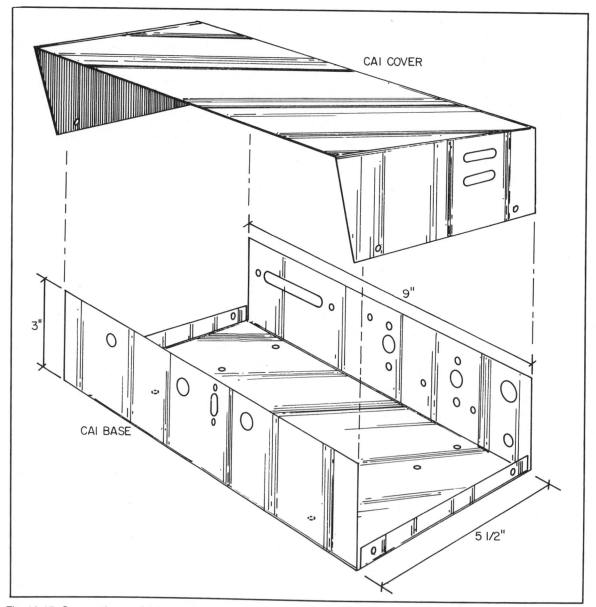

Fig. 10-15. Case and cover fabrication.

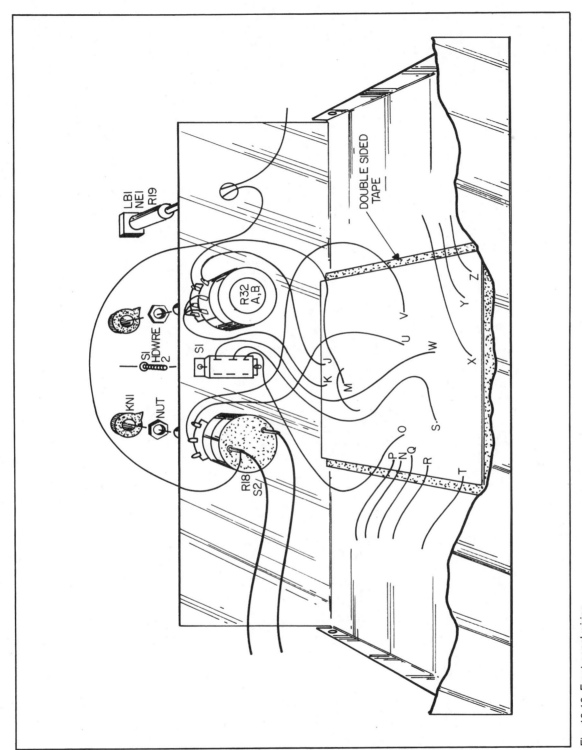

Fig. 10-16. Front panel wiring.

155

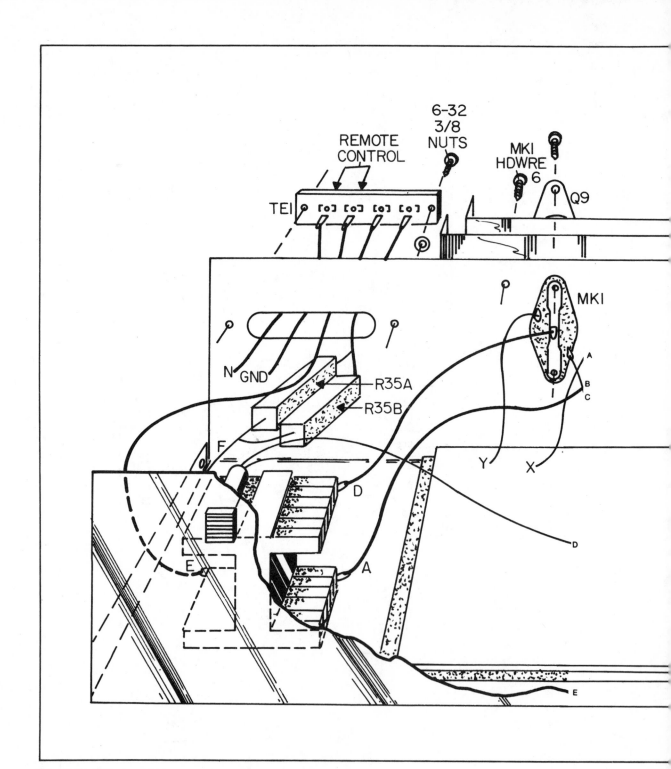

Fig. 10-17. Rear panel wiring.

156

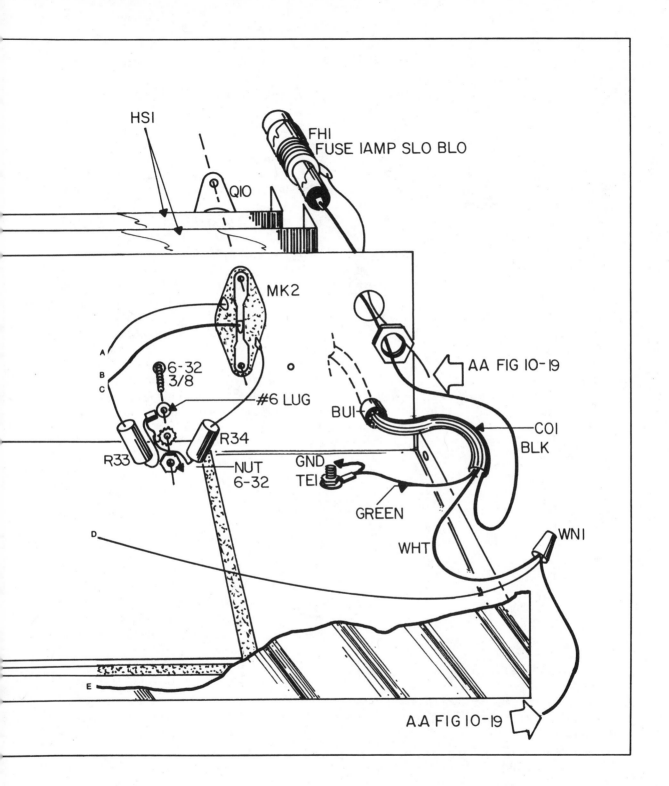

HSI

FHI
FUSE IAMP SLO BLO

QIO

MK2

A
B
C

6-32
3/8

#6 LUG

AA FIG IO-19

BUI

COI
BLK

R34

R33

NUT
6-32

GND
TEI

GREEN

D

WHT

WNI

E

AA FIG IO-19

157

Note the following waveforms at test points Figs. 10-9, 10-10, 10-11, 10-12. Connect external trigger of scope to test point Fig. 10-8 to establish time reference. These waveforms are such *without* x and y being connected to the *final amp*. When connected the voltage should drop from 12 to less than 0.4 on above waveforms at Figs. 10-11 and 10-12.

This stage is a dc amplifier with an inverter stage supplying two positive going symmetrical pulses of current 180 degrees out of phase for driving the final amplifier in a push-pull configuration.

Pulses occurring at B of Q4 switch Q4 and Q6 off and on (saturated). Voltage pulses across R25 are separated and routed to drive transistor Q7 and inverter transistor stage Q5 and driver transistor Q8 providing square waves or voltages 180 degrees of out phase between x and y.

The above described was the most difficult construction and completes the electronic assembly board of this system. This section must be properly operating before interfacing with the remainder of the circuit or damage to the final amp, transducers, etc., can result. The frequency limits described are only an example. Range can be as low as 5 kHz to as high as 25 kHz. Preset at 10 kHz "manual" for remaining assembly.

Final Assembly

It is assumed that the assembly board is completed and set to the frequency limits desired. At this point the builder must decide how many transducers he is going to use, if he is going to mount them in an array, individual placements, etc. The plans and sketches show a console capable of powering four to six remote located transducers. Controls are accessed at the front panel with a terminal strip, fuse, and power cord on the rear panel.

Form CA1 from a piece of #22 gauge galvanized or equivalent thickness in aluminum from a piece of sheet metal, as shown in Fig. 10-15. Note stiffening and cover mounting flanges along bottom. Use this sketch for dimensions. It may also be desired to construct this housing from smaller pieces fastening them together with flanges, brackets, sheet metal screws, etc. Note also that the

cover may slope saving space if desired. Drill holes for controls, mounting of components, heatsinks, etc.

Assemble controls on front panel, fuse, heatsinks, terminals, and bushing on rear panel as shown in Figs. 10-16 and 10-17. Do not attach assembly board to double tape at this point.

Assemble transformer T1 as shown in Fig. 10-18 and wire as shown. Be careful not to break cores—use a thin piece of rubber to mount on. Mount power transformer T2 as shown in Fig. 10-19. Assemble power transistor Q9 and Q10, to heatsink using TO3 (Fig. 10-20) insulating mounting kits. (Check with ohmmeter before applying power.)

Wire as shown in the final amp and power supply schematic (Figs. 10-21 and 10-22). *Note pictorials* Figs. 10-16, 10-19, and 10-17. Apply power and quickly check the waveforms at test points shown Fig. 10-23. These are collectors C of Q9 and Q10 respectively. If waveforms shown are not correct, immediately remove power and check for errors. Externally trigger for time reference if desired. Note that at this time the transducers are not connected to "E" and "F" (T1 unloaded). *Note waveforms* in Fig. 10-24.

V_{ce} must be no more than 0.2 volts with V_C+ being equal to 8.5 V. Waveform must be symmetrical respective to one another with a minimum of ringing and rounding of corners. The adverse conditions will increase as the load is increased. Note that this waveform must be that shown or the transistors may immediately overheat. Also any excessive overshoot could cause breakdown.

Please read the following data pertaining to the final amplifier and transducers. The final amplifier consists of two power transistors connected in a push-push configuration connected to T1 as shown in Fig. 10-21. Note R33 and R34 emitter resistors for current balance between the transistors. Final assembly is shown in Fig. 10-25.

Perform The Following

With the four transducers disconnected observe the waveform across E & F on T1 as shown in Fig. 10-24. (Frequency set at 20 kHz.) Connect four

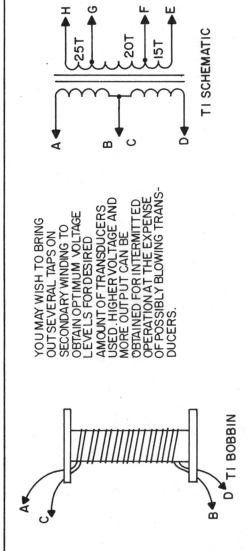

T1 SCHEMATIC

T1 BOBBIN

STEPS FOR WINDING T1 BOBBIN

1. PARALLEL OR BIFILAR WING 15 TURN PAIRS OF #20 ENAMEL COVERED MAGNET (OR VINYL) COVERED WIRE AS SHOWN EVENLY SPACED ALONG ENTIRE BOBBIN LENGTH. THIS FULLY UTILIZES ALL OF THE CORE.

2. SECURE AND SOLDER LEADS, A, B, C, & D TO LUGS AS SHOWN. MAKE SURE THAT THE ENAMEL COVERING IS ENTIRELY REMOVED AND LEADS ARE TINNED BEFORE SOLDERING.

3. WIND A SINGLE LAYER OR TAPE OVER THIS WINDING.

4. WIND 25 TURNS OF #20 PLASTIC COVERED WIRE NEATLY OVER ABOVE AND CONNECT LEADS TO LUGS ON BOBBIN. BRING OUT TAPS, AT 15, 8 20 TURNS FOR SELECTION OF VOLTAGE OUTPUT.

5. CAREFULLY PLACE "E" CORES TO BOBBINS AND TAPE TOGETHER. WHEN THE DESIRED TURNS ARE DETERMINED, YOU MAY WISH TO PLACE THE SINGLE WINDING E8 F FIRST ON THE BOBBIN. NOTE IT IS MORE CONVENIENT TO USE PLASTIC JACKETED WIRE AS THE INSULATION IS EASIER TO REMOVE, WHEN SELECTING THE OPTIMUM TAP ON THE OUTPUT WINDING.

6. ASSEMBLED AS PER FIG. 10-I9.

TURNS ON PRIMARY OF T1 (AB,CD) MAY BE VARIED TO COMPENSATE FOR VALUES OF V_c OBTAINED FROM FIG. 10-20. ADJUST ACCORDING TO REQUIRING OUTPUT.

YOU MAY WISH TO BRING OUT SEVERAL TAPS ON SECONDARY WINDING TO OBTAIN OPTIMUM VOLTAGE LEVELS FOR DESIRED AMOUNT OF TRANSDUCERS USED. HIGHER VOLTAGE AND MORE OUTPUT CAN BE OBTAINED FOR INTERMITTED OPERATION AT THE EXPENSE OF POSSIBLY BLOWING TRANSDUCERS.

Fig. 10-18. T1 transformer winding instructions.

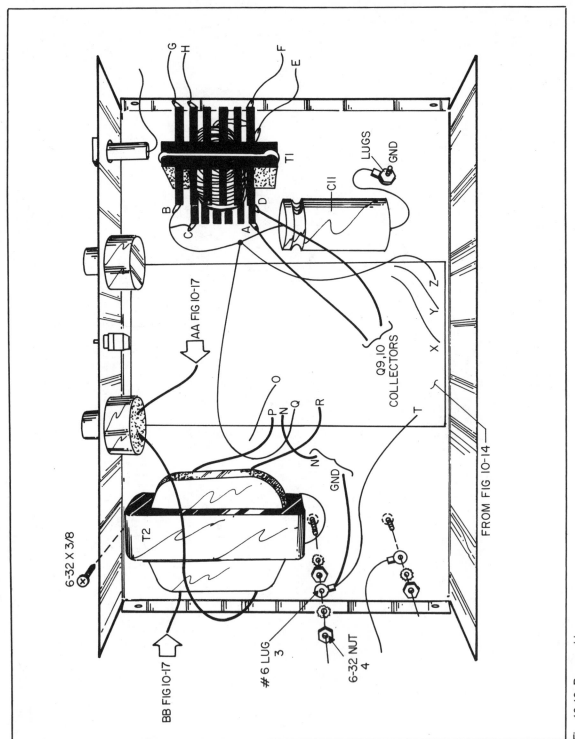

Fig. 10-19. Power wiring.

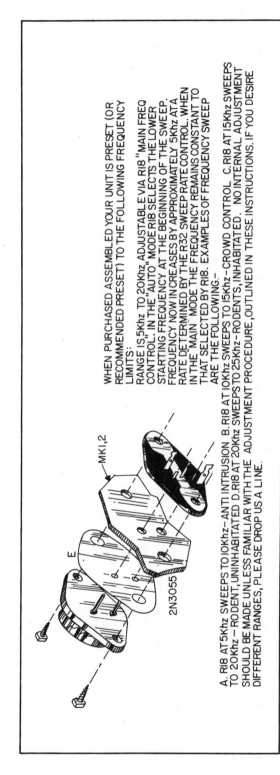

WHEN PURCHASED ASSEMBLED YOUR UNIT IS PRESET (OR RECOMMENDED PRESET) TO THE FOLLOWING FREQUENCY LIMITS:

RANGE IS.5Khz TO 20khz. ADJUSTABLE VIA R18 "MAIN FREQ CONTROL." IN THE "AUTO" MODE R18 SELECTS THE LOWER STARTING FREQUENCY AT THE BEGINNING OF THE SWEEP. FREQUENCY NOW INCREASES BY APPROXIMATELY 5khz AT A RATE DETERMINED BY THE R32 SWEEP RATE CONTROL. WHEN IN THE "MAIN" MODE THE FREQUENCY REMAINS CONSTANT TO THAT SELECTED BY R18. EXAMPLES OF FREQUENCY SWEEP ARE THE FOLLOWING.-

A. R18 AT 5Khz SWEEPS TO 10khz—ANTI INTRUSION B. R18 AT 10Khz SWEEPS TO 20Khz—RODENT, UNINHABITATED D. R18 AT 20khz SWEEPS TO 25Khz—RODENTS, INHABITATED. C. R18 AT 15Khz SWEEPS TO 15Khz—CROWD CONTROL. NO INTERNAL ADJUSTMENT SHOULD BE MADE UNLESS FAMILIAR WITH THE ADJUSTMENT PROCEDURE, OUTLINED IN THESE INSTRUCTIONS. IF YOU DESIRE DIFFERENT RANGES, PLEASE DROP US A LINE.

Fig. 10-20. Q9, 10 MTG method to MK1, 2.

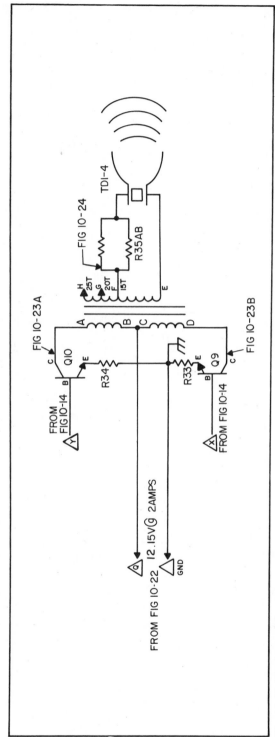

Fig. 10-21. Final amp schematic.

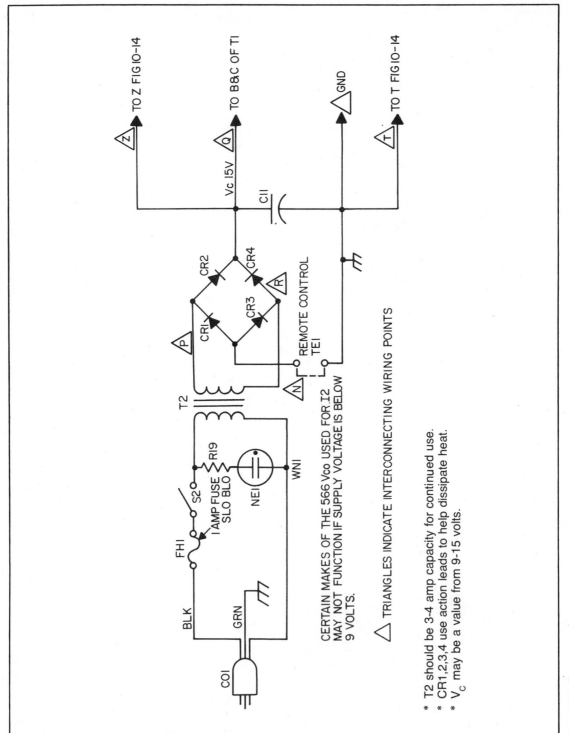

Fig. 10-22. Power supply.

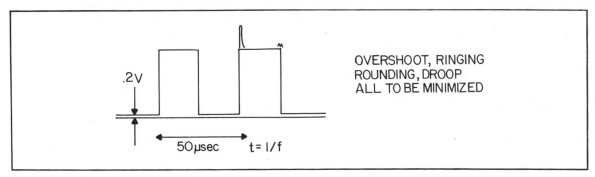

Fig. 10-23A. Waveshape collector of Q10.

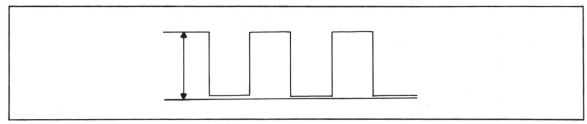

Fig. 10-23B. Waveshape collector of Q9.

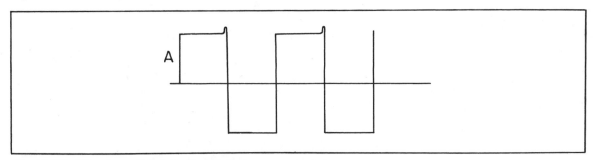

Fig. 10-24A. Waveshape at E & F without transducers connected.

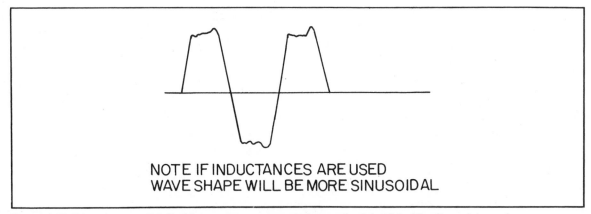

Fig. 10-24B. Waveshape at E & F with transducers connected approximately 30 to 40 volts peak-to-peak.

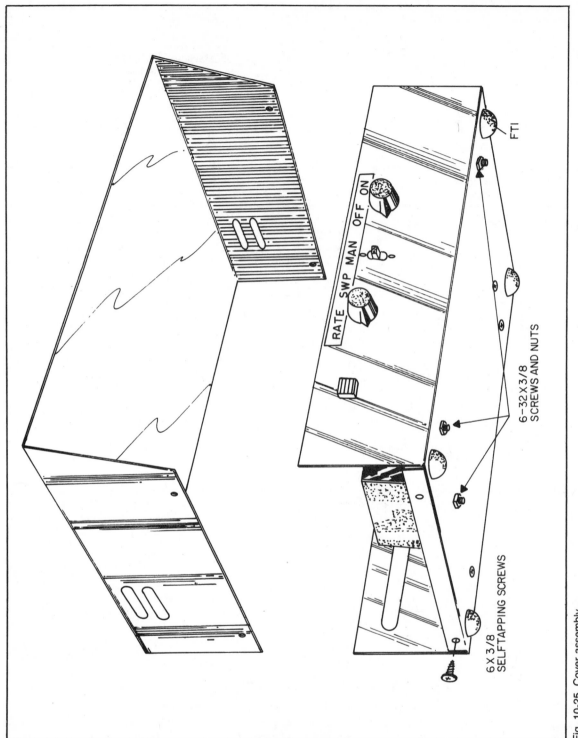

RATE SWP MAN OFF ON

FT1

6–32 x 3/8
SCREWS AND NUTS

6 X 3/8
SELFTAPPING SCREWS

Fig. 10-25. Cover assembly.

164

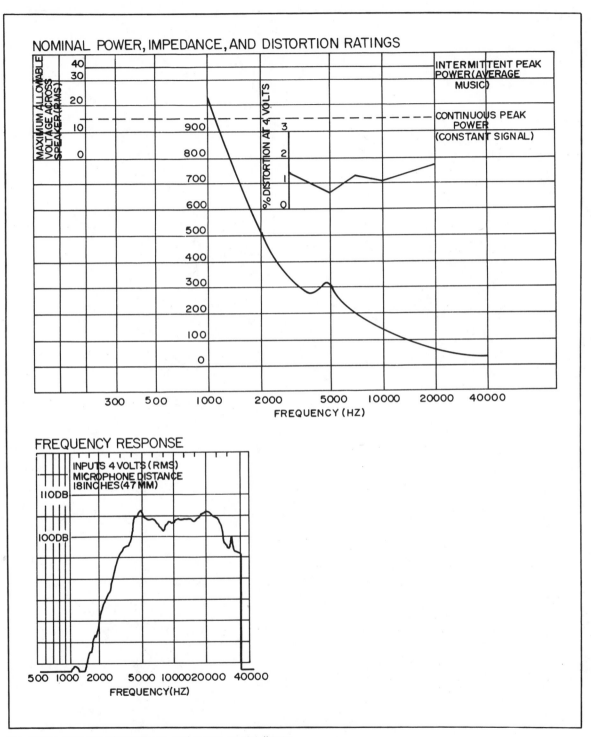

Fig. 10-26. TD1 transducer specs and curves model #6001.

AUTO/SWEEP SWITCH-SELECTS MANUALLY
PRESET TONE OR CONTINUALLY CHANGING TONE.
SWEEP RATE-ADJUSTS RATE OF CHANGING TONE.

MAIN FREQ./ADJ.-SETS LOW FREQUENCY END OF
SWEEP IN "AUTO" OR PRESETS FREQUENCY IN "MAN"
IS APPROX. 5-20KHZ.

I AMP FUSE
POWER TRANSISTORS
HEAT FINS
CONNECTION TERMINALS
TO REMOTE CONTROL-SHORT THESE LUGS
TO OPERATE.

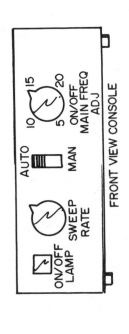

FRONT VIEW CONSOLE

A

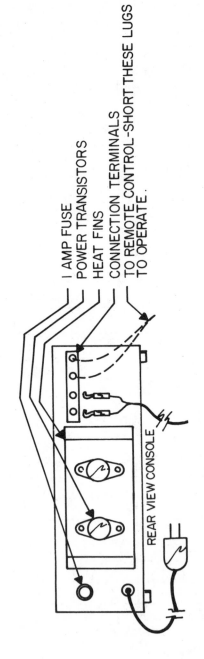

REAR VIEW CONSOLE

B

166

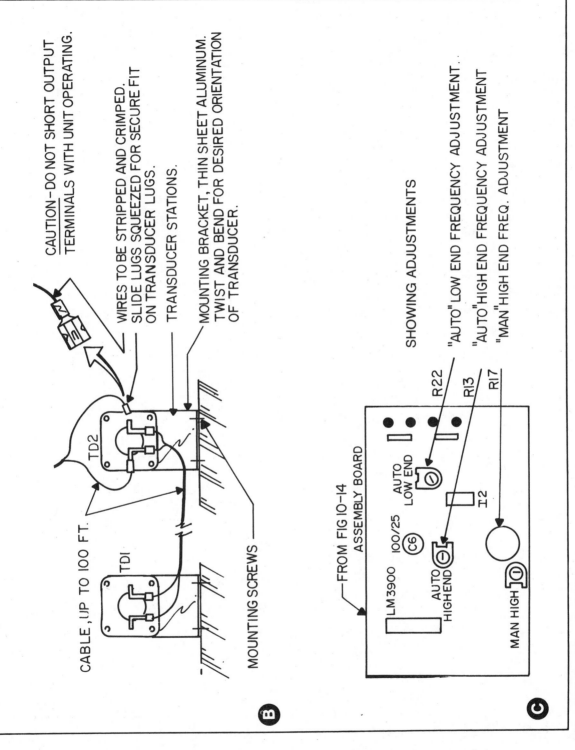

CAUTION–DO NOT SHORT OUTPUT TERMINALS WITH UNIT OPERATING.

WIRES TO BE STRIPPED AND CRIMPED. SLIDE LUGS SQUEEZED FOR SECURE FIT ON TRANSDUCER LUGS.

TRANSDUCER STATIONS.

MOUNTING BRACKET, THIN SHEET ALUMINUM. TWIST AND BEND FOR DESIRED ORIENTATION OF TRANSDUCER.

CABLE, UP TO 100 FT.

TD2

TD1

MOUNTING SCREWS

B

SHOWING ADJUSTMENTS

"AUTO" LOW END FREQUENCY ADJUSTMENT

"AUTO"HIGH END FREQUENCY ADJUSTMENT

"MAN"HIGH END FREQ. ADJUSTMENT

R22

R13

R17

FROM FIG 10-14 ASSEMBLY BOARD

LM3900   100/25   C6

AUTO LOW END

AUTO HIGHEND   I2

MAN HIGH

C

Fig. 10-27. Console and adjustments.

167

EAR DAMAGE	160 db	EXPLOSION, SHELL GRENADES, OUTSIDE DESTRUCTIVE SHOCK-
	150 db	WAVE RADIUS
	140 db	MI6 AUTO RIFLE DOWN RANGE 10'
PAIN	130 db	THRESHOLD OF PAIN
	120 db	PPG1 TO INTRUDERS
	110 db	CIRCULAR SAW, FIRE SIREN
VERY LOUD	100 db	SUBWAY STATION THUNDER
ANNOYING	90 db	BELT SANDER, VACUM CLEANER
	80 db	NORMAL CITY STREET
	70 db	PHONE RING AT 6' NORMAL SPEECH
COMFORTABLE	60 db	AIR CONDITIONER
	50 db	WINDOW FAN
	40 db	NORMAL OFFICE
QUIET	30 db	LIGHT BREEZE
	20 db	RECORDING STUDIO
VERY QUIET	10 db	LOW WHISPER
	0 db	THRESHOLD OF EAR FOR HEARING

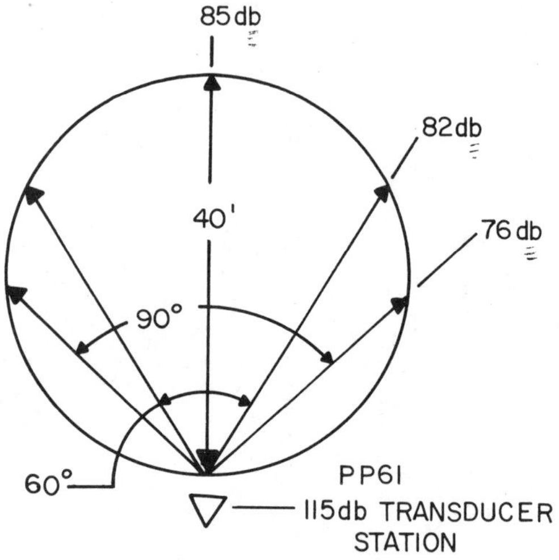

SOUND PRESSURE INTENSITY CHART

DB RATINGS TAKEN AT 18" FROM SOURCES UNLESS OTHERWISE NOTED.

RADIATION PATTERN, SCALE 2.5"=40'
USE A LAYOUT TEMPLATE

Fig. 10-28. Sound pressure chart.

transducers and note the waveform changing from a square wave to a triangular shape. Note voltage value on scope in Fig. 10-24. The power output of this system is about 120 dB measured at 18 inches. It is possible to obtain more power by further selecting the transducers. You will note series resistors R32A and B preventing overheating of the transducers at the high-frequency end. This resistor allows the use of a higher voltage, thus higher output at the lower frequency end curve of transducer (Fig. 10-26).

Obtaining Further Output

It is important to observe safety precautions in performing these following steps. Remember just because you do not hear this energy does not mean it's not affecting you.

It is possible to resonate the inherent capacity of the individual transducers (usually about .15 μF) with an inductance tuning to some desired frequency preferably 20 kHz or the high end of operation. This produces more of a sinusoidal waveshape and is usually more desirable because it allows a higher peak voltage waveform consequently more output without exceeding the rated maximum rms value of the original square wave. A typical value is 500 microhenrys with a frequency peaking at $\cong$ 18.5 kHz. When using inductors watch their current ratings because several hundred milliamperes is possible through these components.

Installation

Connect transducers and wire as shown Fig. 10-27. Note sound pressure chart in Fig. 10-28.

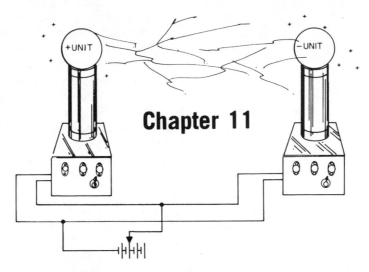

Ultrasonic Pulsed Rodent and Pest Control Device (RAT2)

This product is presently licensed to several manufacturers and is being marketed under the names of Rodent Guard, MSA, Sound Control Systems and also under the name of Resista-Rodent by Sona-Ray. The product as well as controlling rats and other pests without chemicals and messy traps is excellent for supplementing one's income by manufacturing these devices in one's own workshop and marketing via local means.

The following project describes how to construct a device that utilizes the relatively grey area of ultra-high frequency acoustical energy to control certain animals. It has long been known that animals such as dogs are responsive to sounds that the human ear cannot hear. Recent tests done at several universities show how rats when placed in a corner with the only barrier being this sonic energy will starve to death before penetrating it for food!

Bats emit a similar sound used as a type of radar for directing them in flight. The sound given off by the hiss from an air leak or a detuned TV set all affects the behavior of certain animals due to the high-frequency component of this sound. Obvi-

ously, certain types of animals will respond to certain sounds. The ones of interest here are the rodent (Rodentia) family consisting of rats, certain garden pests, squirrels, etc. Bats are a member of the Chiroptera family and may not show the same response to this energy.

It has been found by experimenting that a frequency of approximately 23 kHz creates an annoying, intolerable sound to most of the rodent family. (Obviously, certain species, environment and other conditions will play a part in determining the absolute degree of this effect.) When this sound is momentarily stopped and started the effect on these animals is startling. The author has actually seen certain animals jump into the air at a rate equal to the on/off transitions of this generating device. This method may seem harsh, but is not as cruel as poisoning, shooting, and trapping these animals. This method gives them the choice of vacating the premises to avoid these sounds and causes no permanent injury in anyway except to condition them to stay away.

The accompanying plans show how to con-

Table 11-1. Rat and Rodent Router Parts List.

R1	(1)	Resistor 1.2 M ohm ¼ W
R2	(1)	Resistor 2.7 M ohm ¼ W
R3,13	(2)	Resistor 1.8 k ohm ¼ W
R4	(1)	Resistor 2 k trimpot vert
R5,7	(2)	Resistor 1 k ohm ¼ ohm ¼ W
R6	(1)	Resistor 2.2 k ohm ¼ W
R8	(1)	Resistor 300 ohm 2 W
R9	(1)	Resistor 10 ohm ¼ W
R10	(1)	Resistor 220 ohm ¼ W
R14	(1)	Resistor 39 k ohm ¼ W
C1,3	(2)	Capacitor .01 μF @ 50 V disc
C2	(1)	Capacitor 1.5 μF @ 35 V tant or use 1 μF @ 50 V elect.
C4	(1)	Capacitor .01 μF @ 50 V poly
C5,6	(2)	Capacitor 1 μF @ 50 V EL
C7,8	(2)	Capacitor 100 μF @ 25 V EL
C9	(1)	Capacitor 10 μF @ 25 EL
IC1,2	(2)	555 timer dip 8 pin
Q1	(1)	Npn power tab D40D5
Q2	(1)	Npn plastic semicond PN2222
PC1	(1)	Printed circuit board or use .1 grid perforated board
D1,2	(2)	50 V lamp rect 1N4002
Z1	(1)	15 V zener 1N5245
T1	(1)	12 V @ 100 mA power transformer
L1	(1)	1 k/8 ohm matching transformer 8 Ω winding, not used
CA1	(1)	Enclosure 3¼ × 2⅛ × 4″ metal case or cabinet-fabbed as Fig. 11-5
WN1	(2)	Small wire nuts
TD1	(1)	21-23 kHz transducer
L2	(1)	27 mH inductor
NE1	(1)	Neon lamp short body leads
WR1,2	(2)	4″ length of #24 hook-up wire
PL1	(1)	2 wire 6′ zip cord
SW1	(6)	#6 × ⅜ sheet metal screws
BU2	(1)	1″ ID × ⅛ × ½″ rubber bushing
BU1	(1)	Cord clamp bushing
BU3	(1)	⅜ plastic bushing

Complete kit with optional PC board available through Information Unlimited, Inc., P.O. Box 716, Amherst, N.H. 03031, write or call 1-603-673-4730 for pricing and delivery.

struct a simple electronic device that generates this sonic energy at these frequencies and automatically turns on and off at a preset periodic interval. The plans utilize modern integrated circuits and a final power transistor capable of powering up to three stations that can be located at a considerable distance from the main unit and requires only standard two conductor cords or cables for powering. Each transducer emits a beam of this energy that is semi-circular and therefore directional. When properly installed, one or more of these stations can be used for protecting large and definitive areas.

CIRCUIT DESCRIPTION

IC2 forms a stable oscillator whose frequency and pulse width is determined by the value of R4, R5, R6, and C4 (Fig. 11-1 and Table 11-1). R4 is made adjustable for precise frequency setting. The output of IC2 is pin 3 which is capacitively coupled to the base of Q1. L1 acts as a high impedance choke to the signal while allowing the collector of Q1 to be dc biased. Q1 amplifies the positive pulses from IC2 and step drives the series resonant combination of L2 and TD1. Resistor (R10) serves to broaden the response of this resonant circuit. It should be noted

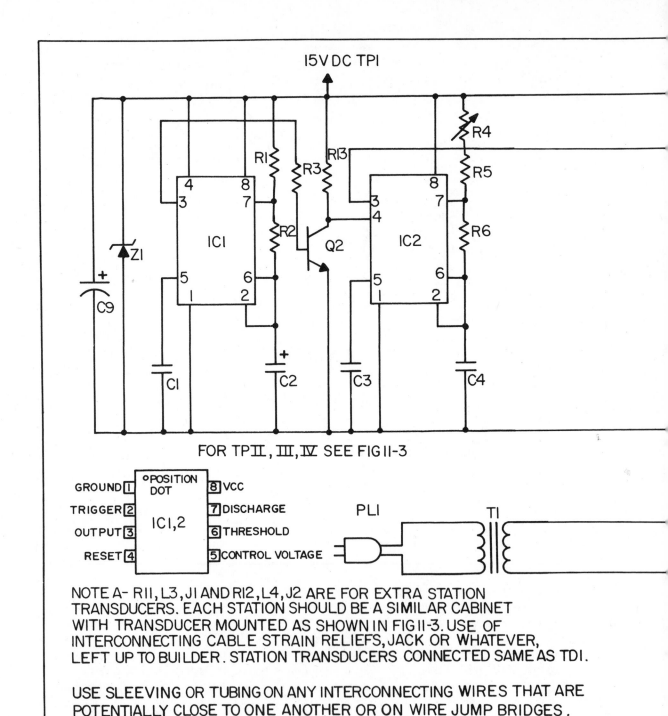

FOR TP II, III, IV SEE FIG 11-3

NOTE A— R11, L3, J1 AND R12, L4, J2 ARE FOR EXTRA STATION
TRANSDUCERS. EACH STATION SHOULD BE A SIMILAR CABINET
WITH TRANSDUCER MOUNTED AS SHOWN IN FIG 11-3. USE OF
INTERCONNECTING CABLE STRAIN RELIEFS, JACK OR WHATEVER,
LEFT UP TO BUILDER. STATION TRANSDUCERS CONNECTED SAME AS TD1.

USE SLEEVING OR TUBING ON ANY INTERCONNECTING WIRES THAT ARE
POTENTIALLY CLOSE TO ONE ANOTHER OR ON WIRE JUMP BRIDGES.
USE COMPONENT LEADS WHEREVER POSSIBLE.

Fig. 11-1. Circuit schematic.

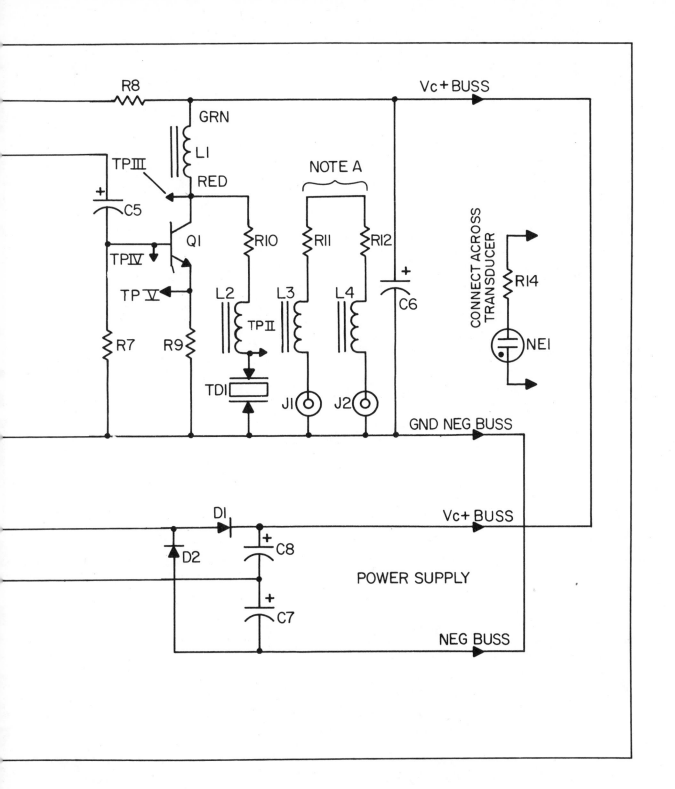

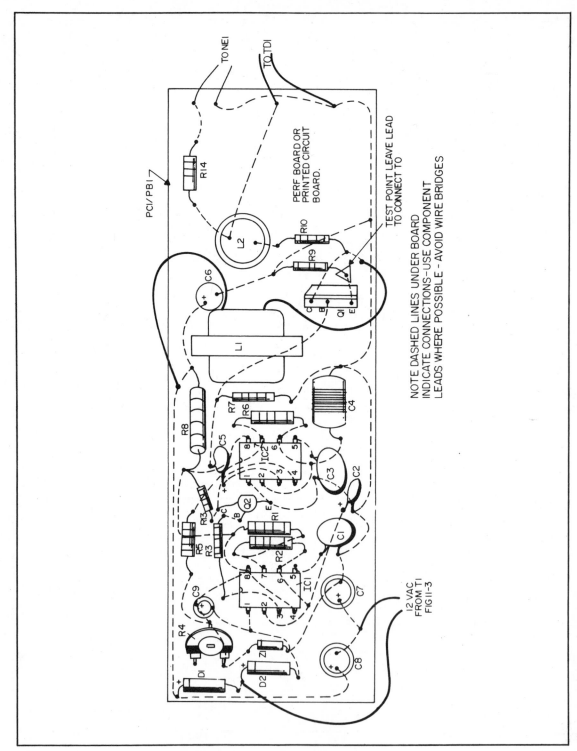

Fig. 11-2. Assembly board layout.

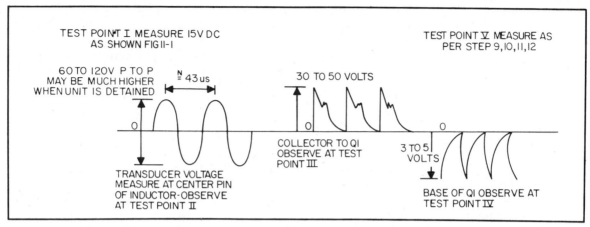

60 TO 120V P TO P
MAY BE MUCH HIGHER
WHEN UNIT IS DETAINED

N ≅ 43 us

30 TO 50 VOLTS

0

0

0

COLLECTOR TO QI
OBSERVE AT TEST
POINT III

3 TO 5
VOLTS

TRANSDUCER VOLTAGE
MEASURE AT CENTER PIN
OF INDUCTOR-OBSERVE
AT TEST POINT II

BASE OF QI OBSERVE AT
TEST POINT IV

Fig. 11-3. Test wavepatterns.

that L2 and the inherent capacity of the transducer (TD1) forms a resonant circuit at around 23 kHz.

It is usually found that most rodents are bothered when the signal is pulsed "on" and "off" with the off time exceeding the on time. This is accomplished via timer (IC1) and time inverter (Q2). IC1 is free running with its period determined via R1, R2, and C2 to approximately 2 seconds off and 3 seconds on, inverted via Q2 and used to gate pin 4 of (IC2) frequency oscillator turning it on for 2 and off for 3 seconds. The power supply is a conventional voltage doubler with a zener regulator for the oscillator voltages.

ASSEMBLY BOARD CONSTRUCTION

1. Identify all components. Separate those that go on the assembly board (Fig. 11-2).

2. Identify board position with that of pictorial. Note corner and mark perfboard to agree for referencing location of components, etc. If PC board is used follow component layout as shown.

3. Drill the following extra holes in perfboard:

☐ 1/16 hole for base lead Q1 (middle lead).

☐ Enlarge holes to 3/32 for mounting tabs of L1.

☐ Holes for the pins of IC1 and IC2. Note that pin spacing is 1/2 that of hole spacing in perfboard. Use a very small drill. (Not necessary with printed circuit board.)

4. Insert components starting from left to right (read notes) and carefully solder and clip excess leads. Be very careful of solder shorts and avoid wire bridges. Note dashed lines indicating connection. Please take note of the following in the assembly steps.

☐ Proper position of IC1 and IC2.
☐ Polarity of D1, D2, and Z1 diodes.

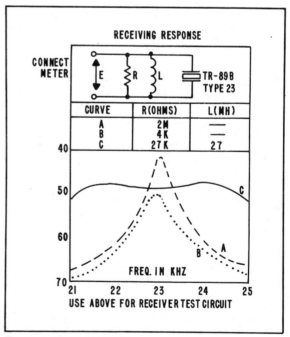

RECEIVING RESPONSE

CONNECT METER

CURVE	R(OHMS)	L(MH)
A	2M	—
B	4K	—
C	27K	27

FREQ. IN KHZ

USE ABOVE FOR RECEIVER TEST CIRCUIT

Fig. 11-4. Receiving circuit.

☐ Polarity of C7, C8, C9 electrolytic capacitors.

☐ Polarity of C2, C5, C6 tantalum or electrolytic capacitors.

☐ Position of Q1 and Q2.

☐ Remove 8-ohm winding of L1. Use the secondary winding as shown. Use ohmmeter for this if not noted on the transformer bearing in mind that the correct winding is the one reading the highest resistance.

☐ Leave excessively long lead on Q1 side of R9 for connection of meter lead in final adjusting.

5. Carefully inspect wiring of assembly board for accuracy, potential shorts, solder joints and other faults. Remember that the ICs may not tolerate a miswire.

6. Test the assembly board as explained below (Figs. 11-1, 11-2, and 11-3).

TEST AND ALIGNMENT STEPS

7. Note test point at R9, Q1 emitter end (excessive lead).

8. Clip in the 12 V winding of transformer T1 to respective terminals on assembly board for temporary power (Fig. 11-2).

9. Insert a voltmeter across R9 set to low range (0-2.5 volts).

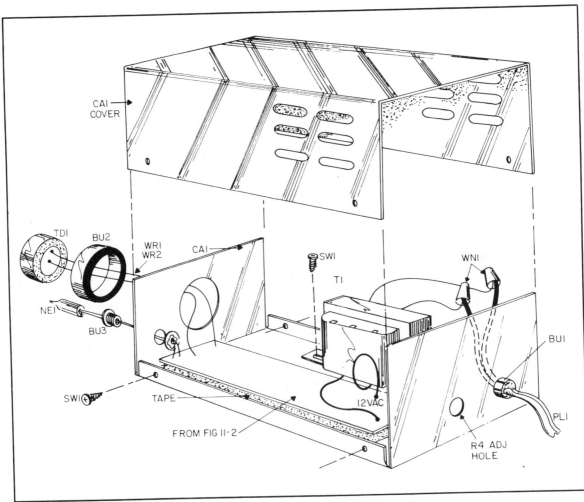

Fig. 11-5. Final assembly.

10. Clip in the transducer to respective terminals on assembly board.

11. Apply power to primary of T1 and quickly adjust R4 for a dip in meter current as indicated on voltmeter. It should sharply dip to 100 to 200 mV or less. Gently touch face of transducer with paper clip and note vibration.

12. Note meter current periodically turning "on" and "off."

13. Note NE1 indicator lamp should ignite with signal "on" time. It may be found that this will only occur with R4 set very slightly off of center of current dip. This is recommended on the cw side of the dip.

14. Note waveshapes at designated test points. Do so only for check purposes.

15. For maximum power output, place a second transducer (referred to as RCVR) about 2 feet from the driven one. Connect to low range on scope or sensitive ac meter. Adjust R4 for a maximum reading on RCVR transducer. Note that this usually occurs at a higher current than the preliminary dip as in the previous step. Current usually will be about 20 mA for a maximum indicating reading at receiver. Please note that the Q1 should at no time be *too hot to touch*. Note schematic of receiver circuit in Fig. 11-4.

16. When driving two or more transducers, it may be necessary to compromise the adjustment of R4 if the individual units vary in operating frequency by much. Note each station should draw an extra 15 mA. Q1 should not overheat. Board is now preliminary tested and is ready to install in the case.

FABRICATING OF CASE AND FINAL ASSEMBLY

17. Center, locate, and punch a 1 7/32″ hole in front panel of the case (Fig. 11-5).

18. Locate and drill a ⅜″ hole in rear panel as shown for the line cord bushing.

19. Locate and drill the two mounting holes for T1 mounted as shown.

20. Locate and drill hole for R4 adjust.

21. Locate and drill ¼″ holes for optional jacks on rear panel to remote transducer stations (if used).

22. Sleeve transducer into large rubber bushing and insert in large hole and secure with RTV, epoxy, etc. Do not push on face of transducer.

23. Mount transformers and optional jacks using screws and nuts.

24. Clamp power cord in place and connect to primary wire of T1. Use tape or small wire nuts. *Caution*—insulate these points well.

25. The same methods of constructing the cases for the optional transducer stations obviously apply.

26. Wire the tested assembly board to its respective connection and retest as per Steps 6 and 7.

27. When proper operation is verified, secure assembly board in place with two-sided tape, or use ingenuity. Note Figs. 11-3 and 11-4, waveshapes and transducer curves for those who desire to check these points.

INSTRUCTION FOR RAT2 UNIT

Best inside performance is obtained with the transducer head placed about three feet above the floor. You will note that the directional intensity radiates in a beam of about 60° and the unit can be placed in many combinations taking advantage of these directional characteristics along with the geometry of the area being protected. The unit should be installed in an area of rodent infestation facing the center of the area. The front of the unit should not be blocked and where there are aisles, it should be aimed down or along partitions, etc. The remote station design permits you to install extra units around walls, counters, partitions or in any closed off area where rodent problems exist. The area one station will cover can vary. Each unit should protect an area of approximately 2000 square feet. After about a week you will notice that there are no longer any signs of rodents or that signs are greatly reduced. Total control can be obtained by installing additional units in the area that still has evidence of rodents. Effectiveness may vary from installation to installation due to the sound absorbing materials in the areas in question. Do not direct the unit to sound absorbing materials. A rule of thumb to follow is that hard and smooth surfaces reflect while soft and porous surfaces absorb the energy. It should be clearly understood that other rodents may visit the area. However, due to the ultrasonic sound, they too will leave in a short period of time. These devices are inexpensive to operate—they use less current than a five watt bulb and require no traps to empty, no poison, no maintenance and they eliminate contamination and smell from dead rats in walls or decaying on the premises.

Some Rat Facts

For years the rat has plagued mankind as being a carrier of such dreaded disease as the bubonic plague that wiped out almost half of Europe. Today, it destroys valuable foodstuffs in the amount of several billion dollars a year, destroys buildings, kills livestock and other animals transmits many diseases through a parasitic flea that makes its home on the rat's hide, and is responsible for over one half of the unexplained fires by gnawing on electric wires and cables. Rats, while being noted for their filth and disease, and also notorious for viciously attacking a person or pet when accidentally cornered in a room or other enclosure. They immediately attack their victim by lunging for the throat and are usually gone before the unsuspecting victim realizes what has happened.

Rats are found anywhere they can find shelter to nest and breed along with food and water for sustenance. Cellars, attics, kitchens, food supply sheds, graineries, barns, dumps, chicken houses, vacated buildings, etc., are prone to their habitation and destruction. Chicken barns and hen houses are a natural for rats where freshly laid eggs become a delicacy in their insatiable greedy diet.

Some Bat Facts

Certain species of bats are close in species to rats, except that they fly! They, like rats, are notorious for spreading disease, rabies, and just being a general nuisance, often invading attics of homes, barns, storehouses, churches, caves, etc. When a heavy infestation takes place in an attic or barn, the usual odor of dead and decaying carcasses becomes annoying and unpleasant as well as unhealthy. Certain species of bats are blood suckers and often prey on livestock and other animals costing the farmer money in feed, veterinary fees, etc. We have all heard the term "blind as a bat" and most of us are aware of this animal's uncanny ability to navigate in tight places by using a type of sonic radar. Some people can hear these high pitched sounds that the bat uses for this purpose, and if we were to emit a sound that would confuse the bat, we would be jamming his ability to navigate and his natural instinct would dictate that he vacate the area of these confusing sounds. This is how the ultrasonic energy emitted from this unit affects these animals.

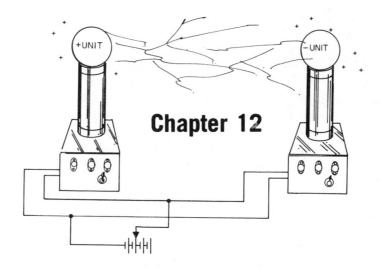

Chapter 12

Ultrasonic Translator Listening Device (HT9)

This project allows one to listen to the world of ultrasonic acoustical sounds that few people realize even exist. It is built in a gun configuration with the barrel housing the necessary circuits. A rear panel contains the jacks and controls. The front of the unit contains the directional receiving transducer. The handle of the device contains the necessary batteries and provides for a convenient method of handling the unit.

One of the most interesting sources of the inaudible sounds receivable with this unit are those that many species of insects generate. On a typical summer day one can easily spend hours detecting and homing in on a particular source of this inaudible sound and find its source to be a certain species of insect. Early evening hours brings the sounds of bats chirping away along with the local insect population. Needless to say, that a whole new world of strange, inaudible sounds emanating from many living creatures is easily detected and listened to. The unit also receives many man-made devices that generates ultrasonic energy. A few of the examples are the following:

☐ Leaking gases; rushing air.

☐ Running water, rolling and beading on a surface generates cracking sounds.

☐ Corona leakage, sparking devices, lightning and other electrical phenomena.

☐ Fires and chemical reactions.

☐ Man-made devices, TV sets, high-frequency oscillators, mechanical bearings, ultrasonic control, rodent devices, TV control, rattle detection, etc.

☐ Acoustical doppler-effect demonstration.

A fun game that can occupy many hours for both youngsters and adults especially at parties is a game of hiding a small, miniature oscillator about ½ the size of a pack of cigarettes and attempting to locate it with the unit. One will quickly master using the device and the challenge quickly reverts to the one who tries to hide the oscillator rather than the one who locates it.

DESCRIPTION OF CIRCUIT

Your unit basically consists of a high gain

amplifier (A1) and an oscillator (A2) combining output into a mixer (A3) where the difference in signal frequency is detected and filtered and further amplified by Q2 for powering external headsets or a speaker (Fig. 12-1 and Table 12-1).

The input amplifier A1 pin 9 is capacitively coupled via C1 to transducer TD1. TD1 is a piezoelectric device that the builder selects within the range of 20 to 30 kHz. TD1 and inductor (L1) are connected via a 10 to 13 inch length of shielded cable WR1. TD1 and L1 must be located away from the remaining circuitry so as to minimize any magnetic coupling that will cause instability problems. L1 is chosen so that it forms a parallel resonant circuit with the inherent capacity of TD1 to its natural frequency. This combination combined with R1 provides a flatter, wider frequency response.

The first two transistors of A1 form a Darlington circuit that is capacitively coupled to the third A1 transistor via C5 and is now fed into A1 fourth transistor via gain control R7 and C3 connected to pin 3. R7 controls the system gain and sensitivity to the desired signals. The fourth transistor is now capacitively coupled to external transistors Q1 via C6. Note that .05 μF coupling capacitors used throughout for attenuating any low-frequency pickup and use of dc bias point determining resistors R4, R8, and R9.

The output of Q1 is taken across collector resistor (R10). The signal at this point is the actual frequency picked up by TD1 only amplified many times. A2 is a 555 timer connected as a conventional astable free-running oscillator whose frequency is determined by trimpot (R19) and poly capacitor (C14). It is this signal that provides the mixer injection frequency necessary for heterodyning with the actual signal in mixer A3. Output of A2 is taken via pin 3 and combines with the actual signal and is injected into A3 via pin 2 through attenuation resistor (R12) and (R13) and decoupling capacitors (C4) and (C8). A3 now mixes these two signals and produces a composite output at pin 6. Resistor R14 and R15 set the gain of A2 to unity. The composite signal consisting of "sum", "difference" and "actual" is filtered and detected to obtain only the

"difference" signal which is what we desire to use here.

Diode D1 integrates the higher frequency component onto C13 while allowing the lower difference component to be capacitively coupled to Q2 via C12 for amplification and impedance matching for output requirements. You will note again for magnetic reasons, the matching transformer T1 is connected in line with the headset leads to reduce magnetic interaction of the circuit. (The headset/transformer combination) is connected to the system via J1. Two batteries are required for system power and are switched via S1. The +9 V is filtered via C15 and is decoupled to A1 via R11. The −9 V is used for mixer A3 and for final amplifier transistor Q2. This reduces power supply interaction.

The circuit is very sensitive to magnetic interaction due to the open magnetic circuit of L1 that is used in the ultra-low-level signal detection section. Grounding is important and should follow the pictorials as also should the circuit layout. The unit is prone to magnetic pickup at power line frequencies and above and therefore must be used accordingly. Many devices powered by ac may fool the user into thinking he is actually listening to an acoustical signal. This is easily distinguished by the low 60 Hz hum or the "doppler test" that involves moving the unit towards the suspected source where the frequency will increase or moving away where the frequency will decrease. Actual acoustical signals will be doppler-shift sensitive. Note that the reverse may occur using the doppler shift if the injection signal is above the detected signal. The tone obviously will be equal to the timer frequency minus the signal frequency. Magnetic shielding of L1 and some ac 60 Hz filtering may be desired by the builder.

CONSTRUCTION STEPS

1. Cut a 4½″ × 2½″ piece of perfboard and drill extra holes for A1.

2. Assemble board as shown in Fig. 12-2 and schematic Fig. 12-1. Start with A1 amplifier. Note A: C5 is mounted on backside of board. (Note B: leads from B1 and B2 are strain-relieved via routing

through holes in PB1. Note twisted leads to R7, not shown twisted, and shielded cable of WR1 and TD1. Chassis ground is via ground lug on J1. Observe polarity of tantalum and electrolytic capacitors, diodes, semiconductors, etc. Carefully check wiring for accuracy and quality of solder joints.

    3. Fabricate PR1 and MP1 per Fig. 12-3.

    4. Assemble and mount R7, S1, J1, BU1, BU2, and TD1 (Fig. 12-4). Note BU1 allows screwdrive adjustment of R19 and eliminates the shorting of screwdriver to metal of RP1. TD1 slides into BU2 for a friction fit. Secure BU2 and TD1 in large hole of MP1 with cement, etc. Interconnect leads to assembly board. Note the only ground connection to the MP1 is via lug under J1.

    5. Fabricate EN1 from an 8″ length of 3½″

Table 12-1. Ultrasonic Listening Device Parts List.

R11,20	(2)	1 k ¼ W resistor
R10,21	(2)	2.2 k ¼ W resistor
R16	(1)	4.7 k ¼ W resistor
R3,5,6	(3)	6.8 k ¼ W resistor
R17	(1)	10 k ¼ W resistor
R1,12	(2)	39 k ¼ W resistor
R9,13,14, 15	(4)	100 k ¼ W resistor
R4,8	(2)	390 k ¼ W resistor
R18	(1)	1 M ¼ W resistor
R2	(1)	5.6 M ¼ W resistor
R7	(1)	5 k pot/switch
R19	(1)	2 k trimpot vert
C15	(1)	100 µF @ 25 V elec cap
C2,3,5,6 13,16	(6)	.05 µF @ 25 V disc caps
C7,17	(2)	.1 µF @ 35 V disc or tantolum caps
C14	(1)	.01 µF @ 50 V poly
C1,4,8,9, 10,11,12	(8)	.01 µF/50 V disc
A2	(1)	555 timer dip 8 pin
A1	(1)	CA3018 Op amp in 12 pin can
A3	(1)	UA741 Op amp in 8 pin can
Q1,2	(2)	PN2222 npn plastic transistor
D1	(1)	1N914 signal diode
PB1	(1)	4½″ × 2½″ perfboard
L1	(1)	27 mH inductor
S1	(1)	Slider switch dpst
KN1	(1)	Plastic knob
J1	(1)	Phono jack
TD1	(1)	23 kHz TR89 transducer
WR1	(10″)	Shielded cable
CL1,2	(2)	6″ snap clips
BU1	(1)	⅜ plastic bushing
BU2	(1)	1″ ID × ⅛ wall × ½″ neoprene rubber bushing
WR3	(4′)	Hook-up wire #24
CA1,2	(2)	Plastic cap 3½″
EN1	(8″)	8 × 3½ PVC SK40
EN2	(5″)	5 × 2 PVC SK40
MP1	(1)	2⅞ × 10⅜ 22 ga mounting plate
RP1	(1)	3¼ × 3¼ 22 ga front panel (fab per Fig. 12-3)
T1		8/1 k matching transformer
B1,2	(2)	9 V ni-cads

    A complete kit is available from Information Unlimited, Inc., P.O. Box 716, Amherst, N.H. 03031, write or call 1-603-673-4730 for price and availability.

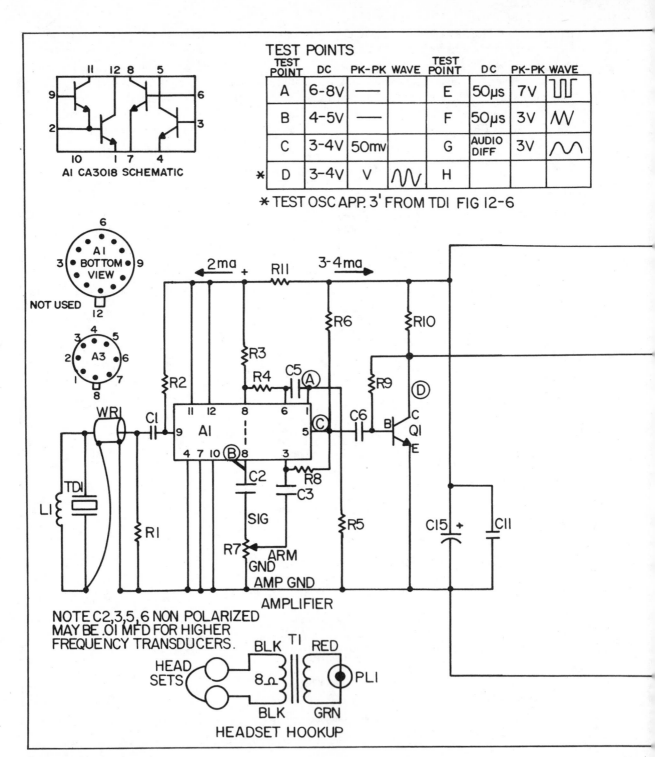

TEST POINTS

TEST POINT	DC	PK-PK	WAVE	TEST POINT	DC	PK-PK	WAVE
A	6-8V	—		E	50µs	7V	⊓⊓⊓
B	4-5V	—		F	50µs	3V	⋀⋀
C	3-4V	50mv		G	AUDIO DIFF	3V	∿
*D	3-4V	V	∿	H			

* TEST OSC APP. 3' FROM TDI FIG 12-6

AI CA3018 SCHEMATIC

NOTE C2,3,5,6 NON POLARIZED MAY BE .01 MFD FOR HIGHER FREQUENCY TRANSDUCERS.

AMPLIFIER

HEADSET HOOKUP

Fig. 12-1. Circuit schematic.

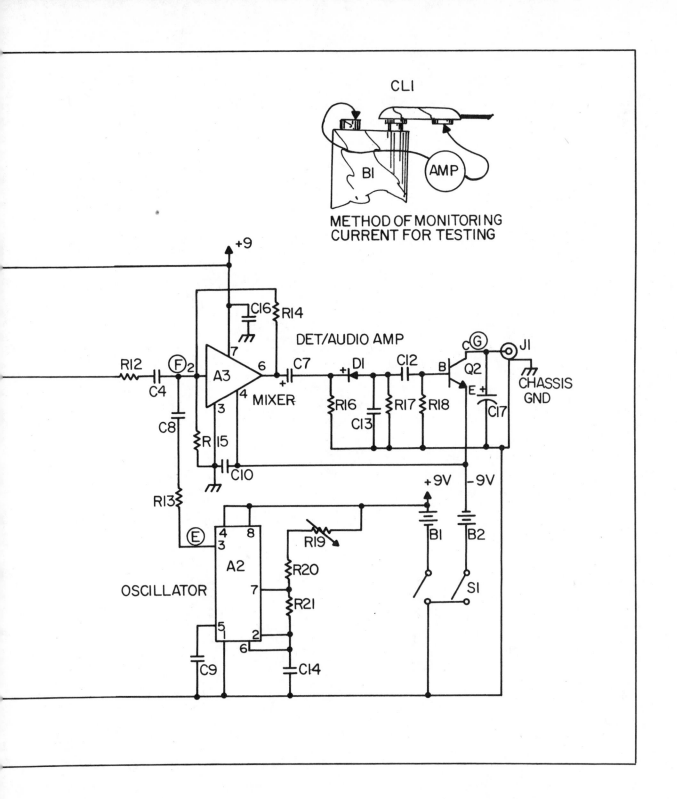

CLI

BI   AMP

METHOD OF MONITORING
CURRENT FOR TESTING

+9

C16 R14

DET/AUDIO AMP

R12 F 2   6 C7   D1   C12   B G   J1
C4   A3   MIXER   Q2   CHASSIS
C8   3   4 +   R16 R17 R18 E +   GND
                                    C13   C17
R 15
        C10

R13

E

+9V   −9V

4   8   R19
3   B1   B2
A2   7   R20
OSCILLATOR   R21
        5   1   S1
        2
        6

C9   C14

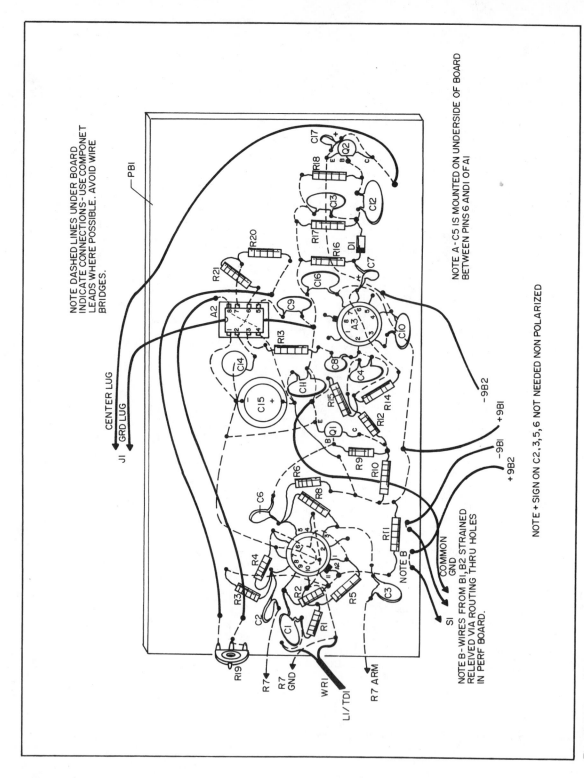

Fig. 12-2. Assembly board.

184

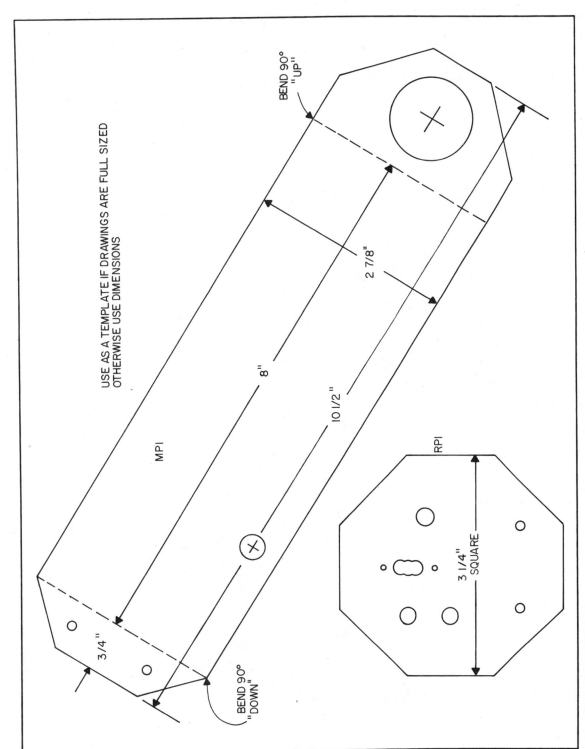

USE AS A TEMPLATE IF DRAWINGS ARE FULL SIZED
OTHERWISE USE DIMENSIONS

BEND 90° "UP"

2 7/8"

8"

10 1/2"

MPI

BEND 90° "DOWN"

3/4"

RPI

3 1/4" SQUARE

Fig. 12-3. Mechanical fabrication layout.

185

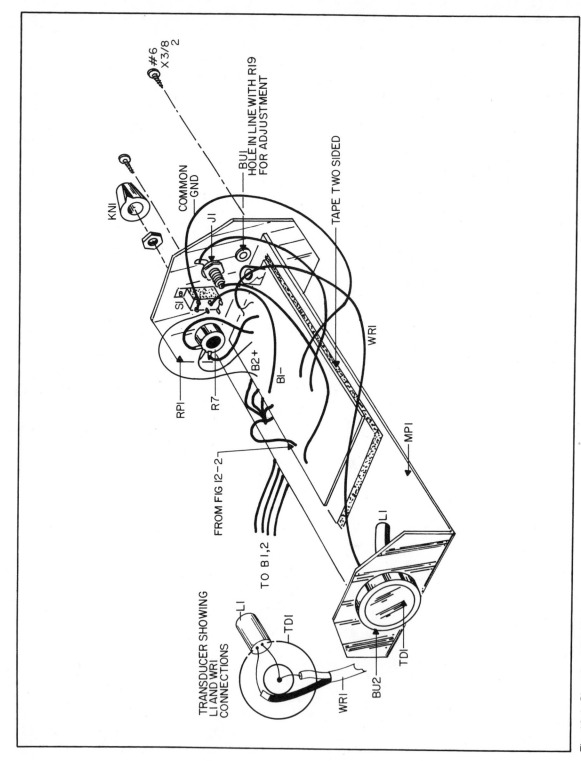

#6
X 3/8
2

BU1
HOLE IN LINE WITH R19
FOR ADJUSTMENT

COMMON
GND

KN1

J1

TAPE TWO SIDED

SI

WRI

RPI

R7

B2+

BI−

MPI

FROM FIG I2−2

LI

TO BI,2

TRANSDUCER SHOWING
LI AND WRI
CONNECTIONS

LI

TDI

WRI

BU2

TDI

Fig. 12-4. Chassis assembly.

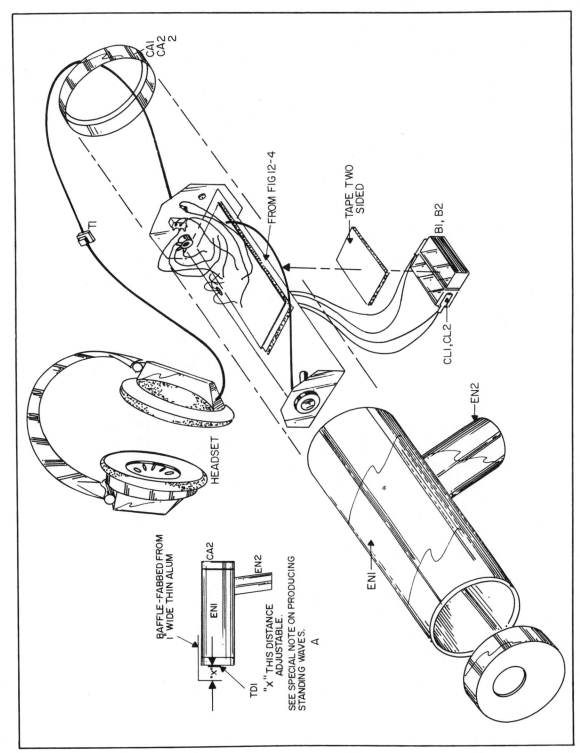

CA1
CA2
2

T1

FROM FIG 12-4

TAPE TWO
SIDED

B1, B2

CL1, CL2

EN2

HEADSET

ENI

BAFFLE - FABBED FROM
1" WIDE THIN ALUM

CA2

EN2

ENI

TD1

"X" THIS DISTANCE
ADJUSTABLE.

SEE SPECIAL NOTE ON PRODUCING
STANDING WAVES.

A

Fig. 12-5. Final assembly.

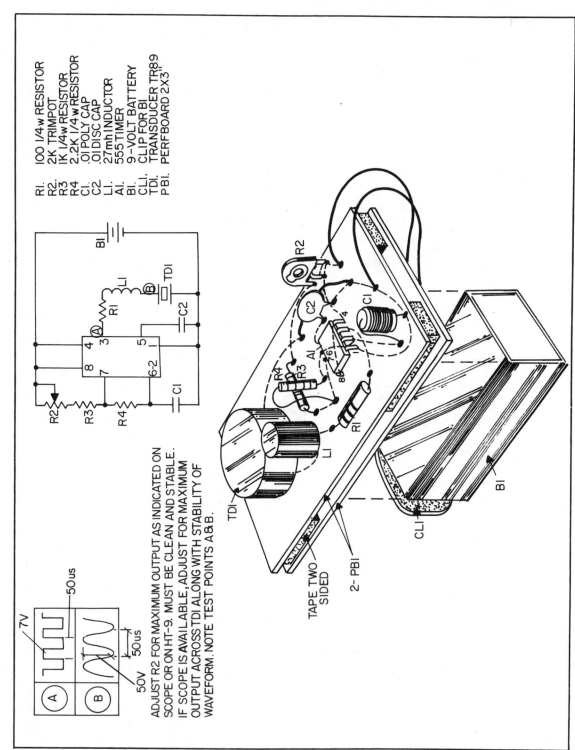

RI.   100 I/4 w RESISTOR
R2.   2K TRIMPOT
R3   IK I/4w RESISTOR
R4   2.2K I/4 w RESISTOR
CI.   .0I POLY CAP
C2   .0I DISC CAP
LI.   27mh INDUCTOR
AI.   555 TIMER
BI.   9-VOLT BATTERY
CLI.   CLIP FOR BI
TDI.   TRANSDUCER TR89
PBI.   PERFBOARD 2x3"

ADJUST R2 FOR MAXIMUM OUTPUT AS INDICATED ON SCOPE OR ON HT-9. MUST BE CLEAN AND STABLE. IF SCOPE IS AVAILABLE, ADJUST FOR MAXIMUM OUTPUT ACROSS TDI ALONG WITH STABILITY OF WAVEFORM. NOTE TEST POINTS A&B.

Fig. 12-6. Test oscillator/hidden transmitter.

OD PVC. Note 1⅞″ hole at slight angle for handle. Use circle saw and firmly secure (Fig. 12-5).

6. Fabricate EN2 from a 5″ length of 2″ OD PVC. Secure EN1 to EN2. Use cement.

7. Fabricate CA1 and CA2 from Alliance #A-3½″ plastic caps as shown in Fig. 12-5.

TEST POINTS AND CIRCUIT DESCRIPTION

8. Connect one terminal of B1 and CL1. Insert 0-100 mA meter in series with other terminals and turn on S1. Note current of 9-10 mA. Remove meter and connect remaining terminal of B1 to CL1 (Fig. 12-1).

9. Repeat with B2 and measure 3 mA, connect remaining terminal of B2 to CL2.

10. Plug in headsets with 1000 ohm/8 ohm matching transformer as per Fig. 12-1. Adjust R7 to hear a hissing sound, usually about ⅛ turn. Too much gain and unit may break into oscillation as indicated by rumbling and squealing in headsets. You should hear a very weak tone. Adjust R19 for lowest note of tone. Noting a tone dip as R19 is adjusted. Keep "in line" transformer T1 for headsets away from L1 inductor.

11. Gently rub TD1 face with finger and note a loud screeching sound. This is a high-frequency sound consisting of many components.

12. Secure assembled board and batteries to MP1 via two-sided tape or RTV.

13. Enclose unit as shown in Fig. 12-5. Note CA1 and CA2 retaining assembly in place.

14. Always operate unit at lowest setting of R7. A true acoustical signal will change tone as the unit is moved relative to the source. This tone change will increase as the device is moved faster. This is commonly known as the "doppler shift" and can be used to determine source direction once you are familiar with it.

SPECIAL NOTE FOR OBTAINING A LITTLE MORE SENSITIVITY

It is possible to produce a standing wave at the face of the transducer TD1 and improve the systems sensitivity. The effect of this can be shown by pointing the device at a steady low-intensity source of ultrasonic energy and carefully adjusting the distance of a flat object "baffle" relative to the transducer face noting an increase in signal. This will occur at ½ wave multiples with the most pronounced effect being those closest to the face. Those desiring to incorporate this feature in their unit should use their own ingenuity in attaching it to EN1. See Fig. 12-5A.

SUPPLEMENTARY APPLICATION OF YOUR HIGH-FREQUENCY TRANSLATOR AND TEST OSCILLATOR

This electronic device can supply hours of fun and entertainment for both young and old alike. A small ultrasonic oscillator (Fig. 12-6) the size of a pack of cigarettes is hidden. The device is now used to locate the hidden box. The highly directional characteristics of the device with its ability to respond to various levels of signal allows one to immediately get a bearing on the hidden box and the fun begins. It should be noted that the system has a range of detecting, seeking, and locating up to ⅛ of a mile. This allows great flexibility in hiding and placing the box. Instead of using a radio transmitter and receiver this system uses an ultrasonic sound generator and detector.

This approach allows a much more defined sense of direction along with a better indication relative to signal strengths. A radio device would require microwave equipment to simulate the desired directions and signal levels obtainable with the ultrasonic approach. Obviously cost and complicated circuitry is a deterrent to this approach. Also using ultrasonics enables the hidden box to be placed on the ground, on a person, inside of boxes, and many other locations that would be inoperable using radio frequencies. The author of these plans, his friends, and family have had many enjoyable hours of playing with this relatively sophisticated toy.

Section III
General Information on
High-Voltage and Scientific Devices

The devices shown range from relatively harmless small and medium Tesla coil generating high-voltage high-frequency currents to a multi-million watt pulser capable of exploding wires, producing EMP or other laboratory uses requiring high energy. A device of this type is considered dangerous in the hands of the inexperienced and is only intended for the serious researcher.

High-voltage dc devices operating from battery or household 120 Vac are intended for research in ion and particle generation, component and dielectric insulation testing, electrostatics, ion propulsion, special effect, X-rays, ignition of spark gap discharge tubes, and many other HV applications. Proper safety is required in handling to avoid a powerful shock.

A magnetic detector device is shown that can indicate the presence of sunspot activity. A two-way communication device uses IR or visible light to produce crystal clear communication for over a mile and provides an excellent high school science-fair project.

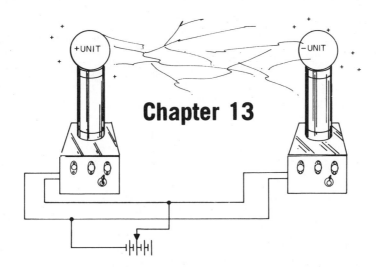

250 kV Tesla Coil/Lightning Generator (BTC3)

A Tesla coil is one of the most fascinating electrical display devices to see in operation. A large unit can produce a continuous spark of a length exceeding the height of the coil. Electric discharges simulating lightning bolts will produce cracks of noise louder than a rifle shot. These sparks as well as being highly impressive and attention getting also can cause bizarre effects to occur in most common materials. For example, wood may explode into splinters or made to glow with an eerie reddish light from *within*. Insulating materials seem to be useless against this energy. Lights light without wires, sparks, and corona in the form of St. Elmo's Fire occurs within proximity of the device. High energy electric and magnetic fields render electronic equipment useless. Phenomena not normally associated with standard HV electricity becomes apparent in the form of many weird and bizarre effects.

GENERAL THEORY

A Tesla coil is a high-frequency resonant transformer. It differs from a conventional trans-

former in that the voltage and current relationships between primary and secondary are independent of turns ratios. A working apparatus basically consists of a secondary (LS1) and a primary (LP1). It is obvious that the primary circuit is capacitance dominant and tuning the primary circuit via taps along the primary coil does not alter the frequency by as much as one would expect. However, this relatively fine tuning of the primary circuit to the secondary is mandatory for proper operation. Force driving the secondary coil (untuned) will produce hot spots and interwinding breakdown.

INTRODUCTION TO YOUR EASY-TO-BUILD TESLA COIL

This project shows how to construct a device using a step-up transformer producing 6000 volts at 20 mA from the 120 Vac line. This voltage current combination can produce a painful shock and all attempts should be made to construct this device with either some personal know-how or experienced help. Safety rules should be followed at all times. The device also produces ozone—therefore, use it in a well ventilated area. Do not use for prolonged periods of time, 30 seconds on at any one time is ample for any demonstration. Avoid eye exposure to spark gap without ample protection such as safety glasses, shielding, etc.

The unit when constructed as shown can develop voltages up to and in excess of 250,000 volts. It will cause a gas discharge lamp such as a regular household fluorescent lamp to glow up to a distance of several feet from the unit. The high-voltage center coil terminal can actually be touched with a piece of metal held securely in the demonstrators hand, creating quite a conversation piece.

The unit can be used for exciting all types of gas discharge tubes, X-ray tubes, plucker tubes, etc. It demonstrates corona, St. Elmo's Fire and the effect of high voltage on many different types of materials. On a good dry day, with the device properly adjusted, it is possible to produce sparks 10″ to 12″ in length. It is designed for the experimenter who wishes to explore the effects of high-frequency and high-voltage energy.

CIRCUIT THEORY

The device consists of a secondary coil LS1 containing approximately 500 turns wound on a PVC form 12″ long (Fig. 13-1 and Table 13-1). This coil possesses an inherent resonant frequency determined by its inductance and capacity. A primary circuit consisting of a drive coil LP1 and capacitor C1 are impulse driven by a spark gap SG1. This primary circuit should also have a resonant frequency equal to that of the secondary coil for maximum performance. It is possible to force the secondary coil with less results. The output voltage of the device is dependent on the ratio of Q between these two coils.

The primary coil has an adjustable tap that allows for fine tuning. It should be noted that it does not take much in the way of added capacitance to the secondary to alter its resonance point. Even a change in the output terminal TER1 may require readjustment of the tap.

Transformer T1 supplies the necessary high voltage. It is rated at 6000 volts ac at 23 mA. A larger capacity transformer will produce more output but may stress the other circuit components. This voltage charges the primary storage capacitor (C1) to a voltage where it fires the spark gap (SG1) producing an impulse of current through the primary inductance (LP1) where oscillations take place. The frequency is determined by the induc-

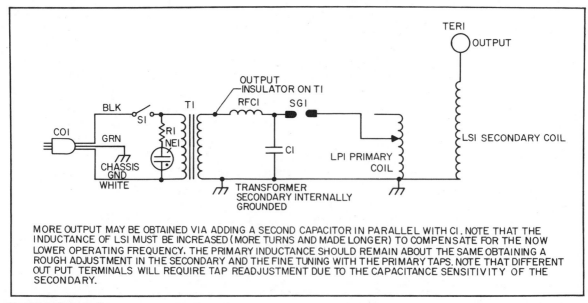

Fig. 13-1. Schematic.

Table 13-1. 250 kV Tesla Coil Parts List.

CTL1	(1)	12″ length of 3 1/2 OD PVC sked 40 wound with 450 turns of #24 wire as construction Step 1
T1	(1)	6000 V @ 23 mA ignition transformer
C1	(1)	.005 μF/6 k Vac capacitor
CO1	(1)	3 wire #18 power cord
WR12	(10′)	#12 covered wire for forming primary
WR13	(500′)	#24 enameled wire
WR7	(6″)	#16 buss wire
WR8	(2′)	1/4″ braid wire
WR17	(1)	3″ #24 precut lead
R1	(1)	33 k 1/4 watt resistor
NE1	(1)	Neon lamp with leads
S1	(1)	Medium toggle switch
RFC1	(1)	2.5 mH rf choke—should have a high voltage rating
CL1	(1)	Alligator clip
BU1	(1)	1/2″ plastic bushing or clamp for line cord
BU2	(1)	3/8″ plastic bushing
WN1	(1)	Small wire nuts #3
FT1	(6)	Rubber feet stick-on
SID1,2	(2)	6 1/2 × 20 1/2 22 gal fab per Fig. 13-4 as 9 × 11 × 5 1/2 box
BOT1	(1)	9″ × 11″ #22 gal fab per Fig. 13-4
BR3	(1)	1″ × 2″ #22 gal for holding C$_1$
MP1	(1)	9 × 11″ finished masonite fab per Fig. 13-4
BL1	(1)	3/8 × 3/8 × 3″ teflon or equivalent fab per Fig. 13-3
SP1-4	(4)	1-1/4 × 3/4 × 1″ insulating blocks fab per Fig. 13-2
TUBE	(1)	3″ × 3/16″ ID poly tube
BR1,2	(2)	Small 3/4 × 3/4 brass bracket Fig. 13-3
CA1,2	(2)	3 1/2 plastic cap
SW3	(4)	6 32 × 1″ PH screw
SW2	(5)	6 32 × 1/2″ PH screw
SW4	(1)	8 × 3/8 sheet metal screw
SW8	(14)	6 × 3/8 self tap type F screw
SW10,20	(2)	1/4 20 × 2″ carriage bolts preferably smooth brass heads
NU1	(11)	6 32 kep nuts
NU3	(4)	10 32 hex nuts
NU4	(2)	1/4 20 hex nuts
TY1	(4)	8″ nylon tyewraps for securing primary to spacers & MP1
TA1		Two-sided tape
TER1	(1)	Smooth brass door knob for terminal (not included in kit)
WA1	(5)	Flat wide shoulder washers
LAB1	(1)	Danger HV warning label

Complete kit with wound and completed secondary coil LS1 available through Information Unlimited, Inc., P.O. Box 716, Amherst, N.H. 03031, write or call 1-603-673-4730 for pricing and delivery.

tance and capacity values of the primary circuit. Voltage output of the secondary coil (LS1) is usually approximately equal to $V_2 = \dfrac{C_1 V_1}{C_2}$ where $C_1 =$ primary storage capacity, $V_1 =$ the spark gap dis-charge voltage, and $C_2 =$ the secondary coil capacitance (usually relatively small). Another way of expressing this relation is that the output voltage is dependent on the input drive voltage times the ratio of primary "Q" to secondary "Q." There are several texts describing some of Tesla's work and these are available through your library.

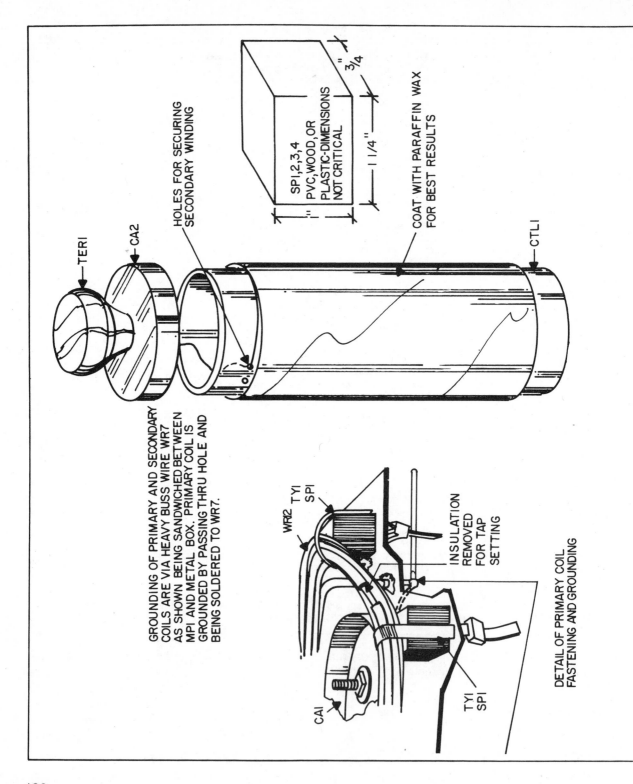

TER1

CA2

HOLES FOR SECURING SECONDARY WINDING

SP1,2,3,4
PVC, WOOD, OR PLASTIC-DIMENSIONS NOT CRITICAL

1"

1 1/4"

3/4"

COAT WITH PARAFFIN WAX FOR BEST RESULTS

CTL1

GROUNDING OF PRIMARY AND SECONDARY COILS ARE VIA HEAVY BUSS WIRE WR7 AS SHOWN BEING SANDWICHED BETWEEN MP1 AND METAL BOX. PRIMARY COIL IS GROUNDED BY PASSING THRU HOLE AND BEING SOLDERED TO WR7.

WR2 TY1 SP1

INSULATION REMOVED FOR TAP SETTING

TY1 SP1

CA1

DETAIL OF PRIMARY COIL FASTENING AND GROUNDING

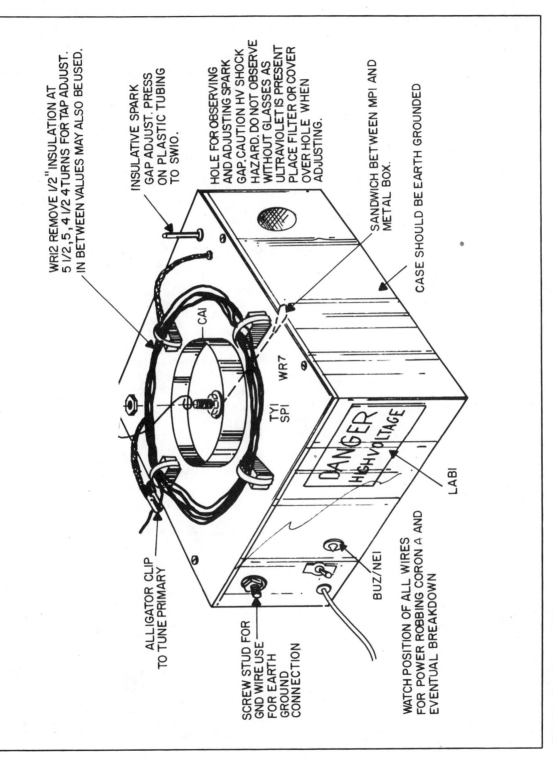

WRl2 REMOVE l/2" INSULATION AT 5 l/2, 5, 4 l/2 4 TURNS FOR TAP ADJUST. IN BETWEEN VALUES MAY ALSO BE USED.

INSULATIVE SPARK GAP ADJUST. PRESS ON PLASTIC TUBING TO SWlO.

HOLE FOR OBSERVING AND ADJUSTING SPARK GAP. CAUTION HV SHOCK HAZARD. DO NOT OBSERVE WITHOUT GLASSES AS ULTRAVIOLET IS PRESENT PLACE FILTER OR COVER OVER HOLE WHEN ADJUSTING.

SANDWICH BETWEEN MPI AND METAL BOX.

CASE SHOULD BE EARTH GROUNDED

CAI

WR7

TYl SPl

DANGER HIGH VOLTAGE

LABl

BUZ/NEI

ALLIGATOR CLIP TO TUNE PRIMARY

SCREW STUD FOR GND WIRE USE FOR EARTH GROUND CONNECTION

WATCH POSITION OF ALL WIRES FOR POWER ROBBING CORON A AND EVENTUAL BREAKDOWN

Fig. 13-2. Complete assembly.

197

CONSTRUCTION STEPS

Please note that substitution of certain parts may be made by the builder, however they may increase or decrease performance.

1. Wind 10″ of #24 magnet wire (WR13) on coil form (CTL1). Leave ¾″ to 1″ on each end (Fig. 13-2). Fasten wires by threading them through small holes in coil form. Leave 12″ leads. Coil form must be dry and clean. Varnish, dope or coat finished winding with paraffin wax several times to seal against moisture and hold in place. Winding must be neat and tight. Avoid kinks and overlaps as this will seriously affect coil performance.

The best way to wind the coil is with a buddy using a broom stick as a shaft and carefully dispensing the wire as needed. Keep wire taut. Winding should be approximately 45 turns per inch for 10 inches using less than 500 feet of wire. Final coil is secured to mounting plate (MP1) via plastic cap (CA1).

2. Cut 10′ of #12 wire and form into a coil of 6 turns with a 6″ diameter as shown Fig. 13-2. Hold together with tape. Fabricate 4 pieces of dry wood or plastic for SP1, 1¼ × ¾ × 1″ for mounting and securing coil primary (LP1). Use tye wraps to secure (SP1) blocks and coil simultaneously to (MP1) mounting plate as shown.

3. To construct spark gap Fig. 13-3 fabricate a 3″ teflon block (BL1) as shown. Note angle brackets (BR1) and (BR2) for attaching (SW10) and (AW20) spark-gap screws. These brackets should be tapped for ¼-20 screws on one end. Leave enough spacing between bottom bracket and chassis to prevent breakdown from occurring. Note flattened shield section with holes for attaching to screws. Note that the bottom screw has nut tightened against the bottom bracket for locking. The top screw is allowed to turn allowing for spark gap adjustment. Attach an insulating shaft of plastic tubing by pressing onto the stud end of this screw. It may be desired to cut out a window for observing the gap. Cover hole with glass, plastic, etc., to protect your eyes from UV. For continued use, it is advised to lock SW10 via locknut to prevent overheating.

4. Capacitor C1 is mounted via bracket as shown on the grounded end (Fig. 13-4). The other end is supported via several pieces of double-sided tape built-up to the correct thickness for proper securing. Only one capacitor is shown. The remaining assembly is illustrated in the drawings with attention being given to proper positioning and spacing of the HV components. MP1 mounting base must be nonmetallic, dimensions of metal enclosure are not critical as long as there is ample spacing. Use sheet metal screws, rivets, etc., for putting enclosure together if home fabricated. Use standard wiring and assembly techniques.

5. A convenient output terminal for the secondary can be a smooth door knob attached via a bolt through CA2. This also looks good and gives the coil a finished look.

6. When firing up your coil, start with the spark gap set at 2 turns from "closed", and adjustment tap at maximum inductance. Note that this lead also contributes to total inductance when routed as shown. (Same direction as turns of primary coil.) It should be possible with this coil to easily obtain 10- to 12-inch streamers when properly adjusted. Careful adjustment of the tap location and gap adjustment will greatly enhance performance. When operating for a prolonged period always note potential breakdown points occurring indicated by heavy corona or premature sparking. These points must be corrected or a burning tracking condition will result.

EXPERIMENTS WITH YOUR COIL

1. Adjustment—suspend a grounded metallic object above the device. Start at about 3″ separation and make adjustments increasing separation until the optimum spark length is obtained. (Note: grounding means connection to metal base.)

2. Effect on human body—use caution as this may cause a reflex secondary reaction even though it is painless. Hold a metal object *tightly* and advance to coil terminal. Note the painless, tingling sensation. Fake out your pals by letting them think it is really painful. This demonstrates the skin or surface effect of high-frequency electricity.

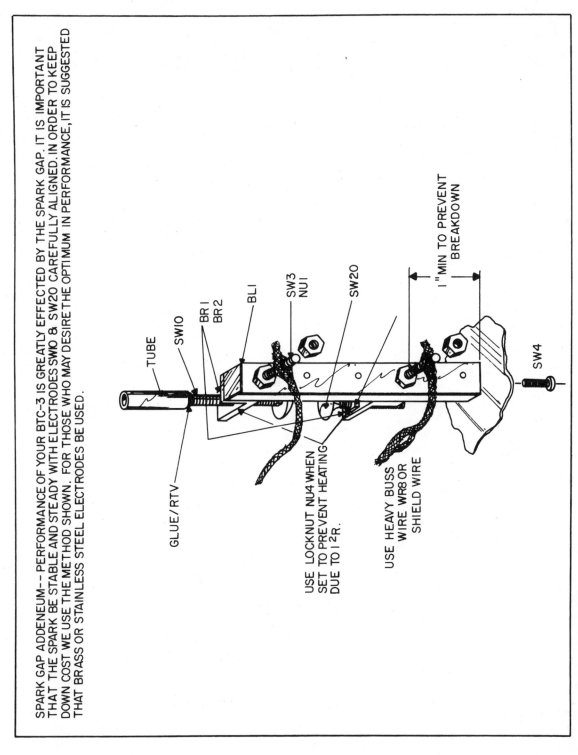

SPARK GAP ADDENEUM-- PERFORMANCE OF YOUR BTC-3 IS GREATLY EFFECTED BY THE SPARK GAP. IT IS IMPORTANT THAT THE SPARK BE STABLE AND STEADY WITH ELECTRODES SW0 & SW20 CAREFULLY ALIGNED. IN ORDER TO KEEP DOWN COST WE USE THE METHOD SHOWN. FOR THOSE WHO MAY DESIRE THE OPTIMUM IN PERFORMANCE, IT IS SUGGESTED THAT BRASS OR STAINLESS STEEL ELECTRODES BE USED.

TUBE

SW10

BR 1
BR 2

BL1

SW3
NU1

SW20

1" MIN TO PREVENT BREAKDOWN

SW4

GLUE / RTV

USE LOCKNUT NU4 WHEN SET TO PREVENT HEATING DUE TO I²R.

USE HEAVY BUSS WIRE WR8 OR SHIELD WIRE

Fig. 13-3. Exploded view of spark gap SG1.

199

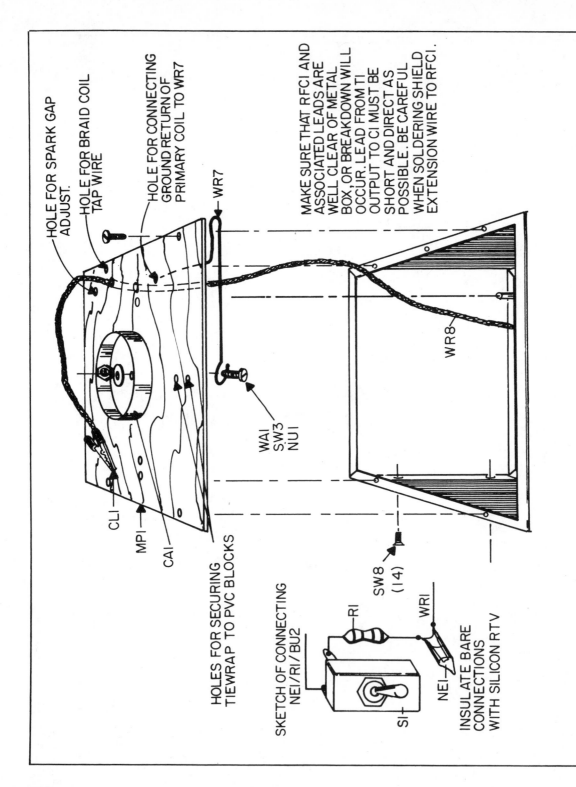

HOLE FOR SPARK GAP ADJUST.

HOLE FOR BRAID COIL TAP WIRE

HOLE FOR CONNECTING GROUND RETURN OF PRIMARY COIL TO WR7

WR7

MAKE SURE THAT RFCI AND ASSOCIATED LEADS ARE WELL CLEAR OF METAL BOX, OR BREAKDOWN WILL OCCUR. LEAD FROM TI OUTPUT TO CI MUST BE SHORT AND DIRECT AS POSSIBLE. BE CAREFUL WHEN SOLDERING SHIELD EXTENSION WIRE TO RFCI.

WR8

WAI
SW3
NUI

CLI

MPI

CAI

HOLES FOR SECURING TIEWRAP TO PVC BLOCKS

SW8
(14)

SKETCH OF CONNECTING NEI/RI/BU2

RI

WRI

SI

NEI

INSULATE BARE CONNECTIONS WITH SILICON RTV

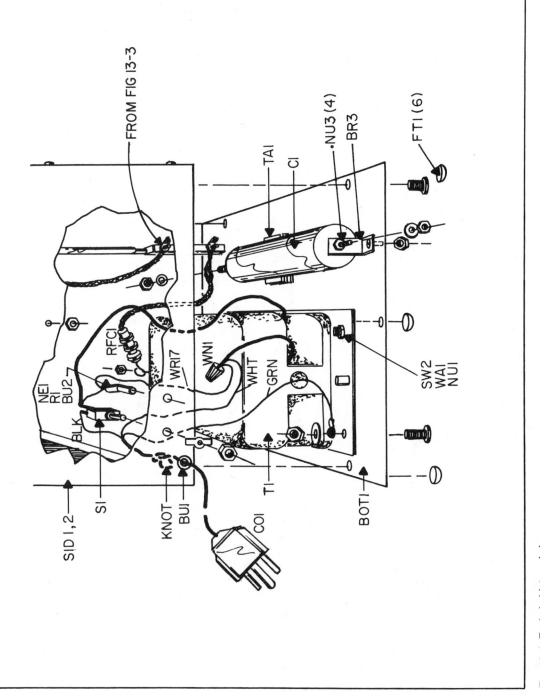

Fig. 13-4. Exploded internal view.

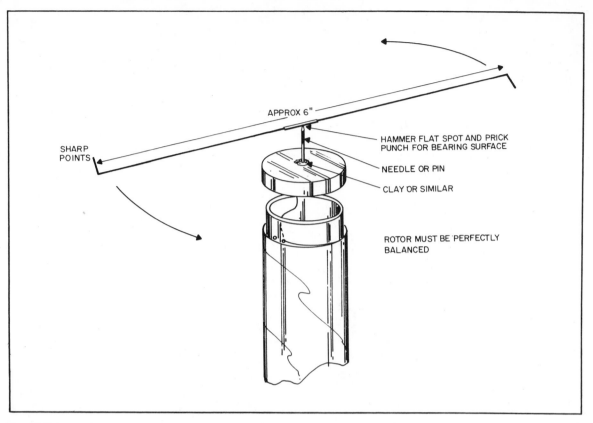

APPROX 6″

SHARP POINTS

HAMMER FLAT SPOT AND PRICK PUNCH FOR BEARING SURFACE

NEEDLE OR PIN

CLAY OR SIMILAR

ROTOR MUST BE PERFECTLY BALANCED

Fig. 13-5. Ion motor.

3. Effect on insulators—place various objects on the top of the coil and note the effects of the high-frequency electricity. The sparks are not stopped by glass or other usual insulators. Experiment using objects such as light bulbs, bottles, glass, etc.

4. Effect on partial insulators—place wood pieces about 12″ × 1″ × 3″ and note red streaks and other bizarre weird phenomena occurring from within the piece.

5. Ionization of gases—obtain a fluorescent lamp and allow it to come within several feet of the device. It will glow and produce light without direct connection, clearly demonstrating the effects of the electric and magnetic fields on the gas. Note the distance from the coil that the lamp will glow. Experiment using a neon lamp. Obtain other lamps and note the colors, distances, and other phenomena.

6. Induction fields—this is demonstrated by obtaining a small filament type lamp such as a flashlight bulb or similar and connecting between a large 1½ to 2′ diameter metal or wire loop. The lamp will now light due to energy coupled by induction. You will note that current is required to light this type of lamp and is entirely different than the radiation field that ionizes and causes the gas lamps to glow.

7. Create special effects of pinwheels, color fires, etc.—by connecting pieces of nichrome wire as shown in sketches. Note: rotor rotates creating a ring of fire. Try different types of rotors.

8. Ion motor—fabricate and carefully balance rotor as shown in Fig. 13-5. Use a piece of #16 or #18 copper buss wire. Rotor will spin at high speeds if carefully balanced, demonstrating ion propulsion.

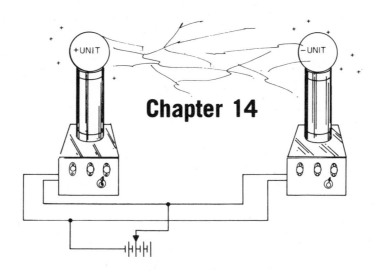

Chapter 14

Ion Ray
Gun/High-Voltage Generator (IOG1)

This project demonstrates the phenomenon of charged particles and their ability to travel distances and charge objects, people, an electroscope, etc. In order to demonstrate this effect, it is necessary to produce voltages of magnitudes that may be at a *hazardous shock potential*. Even though the device is battery operated it must be treated with caution. Use discretion when using it as it is possible for a person wearing insulated shoes to accumulate enough of a charge to produce a moderately painful or irritating shock when he touches a grounded object. The effect can be irritating and *could cause injury* to a person in weak physical condition. Effects depend on many parameters including humidity, leakage sizes and types of objects, proximity, etc.

The device is shown constructed in two ways. When the output of the device is terminated into a smooth large surfaced collector it becomes a useful high-potential source capable of powering particle accelerators and other related devices. It may also be built as a producer of negative or positive ions

demonstrating a phenomenon that is often regarded as a figure of demerit when building and designing high voltage power supplies. The device is now terminated into a sharp point where leakage of positive and negative ions can occur. This will result in corona and the formation of nitric acid via the production of the ozone produced combining with nitrogen and forming, nitrous oxide which with water produces this very strong acid. The production of ions also robs the available current from the supply.

It is well known that high voltage generators usually consist of large smooth surface collectors where leakage is minimized allowing these collective terminals to accumulate high voltages with less current demand. Leakage of a high-voltage point is the result of the repulsion of like charges to the extent that these charges are forced out into the air as ions. The rate of ions produced is a result of the charge density at a certain point. The magnitude of this quantity is a function of a voltage and the reciprocal of the angle of projection of the surface (Fig. 14-1). This is why lightning rods are sharply

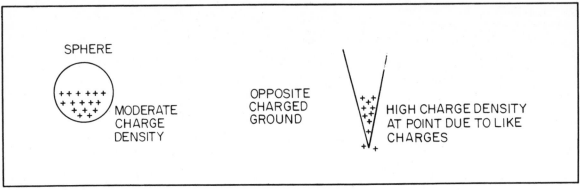

Fig. 14-1, Charge density explanation.

pointed causing the charges to leak off into the air before a voltage can be developed to create the lightning bolt.

It is now evident that to create ions it is necessary to have a high voltage applied to an object such as a needle or other sharp device used as an emitter. Once the ions leave the emitter they possess a certain mobility allowing them to travel moderate distances contacting and charging up other objects by accumulation.

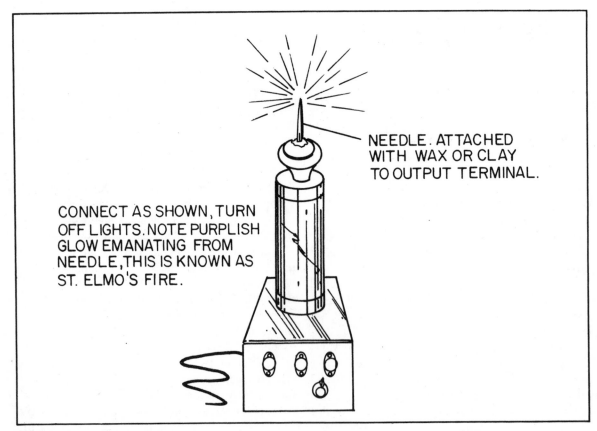

NEEDLE. ATTACHED WITH WAX OR CLAY TO OUTPUT TERMINAL.

CONNECT AS SHOWN, TURN OFF LIGHTS. NOTE PURPLISH GLOW EMANATING FROM NEEDLE, THIS IS KNOWN AS ST. ELMO'S FIRE.

Fig. 14-2. Corona display experiment.

APPLICATIONS
AND EXPERIMENTS USING THE
HIGH-VOLTAGE GENERATOR APPROACH

1. St. Elmo's Fire—this familiar glow discharge occurs during periods of high electrical activity. It is a corona discharge that is brushlike, luminous, and often may be audible, when leaking from charged objects in the atmosphere. It occurs on ship masts, on aircraft propellers, wings, other projecting parts and on objects projecting from high terrain when the atmosphere is charged and a sufficiently strong electrical potential is created between the object and the surrounding air. Aircraft most frequently experience St. Elmo's fire when flying in or near cumulonimbus clouds, thunderstorms, in snow showers, and in dust storms. It is easily artificially produced by using your unit as shown in Fig. 14-2. Also, see Table 14-1.

2. Flashing fluorescent light—this experiment shows the mobility of the ions and their ability to charge up the capacitance in a fluorescent light tube and discharge in the form of a flash. Perform the following: (See Fig. 14-3)

A. Connect a needle or other sharp object so it is pointing in a desired direction. Use clay or wax.

B. Carefully hold a 20 watt fluorescent tube and turn off the lights. Allow eyes to become accustomed to total darkness.

C. Hold end about 3 feet from needle and note lamp flickering. Increase distance and note flicker rate decreasing. Under ideal conditions, and total darkness, the lamp will flicker up to a considerable distance from the source. Use caution in total darkness. Hold lamp by glass envelope and touch end pins to water pipe, metal objects, etc., for best results and brightest flash. CAUTION: Remain clear of device as a painful shock with dangerous secondary reactions may result. Flash time is equivalent to the familiar equation: $T = \dfrac{CV}{I}$ where

"T" is the time between flashes, "V" is the flash breakdown voltage, "C" is the inherent capacity in the tube and "T" is equivalent to the amount of ions reaching the lamp and obviously decreases by the 5/2 power of the distance.

3. Ion charging—this demonstrates the same phenomena as in Experiment 2, but in a different way. Perform the following: (see Fig. 14-4).

A. Set up unit as Experiment 2.

B. Set up objects as shown.

C. Note spark occurring as a result of ion accumulation on sphere. Increase distance and note where spark becomes indistinguishable.

D. Obtain a subject brave enough to stand a moderate electric shock (*Use caution* as a person with a heart condition should not be near this experiment).

E. *Have subject stand on a insulating surface* and then touch a grounded or large metal object. Rubber or similar soled shoes often work to an extent.

4. Ion motor—this dramatically demonstrates Newton's Law of action producing reaction. Escaping ions at high velocity produce reaction. This is a viable means of propulsion for space craft where hypervelocities may approach the speed of light in this frictionless environment. Perform as shown in Fig. 14-5.

A. Form piece of #18 wire as shown. For maximum results carefully balance and provide minimum friction at pivot point. There are many different methods of performing this experiment with far better results. We leave this to the experimenter bearing in mind that a well made and balanced rotor can achieve amazing rpm.

B. Note as rotor spins giving off ions, that one's body hair will bristle, nearby objects will spark and a cold feeling will persist.

5. Demonstrates the destructive shocking power and penetration of a spark discharge. Perform as shown in Fig. 14-6. *Use caution when approaching output terminal.* Rig up a well grounded discharge stick. Use your ingenuity.

6. Lightning generator—Perform as shown in Fig. 14-7.

A. Construct two units but reverse all the diodes in the multiplier stages to produce a negative output in second unit.

B. Obtain some large smooth metal objects that resemble spheres as nearly as possible. Place

Table 14-1. HVM1/IOG1 Parts List.

	IOG1	HVM3	
R1		(1)	0.3 ohm 3-watt pwr resistor
R2,6	(4)	(8)	220 ohm 1 watt in parallel for 55 ohm or use (2) 110 ohm 2 watts
R3,5,7	(1)	(3)	1 k 1/4 watt resistor
R4/S1		(1)	5 k pot and switch
C1		(1)	4700-6800 μF/25 V electrolytic cap
C2,3		(2)	100 μF/25 V electrolytic cap
C4	(1)	(1)	1000 μF 25 V electrolytic cap
D1,2,3, 4,5,6	(2)	(6)	1N4002 1 amp 100 V diodes
Z1		(3)	1N4002 1 amp 100 V diodes/or zener 4.7 V
Q1		(1)	Npn transistor D40D5 pwr Tab Si
Q2,3,4	(2)	(3)	Npn transistor TO3 2N3055 power
T1		(1)	12 V 3 amp transformer
T2	(1)	(1)	TV BW flyback rework Fig. 14-14
CO1		(1)	Power cord 3 wire preferred
MK1,2,3	(2)	(3)	Mtg. kits for Q2,3,4, Fig. 14-17
PB1		(1)	4 × 3 perfboard .1 grid
PB2	(1)	(1)	6 × 2 perfboard .1 grid
S2	(1)		Push-button switch
WR1	(6')	(6')	#18 hook-up for T2 and wiring
WR3	(2')	(2')	#24 hook-up for T2 and wiring
CASE		(1)	7 × 5 × 3" al box
FEET		(4)	Rubber stick-on feet
PLASTIC TUBE	(4")	(4")	3/16" ID flexible plastic tube
TE1	(1)		7 lug terminal strip
BU1		(1)	1/2" bushing plastic
BU2		(1)	Cord clamp bushing
SW3/NU1	(6)	(3)	6 32 × 1 screw and nut
WA1		(1)	Wide shoulder washer
NU2		(4)	4 40 kep nuts
EN1	(1)		8 × 3½ sked 40 PVC Fig. 14-22
EN2	(1)	(1)	8 × 2 ⅜ sked 40 PVC Fig. 14-22
HA1	(1)		6 × 2 sked 40 PVC Fig. 14-22
MP1	(1)		2⅝ × 2⅝ #22 ga Fig. 14-20
CA1,2	(1)	(2)	2⅜" plastic caps
RC1,FC1	(2)		3½" plastic caps
BH1	(1)		4 cell AA holder
LAB1	(1)	(1)	DANGER HV label
C5-10	(6)	(6)	.001/15 kV ceramic cap
CR1-6	(6)	(6)	20 KV HV diodes

The above capacitors and diodes may be added for more output up to 18 stages. (12 are recommended.) Each section will add approximately 10 kV of output.

	IOG1	HVM3	
CU1	(1)		2 3/8 PVC coupling for corona shield
TER1		(1)	Use smooth brass door knob or round ornamental fixture
NL1	(1)		Sharp needle or other similar part for producing ion leakage
WAX			Paraffin wax available in most markets used for canning

Kit of above including six multiplier stages available through Information Unlimited, Inc., P.O. Box 716, Amherst, N.H. 03031. Write or call 603-673-4730 for price and availability.

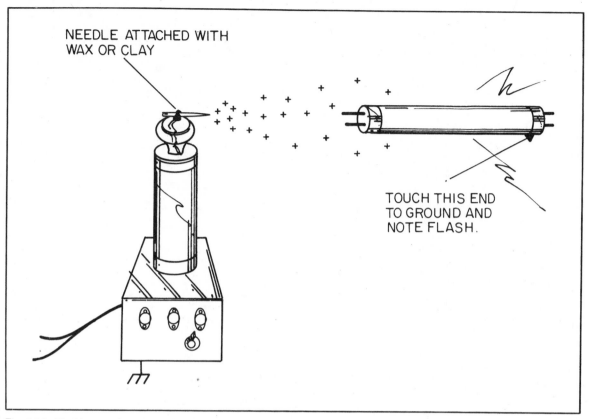

NEEDLE ATTACHED WITH WAX OR CLAY

TOUCH THIS END TO GROUND AND NOTE FLASH.

Fig. 14-3. Flashing gas lamp experiment.

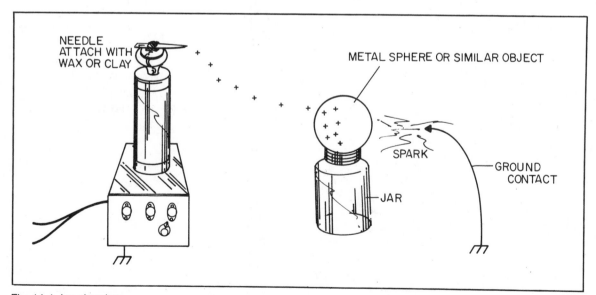

NEEDLE ATTACH WITH WAX OR CLAY

METAL SPHERE OR SIMILAR OBJECT

SPARK

GROUND CONTACT

JAR

Fig. 14-4. Ion charging.

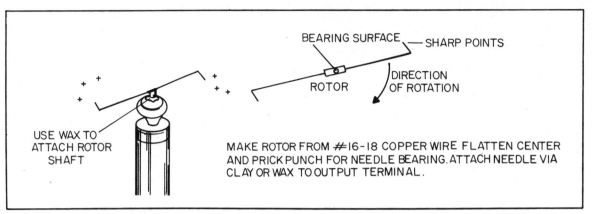

BEARING SURFACE — SHARP POINTS

DIRECTION OF ROTATION

ROTOR

USE WAX TO ATTACH ROTOR SHAFT

MAKE ROTOR FROM #16-18 COPPER WIRE FLATTEN CENTER AND PRICK PUNCH FOR NEEDLE BEARING. ATTACH NEEDLE VIA CLAY OR WAX TO OUTPUT TERMINAL.

Fig. 14-5. Ion motor.

as terminals on units and remove any outward protruding sharp surfaces. Object should be 3″ or more in diameter, use utensils, etc.

C. Locate units as shown.

7. Experiment using your unit in a gun configuration. Capacitor charging performs as shown in Fig. 14-8. The unit generates ions that are accumulated on the insulated spherical object charging it

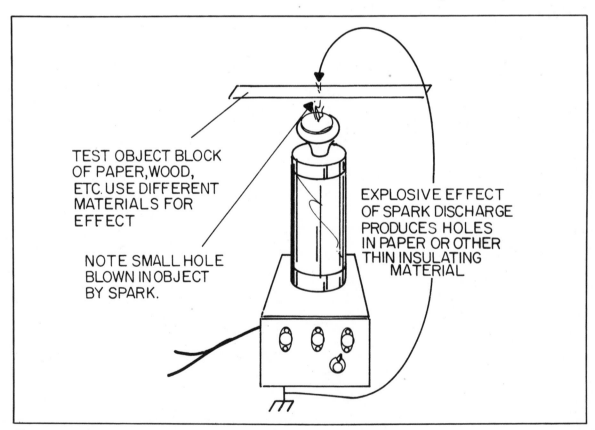

TEST OBJECT BLOCK OF PAPER, WOOD, ETC. USE DIFFERENT MATERIALS FOR EFFECT

NOTE SMALL HOLE BLOWN IN OBJECT BY SPARK.

EXPLOSIVE EFFECT OF SPARK DISCHARGE PRODUCES HOLES IN PAPER OR OTHER THIN INSULATING MATERIAL

Fig. 14-6. Spark discharge.

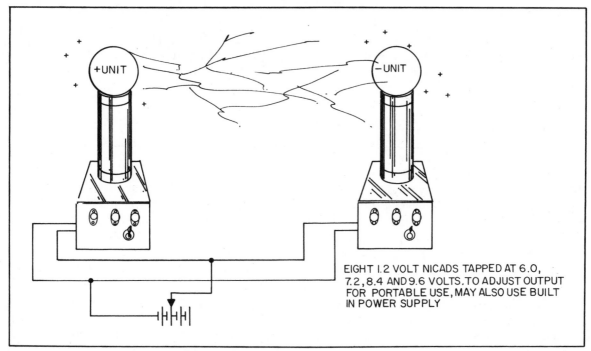

Fig. 14-7. Oppositely charged units experiment.

theoretically to its open circuit potential. (This in practice doesn't occur due to leakage, etc.) The object accumulates a voltage equal to $V = it/c$. Note the unit is also directly grounded to increase this effect by producing the necessary electrical mirror image. The quantity "Q" (Coulombs) of the charge is equal to "CV" where C = capacitance of object and V = voltage charged. The energy "W" (joules) stored is equal to $1/2CV^2$. The capacitance can be calculated by approximating the area of the shadow of the object projected directly beneath it and calculating the mean separation (use inches). The capacitance is now approximately equal to "C" pF = .25 times projected area divided by mean separation.

8. Spark discharge—demonstrates the difference in spark length between a high and a low leakage surface. Perform the following: (see Fig.

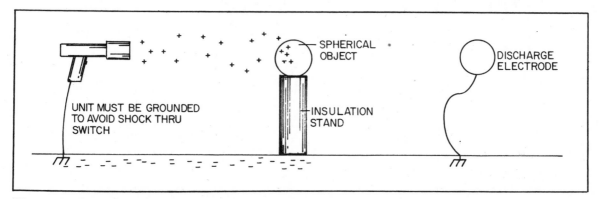

Fig. 14-8. Remote charging by gun configuration experiment.

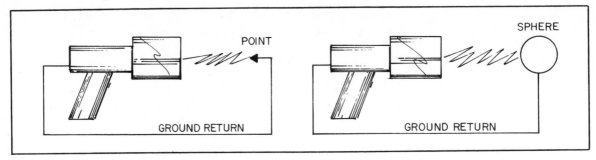

Fig. 14-9. Spark discharge experiment.

14-9). Note that the spark may only be 1″ or so with the pointed object while the smooth surface object produces a 3″ to 4″ spark. This is due to the inability of the device to maintain a voltage under the high leakage conditions with the points while the other produces just enough ions to produce a leader for the main discharge.

9. Demonstrates the transmission of energy via mobile ions. Perform as shown in Fig. 14-10. Objects are round spheres placed on glass bottles used as insulators. One object is grounded using thin insulated wire. Other object discharges to grounded object. Note grounding wire physically jumping at time of discharge, this phenomena is the result of current producing a mechanical force. As unit is brought closer, length of spark discharge and discharge rate will increase. A discharge length of 1″ is easily obtainable with the unit 4 to 5 feet away.

This demonstrates the potential effectiveness of the device.

10. Demonstrates the ability of the device to cause a person to become charged and experience a moderate shock when grounded. Perform as shown in Fig. 14-11. This is similar to Experiment 3 shown in Fig. 14-4. *Caution: do not attempt on individuals unless you know they are in good physical condition.*

The unit is directed towards the victim. An ideal condition is for victim to be standing on insulated floor or wearing rubber or other insulative type shoes. Unit should be grounded for maximum effect, however, some grounding effect will occur through person holding it. Use light switch, pipe, etc., for temporary ground.

Victim charges up with ions and when he touches grounded or large object, a flow of current occurs causing a shock. Please note that the sever-

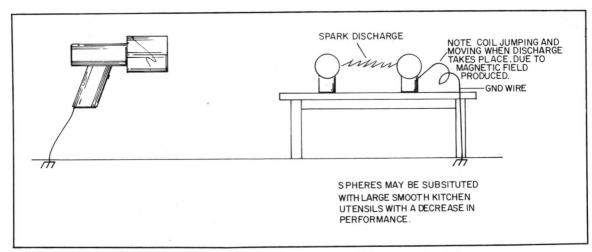

Fig. 14-10. Remote charging experiment.

210

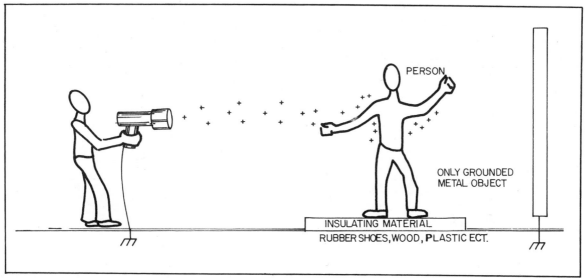

Fig. 14-11. Charging a person experiment.

ity of this shock depends on many factors and can be severe under proper conditions. *Again use caution.*

Other experiments and uses are materials and insulation dielectric breakdown testing, X-ray power supplies, capacitance charging, high voltage work, ignition of gases in tubes and spark gaps, particle acceleration and atom smashers, Kirlian photography, electrostatics and ion generation.

Other related material may easily be obtained on the above subjects.

NEGATIVE ION INFORMATION

In the last two decades a medical controversy has evolved pertaining to the beneficial effects of these minute electrical particles. As with any device that appears to affect people in a beneficial

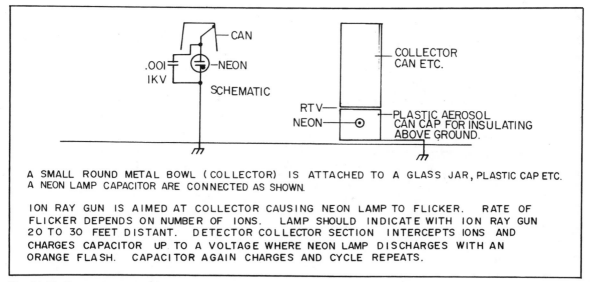

A SMALL ROUND METAL BOWL (COLLECTOR) IS ATTACHED TO A GLASS JAR, PLASTIC CAP ETC.
A NEON LAMP CAPACITOR ARE CONNECTED AS SHOWN.

ION RAY GUN IS AIMED AT COLLECTOR CAUSING NEON LAMP TO FLICKER. RATE OF
FLICKER DEPENDS ON NUMBER OF IONS. LAMP SHOULD INDICATE WITH ION RAY GUN
20 TO 30 FEET DISTANT. DETECTOR COLLECTOR SECTION INTERCEPTS IONS AND
CHARGES CAPACITOR UP TO A VOLTAGE WHERE NEON LAMP DISCHARGES WITH AN
ORANGE FLASH. CAPACITOR AGAIN CHARGES AND CYCLE REPEATS.

Fig. 14-12. Demonstrates a sensitive ion detector.

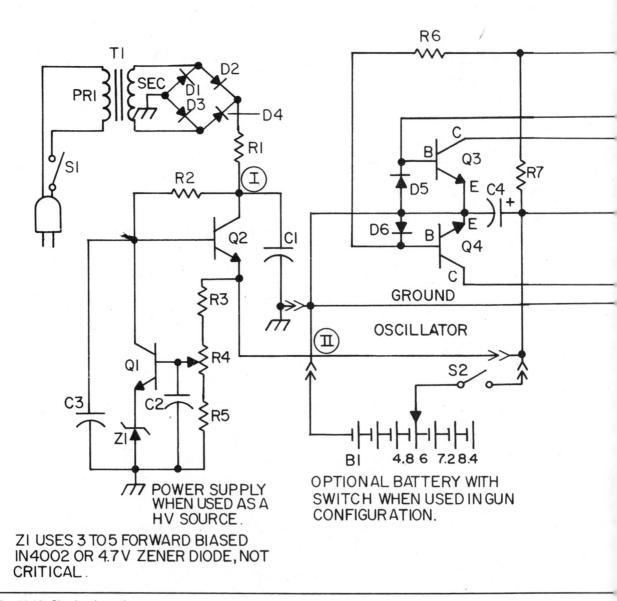

IF OSCILLATOR SECTION DOES NOT WORK TRY
REVERSING FEEDBACK WIRES FROM A&B FROM
T2. AS THE CORRECT PHASE IS REQUIRED.

POWER SUPPLY
WHEN USED AS A
HV SOURCE.

OPTIONAL BATTERY WITH
SWITCH WHEN USED IN GUN
CONFIGURATION.

ZI USES 3 TO 5 FORWARD BIASED
IN4002 OR 4.7V ZENER DIODE, NOT
CRITICAL.

Fig. 14-13. Circuit schematic.

NOTE ONLY THREE SECTIONS OF VOLTAGE MULTIPLICATION
ARE SHOWN. YOU MAY USE UP TO SIX WITH GOOD RESULTS.
MORE STAGES WILL REQUIRE BETTER INSULATING
MATERIALS.

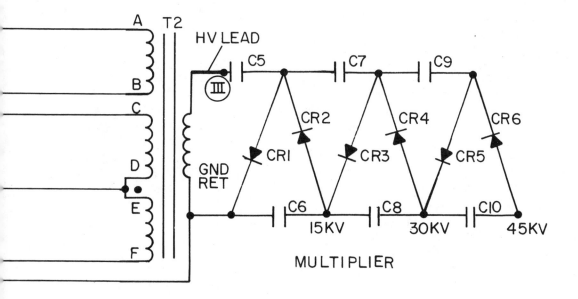

MULTIPLIER

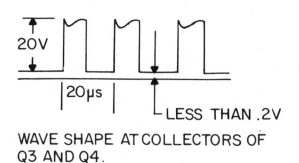

LESS THAN .2V

WAVE SHAPE AT COLLECTORS OF
Q3 AND Q4.

sense there are those who sensationalize and exaggerate these claims as a cure for all ailments and ills. Such people manufacture and market these devices under these false pretences, and consequently give the products a diverse name. The Food and Drug Administration now steps in on these claims and the product along with its beneficial facets goes down the tubes.

The builder may wish to obtain the following articles: *Negative Ion's Affect Health* - Oct. 22, 1976, Int'l. Press Release; Ions - Oct. 1960 Rotarian, Oct. 1960 Readers Digest; *Negative Ions*, Popular Electronics, Sept. 1961.

People are affected by negative ions from the property of these particles to increase the rate of activity by cilia thus enhancing (whose property is to keep the tracheas clean from foreign objects) oxygen intake and increasing the flow of mucous. This property neutralizes the effects of cigarette smoking that slows down the activity of the cilia. Hay fever and bronchial asthma victims are greatly relieved by these particles. Burns and surgery patients are relieved of pain and heal faster. Tiredness, lethargy and the general dragged out feeling are replaced by a sense of well being and renewed energy. Negative ions destroy bacteria and purify the air with a country air freshness. They cheer people up by decreasing the serotonin content of the blood. As can be seen in countless articles and technical writings, negative ions are a benefit to man and his environment.

Negative ions occur naturally from static electricity, certain winds, waterfalls, crashing surf, cosmic radiation, radioactivity and ultraviolet radiation. Positive ions are also produced from some of the above phenomena and usually neutralize each other as a natural statistical occurrence. However, many man-made objects and devices have a tendency to neutralize the negative ions, thus leaving an abundance of positive ions which create sluggishness and most of the opposite physiological effect of its negative counterpart.

One method of producing negative ions is obtaining a radioactive source rich in beta radiations (electrons are negative). Alpha and gamma emission from this source produces positive ions that are neutralized electrically. The resulting negative ions are electrostatically directed to the output exit of the device and further dispersed by the action of a fan (this method has recently come under attack by the Bureau of Radiological Health and Welfare for the use of tritium or other radioactive salts). This approach appears to be the more hazardous of the two according to the Product Consumer Safety people.

A more accepted method is to place a small tuft of stainless steel wool as the Ion Emitter at the output terminal of a negative HV dc power supply. The hair like property of the stainless steel wool allows ions to be produced at relatively low voltage yet reducing ozone output. Ions are produced by leakage of the particles charging air molecules in the immediate vicinity of the steel wool emitter. The unit should be operated below 14 kV as overvoltage can produce substantial amounts of ozone that can mask the beneficial effect of the increased ions obtained. A suitable and effective ion detector capable of indicating relative amounts of ion flux is shown in Fig. 14-12.

CIRCUIT DESCRIPTION

Q3 and Q4 comprise a high frequency oscillator alternately driving the core of T2 via winding C, D, E, and F at a frequency determined by its magnetic properties. Usually about 20 kHz for the components shown (Fig. 14-13). Feedback for the transistors is via winding A and B with base current limiting resistor (R6). Diodes D5 and D6 provide the return path for this current. R7 provides the circuit imbalance necessary to start oscillation. T2 produces about 10 to 15 kV at its output and is fed into a Cockcroft-Walton multiplier where it may be rectified and increased to nearly 100 kV via diodes and capacitors. This output may be converted to directional ions via a sharp pointed needle at the output or can be terminated into a large smooth terminal to produce a high potential. Please note that the diodes are shown connected for positive ions, however, they can be reversed for generation of negative ions producing the usual beneficial results attributed to this healthful phenomenon.

When the device is used as a high-voltage power supply, power is obtained from T1, rectified by diodes D1, 2, 3, 4 and controlled by *pass* transistor Q2. Voltage output is determined by the setting of R4 and is adjustable from 5 to 12 volts. Z1 is a voltage reference for Q1 that controls the voltage at the base of Q2, consequently the voltage output at the emitter. Regulation feedback is via the loop divider consisting of R3, R4, and R5. This adjustment allows the output to be varied over a considerable range and is a definite advantage for many experiments.

When used as the Ion Ray Gun this section consisting of the power supply is eliminated and batteries are used. Proper grounding of the chassis and mechanical parts of the device are necessary to avoid possible electric shock. Grounds should be at identical potential with the ground on which the operator is standing regarded as the common point.

CONSTRUCTION HV DC
SUPPLY CONFIGURATION (HVM3)

1. Rework T2 flyback transformer as per Fig. 14-14.

2. Assemble board as shown in Fig. 14-15. Note polarity of the diodes and capacitors. See note on R2, R6, and Z1 zener on substitution of other components. Board should have #18 leads to collector of Q2, emitter of Q2, ground return T2, and the emitter of Q3, Q4. It also should have #24 leads to base of Q3 and Q4.

3. Assemble multiplier board as shown in Fig. 14-16.

4. Fabricate case as shown for mounting of components. Mount transistors with mounting kits as shown and wire as shown in Fig. 14-17. Interconnect wires from assembly board. Wire in T1 and T2. Make sure T2 high voltage coil has adequate clearance from wires or metallic objects. This is important or power robbing corona will result. Note method of securing T2.

5. Carefully check wiring to this point and layout as shown in Fig. 14-18. Note case cover and HV-multiplier section are not connected at this point.

6. Connect to 120 Vac and turn R4 fully ccw. Measure 15 Vdc at TPI and 5 volts at TPII. Rotate R4 full cw and measure 11 to 13 V at TPII. Touch TPIII with an insulated metal screwdriver and note

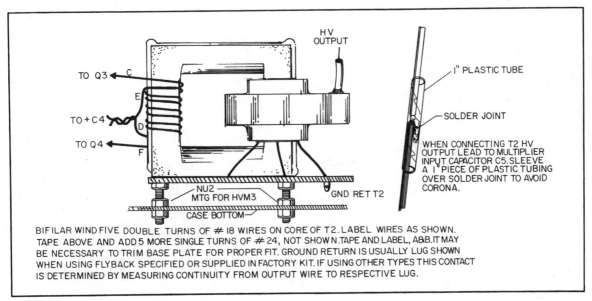

BIFILAR WIND FIVE DOUBLE TURNS OF # 18 WIRES ON CORE OF T2. LABEL WIRES AS SHOWN. TAPE ABOVE AND ADD 5 MORE SINGLE TURNS OF # 24, NOT SHOWN. TAPE AND LABEL, A&B. IT MAY BE NECESSARY TO TRIM BASE PLATE FOR PROPER FIT. GROUND RETURN IS USUALLY LUG SHOWN WHEN USING FLYBACK SPECIFIED OR SUPPLIED IN FACTORY KIT. IF USING OTHER TYPES THIS CONTACT IS DETERMINED BY MEASURING CONTINUITY FROM OUTPUT WIRE TO RESPECTIVE LUG.

Fig. 14-14. T2 transformer.

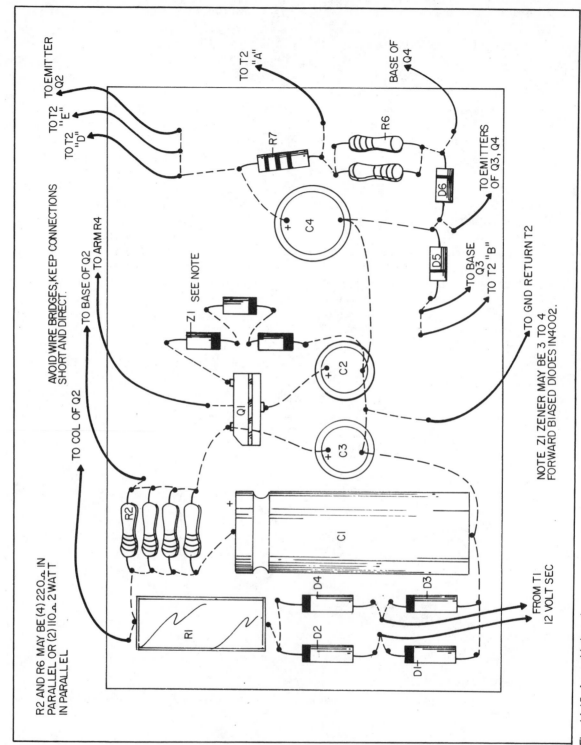

Fig. 14-15. Assembly board.

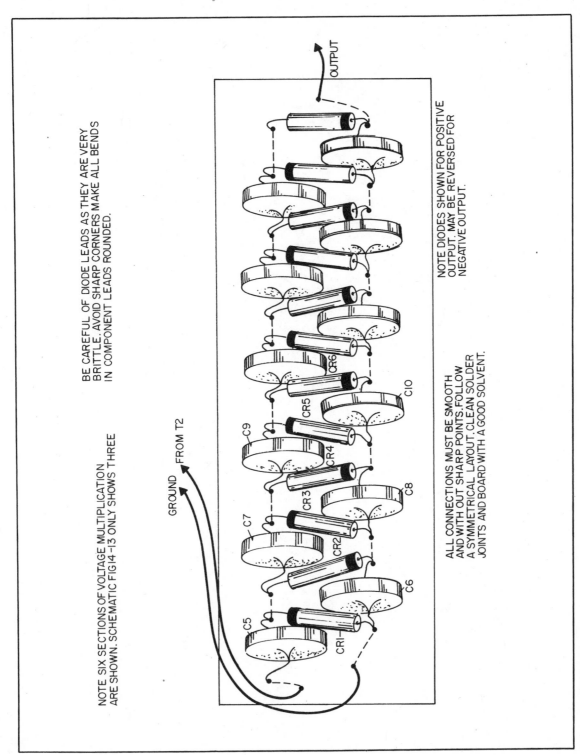

BE CAREFUL OF DIODE LEADS AS THEY ARE VERY BRITTLE. AVOID SHARP CORNERS MAKE ALL BENDS IN COMPONENT LEADS ROUNDED.

NOTE SIX SECTIONS OF VOLTAGE MULTIPLICATION ARE SHOWN. SCHEMATIC FIG14-13 ONLY SHOWS THREE

GROUND

FROM T2

ALL CONNECTIONS MUST BE SMOOTH AND WITH OUT SHARP POINTS. FOLLOW A SYMMETRICAL LAYOUT. CLEAN SOLDER JOINT'S AND BOARD WITH A GOOD SOLVENT.

NOTE DIODES SHOWN FOR POSITIVE OUTPUT. MAY BE REVERSED FOR NEGATIVE OUTPUT.

OUTPUT

C5

C7

C9

CR2

CR3

CR4

CR5

CR6

CR1

C6

C8

C10

Fig. 14-16, Multiplier board.

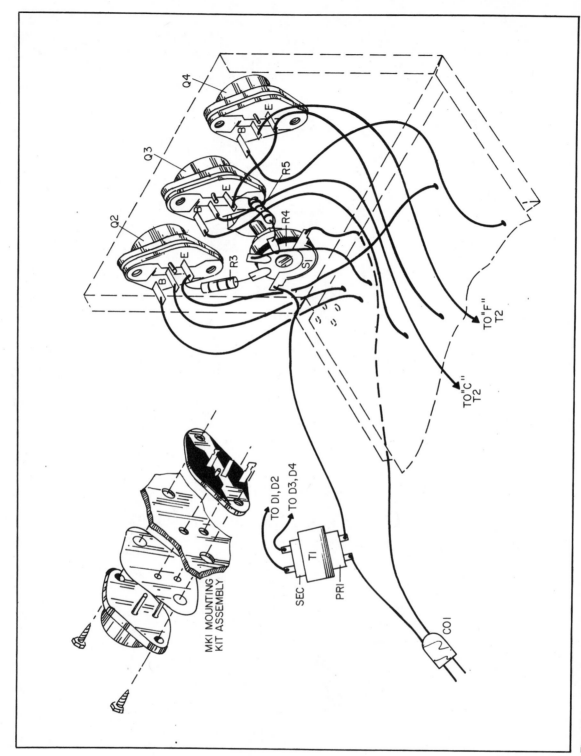

Fig. 14-17. Transistor and power connections.

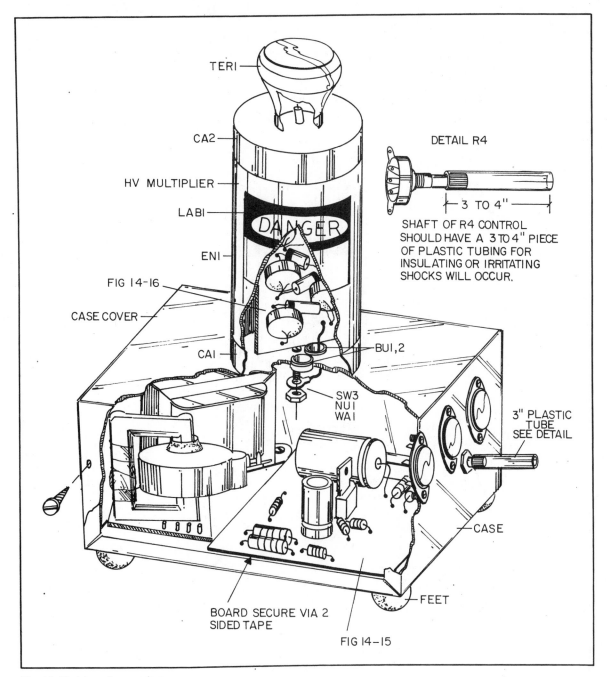

TER1

CA2

HV MULTIPLIER

LAB1

DANGER

EN1

FIG 14-16

CASE COVER

CA1

BU1,2

SW3
NU1
WA1

DETAIL R4

3 TO 4"

SHAFT OF R4 CONTROL
SHOULD HAVE A 3 TO 4" PIECE
OF PLASTIC TUBING FOR
INSULATING OR IRRITATING
SHOCKS WILL OCCUR.

3" PLASTIC
TUBE
SEE DETAIL

CASE

BOARD SECURE VIA 2
SIDED TAPE

FEET

FIG 14-15

Fig. 14-18. Internal x-ray view.

a healthy ½″ purplish arc. Do not ground. Check collectors of Q3 and Q4 and note waveshape as shown in Fig. 14-13.

7. Temporarily connect the multiplier section and place on an insulating surface to avoid shorting the connections on the board. Apply power

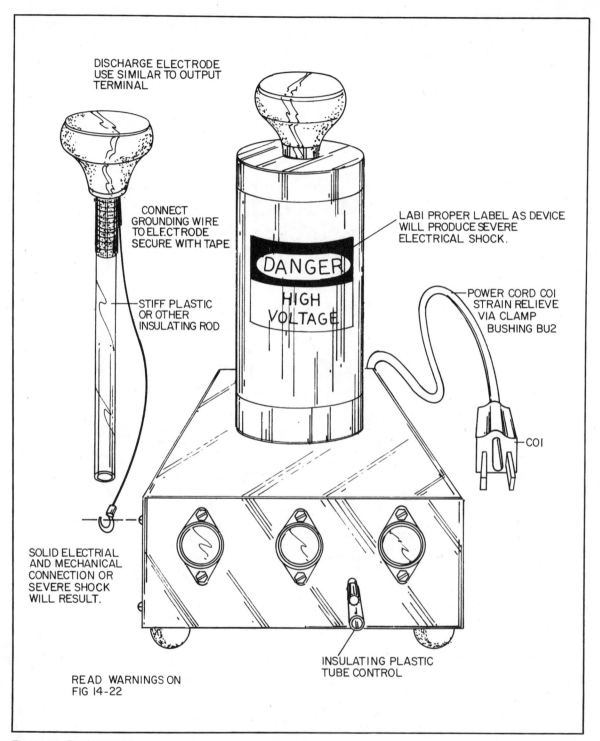

DISCHARGE ELECTRODE
USE SIMILAR TO OUTPUT
TERMINAL

CONNECT
GROUNDING WIRE
TO ELECTRODE
SECURE WITH TAPE

LABI PROPER LABEL AS DEVICE
WILL PRODUCE SEVERE
ELECTRICAL SHOCK.

DANGER
HIGH
VOLTAGE

POWER CORD COI
STRAIN RELIEVE
VIA CLAMP
BUSHING BU2

STIFF PLASTIC
OR OTHER
INSULATING ROD

COI

SOLID ELECTRIAL
AND MECHANICAL
CONNECTION OR
SEVERE SHOCK
WILL RESULT.

INSULATING PLASTIC
TUBE CONTROL

READ WARNINGS ON
FIG 14-22

Fig. 14-19. Final assembly.

220

at lowest setting of R4 and note a HV spark or ¾" to 1" or more. Turn off lights and note any points where corona may be occurring. Correct if any and advance R4 to mid range. Note spark distance increasing to 1½" or more. Contact electrode must be a large well grounded smooth object or leakage will occur at a faster rate than the supply can deliver. *Do not at this point apply full power* as HV multiplier section must be potted to withstand the high potentials generated and flashover may occur destroying a capacitor or diode.

8. Fabricate case cover as shown in Fig. 14-18. Note two holes for each wire from multiplier section. The hole for lead to C5 from T2 must have a plastic bushing as there is high voltage at high frequency at this point and corona must be kept minimized. These holes are spaced 1½ from one another. A hole for screw S1 to secure CA1 plastic cap via washer WA1 is between these two larger holes. This plastic cap secures the HV multiplier section in place with the retaining screw also being used to connect the ground input of the HV multiplier.

9. EN1 is a 6" length of 2⅜" sked 40 PVC tube. The multiplier board output lead is attached to the output terminal via a convenient means. This terminal is also secured to plastic cap CA2 and should be liquid-wax tight to avoid leaking with potting. Since there are many different variations in output terminals the exact method of securing is left to the ingenuity of the builder. The important part is that the multiplier output lead must be direct and as short as possible to the output terminal and the seal should be liquid-wax tight as the wax is poured into the input end.

When melting paraffin wax always use a container with water to prevent the wax from overheating. Paraffin wax may be replaced by other potting compounds that are better, however, paraffin is forgiving and can easily be remelted and reused over should the builder make an error or experience a breakdown.

10. Attach HV multiplier section as shown in Fig. 14-18 and wire to T2 as shown in Fig. 14-14. Note plastic tubing for shield of actual solder joint.

11. Assemble the discharge electrode as shown in Fig. 14-19. Attach a section of insulating plastic tube to shaft of R4 as shown in Fig. 14-18. Flexible 3/16 ID vinyl works well and can be pressed on the shaft of R4.

12. To test unit apply power and draw spark with the *grounded* discharge electrode. Spark should be 3" or more or equivalent to nearly 100 kV with the six-section multiplier shown in Fig. 14-16. Note that a *very powerful electric shock* is possible from this unit even when power is removed due to the storing capability of the capacitors in the multiplier.

CONSTRUCTION ION RAY GUN CONFIGURATION (IOG1)

1. Fabricate MP1 plate as shown in Fig. 14-20 from a 2⅝ × 2⅝ square. Trim corners and fabricate holes for components. Note Fig. 14-17 for mounting of transistors Q3 and 4.

2. Assemble T2 from Fig. 14-14.

3. Fabricate EN1 from a 8" length of 3½ OD sked 40 PVC. Drill a 1⅞" hole at an angle for HA1 handle. Use holesaw and firmly secure in vise.

4. Fabricate EN2 from a 8" length of 2⅜ OD sked 40 PVC.

5. Fabricate HA1 from a 6" length of 2" OD sked 40 PVC.

6. Assemble and wire T2 to mounting plate as shown in Figs. 14-20 and 14-21.

7. Connect a 6-volt battery and note a current draw of less than 1 amp. You should be able to obtain a purplish arc from the output of T2. If not, reverse feedback wires A and B and check for errors. Note waveform sketch in Fig. 14-13. (See note in Fig. 14-22 on batteries.)

8. Assemble multiplier board as shown in Fig. 14-16.

9. Final assembly as shown in Figs. 14-21 and 14-22, use potting instructions given for HVM3.

10. Assembled unit will produce a high field of ions capable of causing irritating electric shocks. **Do not use in an explosive environment. Do not use near sensitive electrical equipment such as a computer, etc. Do not use without a positive ground or severe shock may result to the user via switch S2.**

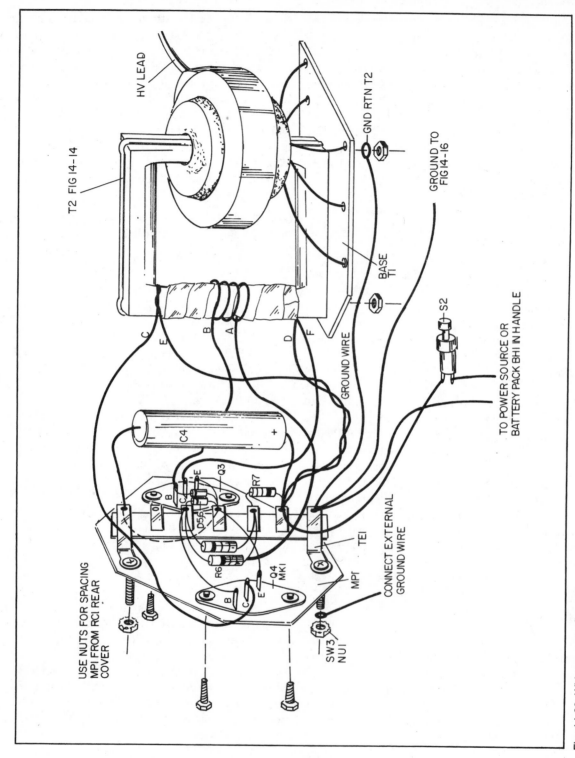

Fig. 14-20. Wiring—gun configuration.

222

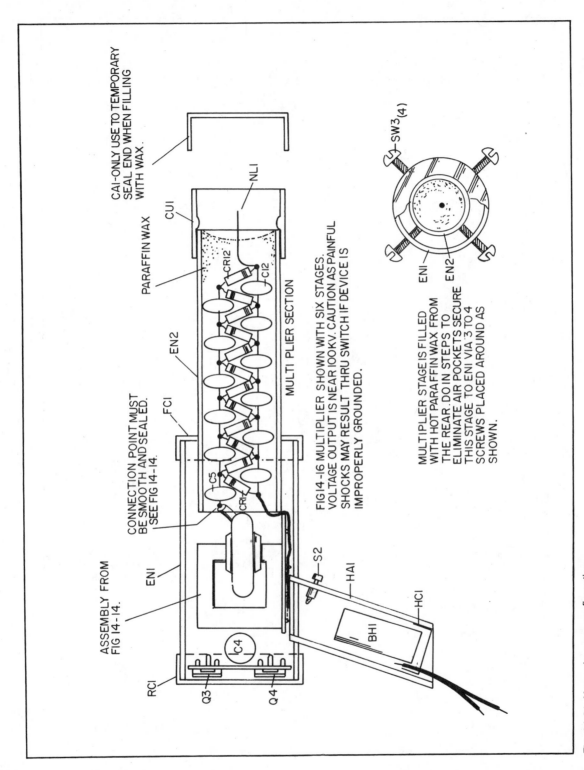

CA1-ONLY USE TO TEMPORARY SEAL END WHEN FILLING WITH WAX.

PARAFFIN WAX

CU1

NL1

CR12

C12

EN2

MULTI PLIER SECTION

FC1

FIG14-16 MULTIPLIER SHOWN WITH SIX STAGES. VOLTAGE OUTPUT IS NEAR 100KV. CAUTION AS PAINFUL SHOCKS MAY RESULT THRU SWITCH IF DEVICE IS IMPROPERLY GROUNDED.

CONNECTION POINT MUST BE SMOOTH AND SEALED. SEE FIG 14-14.

C5

CR12

ASSEMBLY FROM FIG 14-14.

EN1

S2

HA1

BH1

HC1

C4

RC1

Q3

Q4

SW3 (4)

EN1

EN2

MULTIPLIER STAGE IS FILLED WITH HOT PARAFFIN WAX FROM THE REAR DO IN STEPS TO ELIMINATE AIR POCKETS SECURE THIS STAGE TO EN1 VIA 3 TO 4 SCREWS PLACED AROUND AS SHOWN.

Fig. 14-21. X-ray view gun configuration.

223

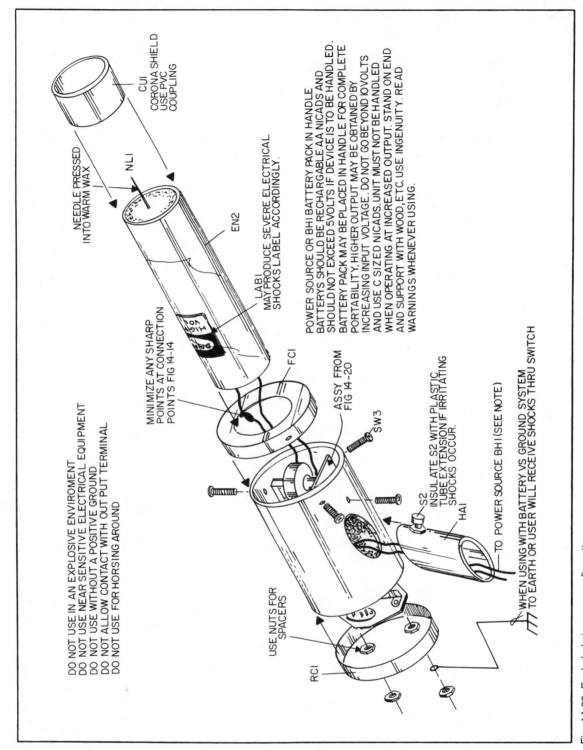

DO NOT USE IN AN EXPLOSIVE ENVIROMENT
DO NOT USE NEAR SENSITIVE ELECTRICAL EQUIPMENT
DO NOT USE WITHOUT A POSITIVE GROUND
DO NOT ALLOW CONTACT WITH OUT PUT TERMINAL
DO NOT USE FOR HORSING AROUND

MINIMIZE ANY SHARP
POINTS AT CONNECTION
POINTS FIG I4-I4

USE NUTS FOR
SPACERS

RCI

NEEDLE PRESSED
INTO WARM WAX

NLI

CUI
CORONA SHIELD
USE PVC
COUPLING

EN2

LABI
MAY PRODUCE SEVERE ELECTRICAL
SHOCKS LABEL ACCORDINGLY.

POWER SOURCE OR BHI BATTERY PACK IN HANDLE
BATTERYS SHOULD BE RECHARGABLE AA NICADS AND
SHOULD NOT EXCEED 5VOLTS IF DEVICE IS TO BE HANDLED.
BATTERY PACK MAY BE PLACED IN HANDLE FOR COMPLETE
PORTABILITY. HIGHER OUTPUT MAY BE OBTAINED BY
INCREASING INPUT VOLTAGE. DO NOT GO BEYOND I0VOLTS
AND USE C SIZED NICADS. UNIT MUST NOT BE HANDLED
WHEN OPERATING AT INCREASED OUTPUT. STAND ON END
AND SUPPORT WITH WOOD, ETC. USE INGENUITY. READ
WARNINGS WHENEVER USING.

FCI

ASSY FROM
FIG I4-20

SW3

S2
INSULATE S2 WITH PLASTIC
TUBE EXTENSION IF IRRITATING
SHOCKS OCCUR.

HAI

TO POWER SOURCE BHI (SEE NOTE)

WHEN USING WITH BATTERY VS GROUND SYSTEM
TO EARTH OR USER WILL RECEIVE SHOCKS THRU SWITCH

Fig. 14-22. Exploded view gun configuration.

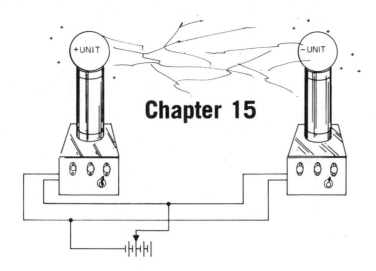

Chapter 15

High-Energy Pulsed Source (HEG1)

Caution!—Caution!—Caution This is a laboratory device. It serves no useful purpose to those not versed in the fields of high-energy circuits. Do not attempt unless thoroughly familiar with all HV safety precautions and procedures. This device is a highly dangerous electrical device. It can kill by electrocution. We have designed the power pack to operate from a rechargeable 12-volt gel cell. This by no means suggests the relaxation of safety procedures but does minimize dangerous potential ground currents through power lines, etc. The entire device should be completely electrically isolated from any earth grounds. The operator should always keep one hand in his pocket when operating and use the discharge rod as indicated at all times. An explosion danger exists in the test chamber and adequate shielding must be provided for all personnel within operating range. Many materials will produce exothermic reactions creating shrapnel from a detonation and brisance effect.

HIGH ENERGY PULSER

The following plans show how to construct an electrical device capable of producing multi-million watt pulses of instantaneous power. (Note the references made to Chapter 4.) It can be used for a variety of advanced scientific and technical functions and is to be considered a research tool. Uses for this device include supplying the necessary pulse of energy for laser flash tubes, producing electromagnetic pulses known as EMP, high-current magnetizing, exploding wires, destructive testing of components and materials, pulse welding, high speed photography, nuclear physics, acoustics, and experimental testing by super heating chemicals, and using the propulsion effects for producing hyper velocities. Research and testing on many materials makes this unit a valuable laboratory tool for advanced technicians.

CIRCUIT THEORY

The pulser unit described in these plans (Fig. 15-1 and Table 15-1) utilizes a portable energy storage charging power supply using a highly efficient transistor-switching oscillator circuit and ferrite transformer (T1). The switching frequency of

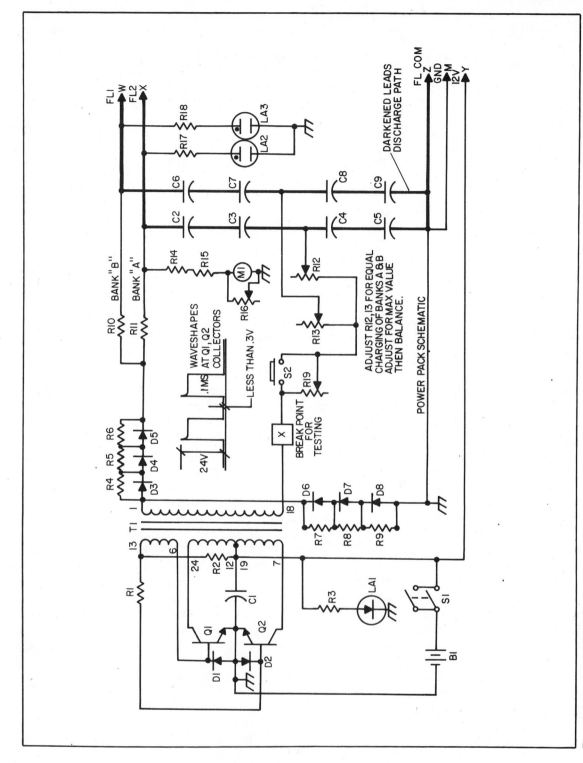

Fig. 15-1. Power pack/energy storage schematics.

226

Table 15-1. Spark-gap Switch and Test Chamber.

Note: This system uses the RUB3 power pack and energy storage system described in Chapter 4 with modifications as outlined in the text.

TC1	(1)	Test chamber - plastic box with cover 6 × 4 × 2 or similar
BASE	(1)	8 × 13 × 1 hardwood base—varnished
BR1,2	(2)	Electrode holder brackets fab as shown Fig. 15-3
SPH1,2,3	(3)	1″ solid brass spheres tapped for ¼ 20
BLK1	(1)	⅜ × ⅜ × 3″ teflon or plastic block
INT1	(1)	Integrator for housing R20, 21, C10, D9 in plastic cap potted with paraffin wax Fig. 15-7
S3	(1)	Push-button switch
WR5	(12″)	#22 plastic covered wire
BC12		Battery charger—optional
HOL1,2	(2)	Sample holders fab as Fig. 15-5
IGN1	(1)	Ignitor constructed per Fig. 4-10 and 4-21
BU1	(2)	½″ plastic bushing
CL1	(1)	Nylon clamp
LAB1,2	(2)	HV danger decals
WN1,2	(2)	Small wire nuts
NED1	(1)	2″ stainless steel needle
SW4	(7)	#8 × ⅜ sheet metal screws
DISROR	(1)	Discharge rod 1″ sphere with ¼ 20 threaded shaft and insulating handle as shown Fig. 15-9
R20,21	(2)	22 meg resistors
D9	(1)	10 kV diode
C10	(1)	.001 μF/15 kV capacitors
CAP1	(1)	Plastic cap for INT1
LAB1,2,3,4	(4)	HV Labels
BC12		Optional Battery Charger

Complete kit of above is available through Information Unlimited, Inc., P.O. Box 716, Amherst, N.H. 03031, write or call 1-603-673-4730 for pricing and delivery.

these transistors, (Q1 and Q2) is determined by the magnetics of T1 and is about 20 kHz for the values shown in these plans. Resistor (R1) limits the base drive to Q1 and Q2 obtained via the feedback winding on T1. Diodes (D1 and D2) form the return path for the base drive of the opposite transistor. Resistor (R2) provides the dc unbalance necessary to initiate oscillator switching. Power to the circuit is obtained from a 12-volt gel cell (B1) and is controlled via switch (S1). Resistor (R3) limits the current to the power "on" lamp (LA1). High voltage square-wave pulses are obtained from the secondary of T1 and are rectified and voltage doubled via diodes (D3-D8). Resistors (R4-R9) divide the reverse voltage across the diodes during their non-conducting period. The rectified dc pulses are now integrated into the energy storage capacitors (C2-C9).

The circuit shows the dual capacitor bank option for obtaining 900 joules of energy. The builder may decide only to use one section for the 450 joule configuration, eliminating (Bank B) C7-C9 and surge separating resistors (R10 and R11). These resistors are only required for even peak-power balancing between the two energy banks. Metering of the energy bank voltage is done by monitoring only (Bank A) C2-C5. Resistors (R14, R15) determine the meter current while (R16) shunts this current for meter calibration. LA2 and LA3 are neon lamps with current-limiting resistors (R17 and R18) and give the operator indication that the energy bank still maintains a potentially hazardous energy charge at the time.

Whenever charging any capacitor bank in this manner a 50% loss of energy is always encountered due to the lumped value equivalent of a series current-limiting charge resistor. The circuit shown contains several circuit resistances such as transformer secondary resistance, forward diode resistances, etc. These are too low to properly limit this initial charging current to a value that can be safely handled by the transistor switching circuit. Hence, resistors R12, R13, and R19 provide this necessary resistance and may be slider types adjustables of 10 watts or more. These are set to a value for obtaining the fastest charge possible without circuit overloading. S2 shorts out R19 speeding up the charging time only after some value of initial charge is already accumulated. These values and parameters will vary from circuit to circuit and may require final adjustment for optimum performance.

When the charges on both Banks A and B are sufficient (usually somewhere between 1500 and 1800 volts) the spark gap is ready to be fired. When ignition occurs and the gap conducts a pulse of current occurs that is equal to the charge voltage divided by the circuit resistance. This consists of leads, spark-gap dynamic resistance, capacitors, and the resistance of the test load. It is obvious that to transfer as much energy to the load requires that all other resistances must be comparatively low. Inductance due to lead length also contributes much to the decreasing usable power by the familiar relation:

$$e = L \frac{d_i}{d_t}$$

The ignitor (Fig. 4-10) is powered by leeching a small amount of power from both energy banks simultaneously to maintain charge balance. These are line FL1 and FL2 and charge the ignitor (IGN).

A dump capacitor (C1) is charged via divider resistors (R1), (R2), (R3), and (R4) and divided down by R5. Diodes (D1) and (D2) are isolation diodes. A voltage developed across R5 now charges C1 to 300-400 volts through the primary of the HV pulse transformer (T1). SCR1 is triggered "on" and shorts capacitor C1 across the primary of T1 and dumps its energy producing a high-voltage pulse at the output of the secondary winding. When C1 is dumped in the primary inductance of T1 a ringing wave is produced whose frequency is equal to the resonance of these components. Diode (D3) recovers the negative part of this energy in this waveform. SCR1 is triggered "on" by the UJT (Q1 connected as a free-running relaxation oscillator whose repetition rate is determined by R7 and C2. This circuit is activated via trigger switch S3 that energizes the UJT circuit. C2 now charges up and eventually fires the UJT turning "on" SCR1 producing the HV trigger pulse. This pulse is 6 to 10 kV and triggers the spark gap of the pulser system, that supplies the necessary energy to the sample under test in the chamber TC1.

The spark-gap switch consists of round electrodes (SPH1,2) being spaced at the point before breakdown occurs at the charge voltage of the energy bank capacitors. A stainless steel needle (NED1) is positioned in relation to the gap electrodes so as to initiate breakdown by ionizing the air in the gap. The needle is charged by integrating the high-voltage dc pulses onto capacitor (C10) through diode (D9) from the HV pulsed output of the ignitor (IGN1). Resistors (R20,21) serve as a HV return path for the ignition pulse. The dynamic resistance

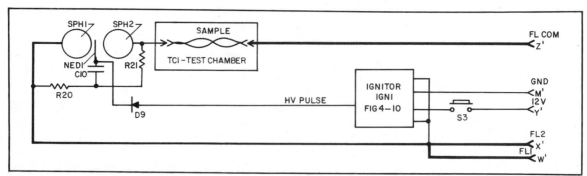

Fig. 15-2. Spark-gap switch and test chamber schematic.

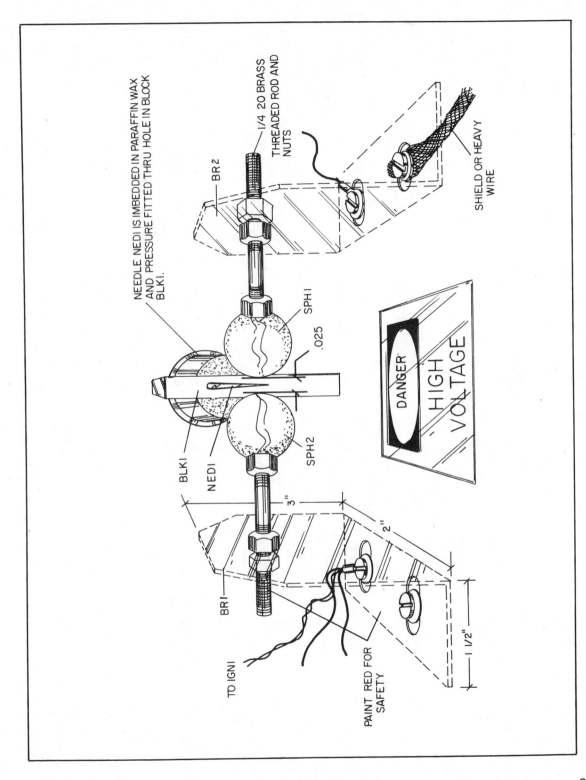

NEEDLE NED1 IS IMBEDDED IN PARAFFIN WAX
AND PRESSURE FITTED THRU HOLE IN BLOCK
BLK1.

1/4 20 BRASS
THREADED ROD AND
NUTS

BR2

SHIELD OR HEAVY
WIRE

SPH1

.025

DANGER
HIGH
VOLTAGE

BLK1

NED1

SPH2

3"

BR1

2"

1 1/2"

TO IGN1

PAINT RED FOR
SAFETY

Fig. 15-3. Spark-gap switch.

229

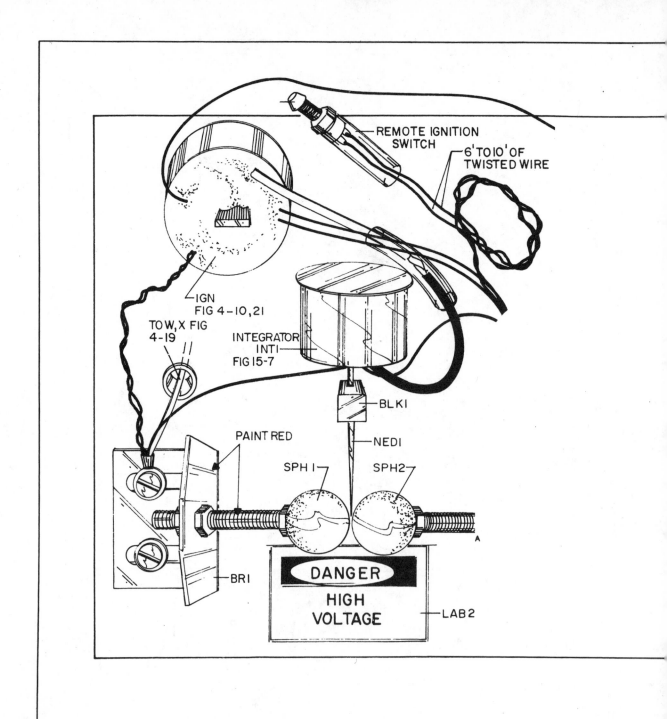

REMOTE IGNITION SWITCH

6' TO 10' OF TWISTED WIRE

IGN FIG 4-10,21

TO W,X FIG 4-19

INTEGRATOR INT1 FIG 15-7

BLK1

PAINT RED

NEDI

SPH1

SPH2

DANGER

HIGH VOLTAGE

LAB2

BR1

Fig. 15-4. Spark-gap switch and test chamber wiring.

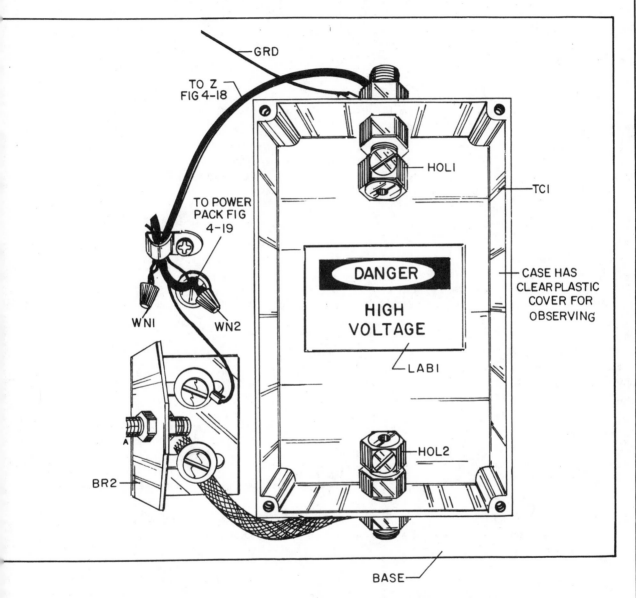

GRD

TO Z
FIG 4-18

TO POWER
PACK FIG
4-19

WN1

WN2

BR2

A

HOL1

TC1

DANGER

HIGH
VOLTAGE

LAB1

CASE HAS
CLEAR PLASTIC
COVER FOR
OBSERVING

HOL2

BASE

NOTE "W" "X" AND "Z" ARE ROUTED AS A SINGLE HEAVY
#14 WIRE TO CONNECTION POINTS AS SHOWN FIG 4-18,19
WIRES CONNECT TO POINTS DIRECTLY AND DO NOT GO THRU HOLE IN CASE.

of the spark gap is usually in the milliohms which for all practical purposes relative to the other circuit parameters is equivalent to the closing of a switch consequently *spark-gap switch*. An excellent text on spark gaps and high speed pulse technology is available through Information Unlimited, Inc.

CONSTRUCTION STEPS

The HEG1 pulser is shown constructed in two sections. These are the power pack/energy storage stage and the spark-gap switch and test chamber stage. The power pack section is almost identical to the RUB3 Ruby Laser Power Pack and it is suggested that the builder construct this section per the "construction steps" one through thirteen omitting the inductor L1 in step 5 and bringing out the wires termed "umbilical" in steps 12, 13 directly to the spark-gap section as shown in Fig. 15-2.

The HEG1 schematic Fig. 15-1 is similar to the RUB3 schematic Fig. 4-9 with the exception of L1 being omitted and the "laser-head section" being replaced by the "spark-gap section". It should be noted that the functions of the spark-gap section and the laser-head section are similar in operation in that they both require the ignitor IGN1 for providing the ignitor of the flash tubes RUB3 or the spark gap switch HEG1 for discharging the energy storage capacitors. The testing of the power pack section can be completed using the RUB3 steps one through eleven. The assembly on the ignitor IGN1 Figs. 4-10, 4-21 is assembled as instructed.

SPARK-GAP SWITCH AND TEST CHAMBER

1. Obtain a hardwood base of 8″ × 13″ × 1″ thickness sand smooth and varnish for looks.
2. Fabricate brackets for spark-gap spheres SPH1, 2, as shown in Fig. 15-3.
3. Construct integrator INT1 as shown in Figs. 15-4 and 15-7.
4. Fabricate sample holder as shown in Figs. 15-5 and 15-6.
5. Mount components on base as shown (Fig. 15-8). Note BR1 shows slots for a sliding adjustment.

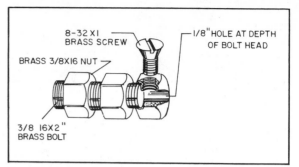

Fig. 15-5. Detail of screws and holder.

6. Screw in ¼ brass rod into SPH1, 2 and use ¼-20 brass nuts as shown.
7. Assemble discharge rod as shown in Fig. 15-9.

ADJUSTING AND USING THE HEG1

Again we caution the user/builder of the dangerous potentials that are present in this device. Always disconnect the battery and discharge both spheres SP1, 2 with the discharge rod shown in Fig. 15-9 before touching any part of the system for sample changing or adjustment. It is suggested to paint or mark all high voltage carrying components with red paint or dope.

It is assumed that the Power Pack Section as outlined in the previous instructions is performing correctly.

1. Connect all the appropriate wires to the spark-gap switch section as shown Figs. 15-1, 15-2, and 15-4. Note heavy #14 wires used for connecting W, X, and Z. These must be direct and short as possible. The other connecting wires can be #22 gauge.
2. *Use Discharge Rod* Fig. 15-9 before connecting a 100-1000 ohm 50 watt resistor into the test chamber electrode holders (HOL1,2).
3. Adjust the spacing between the electrodes spheres to .025″ (use a feeler gauge).
4. Very carefully position needle NED1 so that its point is almost midway between the sphere. The adjustment of this needle will determine the ignition point. The needle must *not* be in the path of the direct discharge or its point will be blown away.

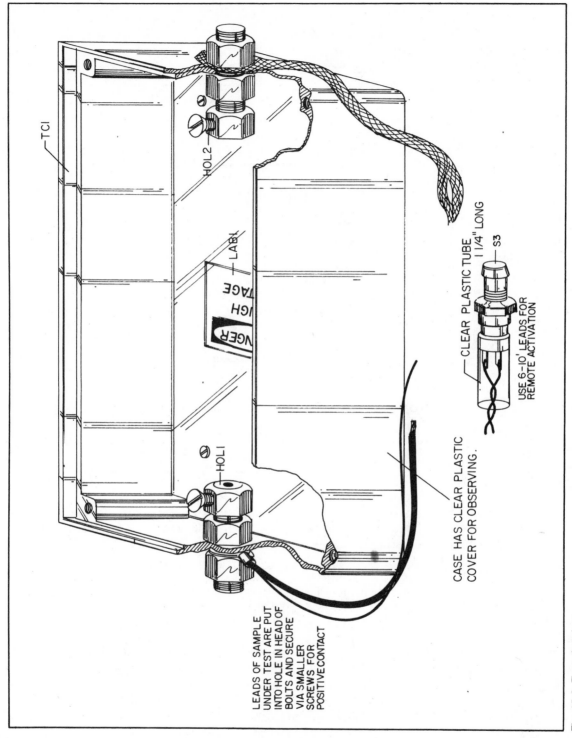

TC1

HOL2

LAB1

DANGER HIGH VOLTAGE

HOL1

CLEAR PLASTIC TUBE
1 1/4" LONG

S3

USE 6–10' LEADS FOR
REMOTE ACTIVATION

CASE HAS CLEAR PLASTIC
COVER FOR OBSERVING.

LEADS OF SAMPLE
UNDER TEST ARE PUT
INTO HOLE IN HEAD OF
BOLTS AND SECURE
VIA SMALLER
SCREWS FOR
POSITIVE CONTACT

Fig. 15-6. Test chamber.

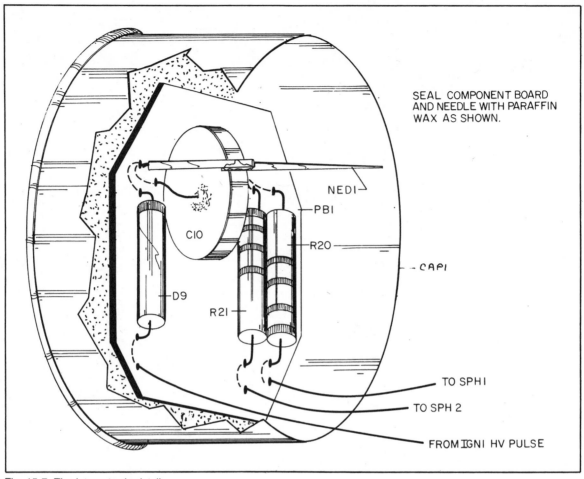

SEAL COMPONENT BOARD
AND NEEDLE WITH PARAFFIN
WAX AS SHOWN.

NEDI

PBI

R20

CAPI

D9

R21

TO SPHI

TO SPH 2

FROM IGNI HV PULSE

CIO

Fig. 15-7. The integrator in detail.

5. Connect battery and allow system to charge up to 1000 volts. Check action of ignitor by depressing S3 causing proper discharge between the spheres. This should be a quiet flash causing the storage voltage to fall to almost zero indicating near total energy usage through the test resistor. Discharge any residual voltage with *Discharge Rod* before proceeding. The test resistor should be warm. You will note that the actual energy dissipated between the switching spheres is only a fraction of that delivered to the load hence the relatively small ineffective spark discharge between the spheres. **Do not be fooled by this as the peak power produced in the resistor no doubt was in the kilowatts.**

The spark-gap dynamic resistance is less than 50 milliohms. If the load in the test chamber for example is 500 ohms then the usable power is $(\frac{500}{5 \times 10^{-3}}) = 100,000$ times more than the energy in the spark gap between the electrode spheres. As the resistance of the test chamber load decreases to a low value, the power now begins to redistribute into the spark gap as it becomes more significant. A dramatic example is to obtain a 100 ohm 2 watt carbon resistor and place it in the test chamber. Place the cover on and allow to charge to 1000 volts. Shield yourself between some barrier and the test chamber and trigger the system with the remote switch. The resistor should blow up with the force

234

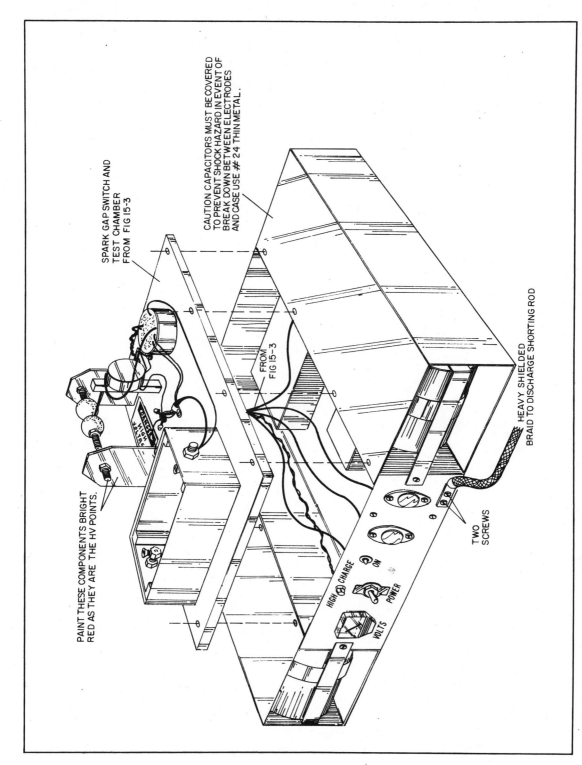

SPARK GAP SWITCH AND TEST CHAMBER FROM FIG 15-3

CAUTION CAPACITORS MUST BE COVERED TO PREVENT SHOCK HAZARD IN EVENT OF BREAK DOWN BETWEEN ELECTRODES AND CASE USE # 24 THIN METAL.

PAINT THESE COMPONENTS BRIGHT RED AS THEY ARE THE HV POINTS.

FROM FIG 15-3

HEAVY SHIELDED BRAID TO DISCHARGE SHORTING ROD

TWO SCREWS

HIGH CHARGE

ON

POWER

VOLTS

Fig. 15-8. Final power pack and spark gap assembly.

235

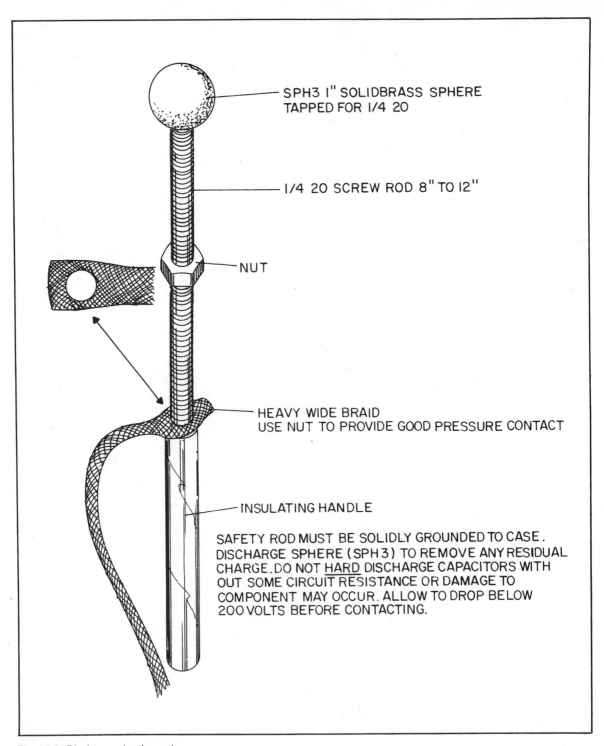

SPH3 I" SOLIDBRASS SPHERE
TAPPED FOR 1/4 20

1/4 20 SCREW ROD 8" TO 12"

NUT

HEAVY WIDE BRAID
USE NUT TO PROVIDE GOOD PRESSURE CONTACT

INSULATING HANDLE

SAFETY ROD MUST BE SOLIDLY GROUNDED TO CASE.
DISCHARGE SPHERE (SPH 3) TO REMOVE ANY RESIDUAL
CHARGE. DO NOT HARD DISCHARGE CAPACITORS WITH
OUT SOME CIRCUIT RESISTANCE OR DAMAGE TO
COMPONENT MAY OCCUR. ALLOW TO DROP BELOW
200 VOLTS BEFORE CONTACTING.

Fig. 15-9. Discharge shorting rod.

of a small explosive charge causing the lid to fly off and pieces to fly around like shrapnel. This demonstrates the peak power obtainable. Again remember to short out electrodes with *Discharge Rod* before inspecting or reloading test chamber.

The unit is capable of charging safely to 1500 volts and intermittedly to 2000. The energy discharge at this maximum value is in excess of 1000 joules. If the load in the test chamber is 10 ohms the peak power pulse is approximately equal to $E^2/R = \dfrac{4 \times 10^6}{10} = 400{,}000$ watts. The pulse duration for 70% is approximately 6 milliseconds.

The many applications of this device are left to the user bearing in mind that it has the capability of switching moderately high energy levels in reasonably short times. Stray circuit inductance and high inherent resistances limit this device from producing the extremely ultra-short-level high-powered pulses sometimes required for certain applications in nuclear physics, etc. The builder may substitute any reasonably similar power supply and energy storage system to work in conjunction with the spark switch. The objective with the battery is to allow portable operation and lessen the danger of this unit.

Experiments producing an electromagnetic pulse effect (EMP) can be produced by this unit to a limited degree. Discharging into a large coil of heavy wire can produce highly inductive fields capable of inducing high voltage into various objects located within its induction field. Magnetic fields can be produced capable of magnetizing many objects and materials.

Experiments involving the discharging of this energy into certain chemicals and materials can produce interesting results generating propulsion and explosive effects. The many applications are left to the user as this device serves no useful purpose to those who are not versed in this type of equipment and should be considered for educational use only.

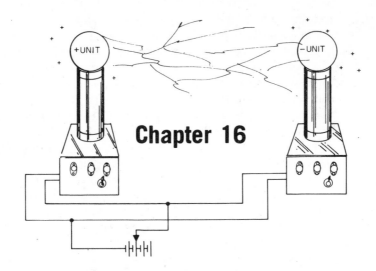

Chapter 16

Magnetic Field Distortion Detector (MGD1)

This project shows how to construct an ultrasensitive magnetic field distortion detection system. The device when constructed is easily capable of detecting slight solar storms as a result of normal solar activity. Major magnetic disturbances caused by heavy solar activity are therefore, obviously easily detected.

Magnetic fields produced by many devices both man-made and natural are detectable by this system. Interesting results are produced when this unit is placed near anything electrical or magnetic such as a car, motor, TV, etc. Sensitivity is impressive when one sees the results of a pulse of one amp of electrical current through a wire located one meter away from the unit (this corresponds to a milligauss). The earth's magnetic field is 0.5 gauss. When the system is properly installed it will detect passing automobiles, aircraft or any moving ferrous source that will cause the terrestrial magnetic field to distort. The unit is an excellent science fair project and can supply the amateur scientist with hours of knowledgeable entertainment listening

and detecting magnetic fields and their abnormalities (see Table 16-1).

The unit shown is constructed to be as flexible as possible. It is built in two sections: the sensing head and the control box that are interconnected via a length of shielded mike cable. Maximum usable sensitivity requires that the sensing head to be placed away from power lines and other potential disturbances.

The sensing head (Fig. 16-1A) must be securely mounted as mechanical movement due to wind and vibration can cause erroneous signals as it shifts through the earth's magnetic field. The unit can be made as a sensitive magnetic probe by installing the sensing head and the control box together bearing in mind that the usable sensitivity may be decreased due to erroneous movement near magnetic materials when in use. Detection of field changes produce moderately loud audible alarms as well as giving a relative indication on a meter (M1) located on the panel. An optional jack (J2) is used for a chart recorder, if desired. Front panel control

Table 16-1. Magnetic Field Distortion Meter (MGD1)

R1,3	(2)	1.5 M 1/4 watt resistor
R2	(1)	5.6 M 1/4 watt resistor
R4	(1)	10 M 1/4 watt resistor
R5	(1)	1 M 1/4 watt resistor
R6,10	(2)	4.7 k 1/4 watt resistor
R7	(1)	1 k 1/5 watt resistor
R8	(1)	10 k 1/4 watt resistor
R9	(1)	100 ohm 1/4 watt resistor
R11,14	(2)	100 k trimpot
R12,13	(2)	5 k pot & switch
C1	(1)	4.7 M/25 V E1 capacitor
C2	(1)	1 μF 50 V E1 capacitor
C3,4,5, 6,7,8	(6)	.1 μF/25 V disc
A1	(1)	LM4250 nano-amp dc amplifier (can or dip)
A2	(1)	UA741 Op-amp IC
Q1	(1)	PN2907 pnp transistor
Q2,3	(2)	PN2222 npn transistor
D1,2,3,4	(4)	1N914 signal diode
T1	(1)	1 k/8 ohm matching transformer
P1	(2)	Phono plug
J1,2	(2)	Phono jack
BH1,2	(2)	Plastic 4 "AA" holder
BH3	(1)	Dual "AA" holder metal
CA1	(1)	7 × 5 × 3" al box
CL1,2	(2)	6" snap clips
SP1	(1)	2" 8 ohm speaker
M1	(1)	100 μa panel meter
WN1,2	(2)	Small wire nuts
WR3	(36")	#24 red plastic stranded
WR4	(36")	#24 black plastic stranded
WR15	(12")	#24 solid buss wire
WR10	(10')	Shielded mike cable
CP1,2	(2)	3 1/2" plastic cap
BU1	(1)	Plastic bushing 3/8"
KN1,2	(2)	Small knob 1/4"
SW2	(3)	6 32 × 1/2 screw
NU1	(3)	6 32 kep nuts
PB2	(1)	4 × 1 3/4" .1 × .1 perf board
PB1	(1)	6 1/4 × 2 1/2 .1 × .1 perf board
EN1	(1)	8" × 3 1/2" OD PVC sked 40
TA1	(10")	2 × 1 × 1/8" tape
B1-10	(10)	AA cells not included in kit

Complete kit available through Information Unlimited, Inc., P.O. Box 716, Amherst, N.H. 03031. Write or call 1-603-673-4730 for pricing and delivery. (Note: Mounting hardware, Coil L1 block BL1, and Core CR1 are not included in kit as these are selectable by the builder.)

(R13) and switch (S1) control power and threshold activation for the alarm section. R12 provides meter sensitivity and also controls the output to J2 for the chart recorder. Note these controls are independent of one another for convenience in operating.

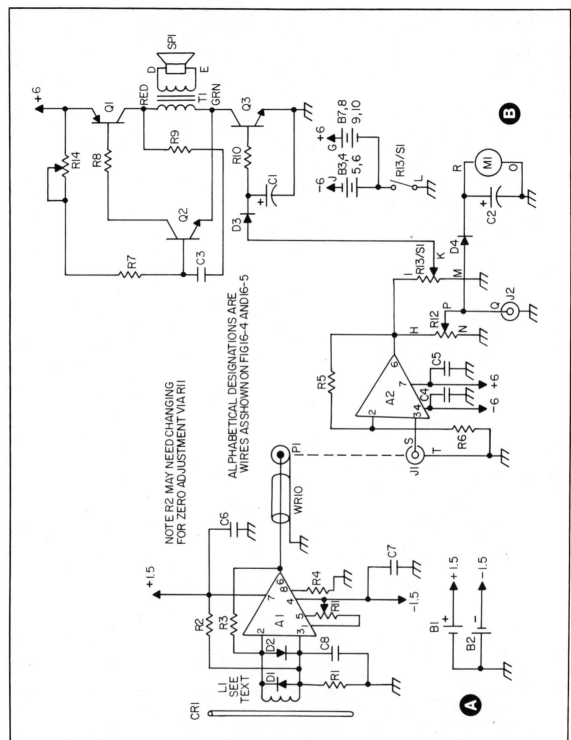

Fig. 16-1. Sensing head and control box.

CIRCUIT DESCRIPTION

A coil consisting of many turns is wound over a soft iron core of about 2 feet to 3 feet in length (Figs. 16-1A and B). Any magnetic changes induce small voltages to occur at the output of the coil (L1) that is fed into a nanoamp-sensitive dc amplifier (A1), providing a current gain of ×500. The iron core of L1 serves to concentrate the magnetic flux lines by offering a lower reluctance path than that of air. The input of A1 is protected from overvoltage and transients via clipping diodes (D1) and (D2). The output of A1 is forced to zero via balance compensating resistor (R2) and (R11). It should remain at zero as long as no field change is occurring. It may be found that R2 is not needed if zero offset can be obtained with R11. Or again, R2 may be a different value than that specified due to other variables. Power for A1 consists of B1 and B2 that are left in the circuit permanently as current drain is very low. It should be noted that if the sensing head is to be located in a hard to get at area that the builder use good quality D-size cells as these will last that much longer. The sense head output is fed to the control box via a length of shielded cable (WR10). As was mentioned before, the sensing head should be located away from any potential magnetic interfering devices such as anything electrical and any large ferrous objects.

The sensing head and control box may be assembled together and used as a probe. This configuration may reduce system sensitivity due to erroneous movements and interacting of the circuits such as magnetic speakers, meters, etc.

The control box provides another dc amplifier consisting of A2 and having a gain of ×200. This gives the system an overall gain of 100,000. Full scale meter deflection of 100 μa on M1 occurs with a current of 1 nanoamp (10^{-9}A) from L1. It is obvious that with the many thousands of turns on L1 that a detectable signal can occur with very minute flux changes such as changes measured in microgauss. *It should be noted that the earth's magnetic field is approximately 0.5 gauss.*

A1 is connected as a conventional dc operational amplifier with its output connected to R12 and R13. You will note that R13 also is ganged with

S1 that controls power to this section. R12 controls the deflection of meter M1 with D4 and C2 serving to smooth out and average the signal to M1. J2 is also controlled by R12 and is an auxiliary jack intended for a chart or other recorder. R13 controls transistor switch Q3 again with D3 and C1 being used to smooth out the signal. Q3 controls the audible alarm by clamping the negative power returning to ground. R13 is set just to the point where the alarm starts to sound and then backed off. This is the maximum sensitivity for audible indication. The alarm is nothing more than a reflex-type oscillator consisting of Q1 and Q2 with its feedback paths consisting of R9 and C3. R14 is set for a reliable clean sounding signal.

CONSTRUCTION STEPS

Construction consists of two parts, the sensing head (Figs. 16-1A and 16-2) and the control box (Figs. 16-1B and 16-3). Note that the dashed lines in Figs. 16-2 and 16-4 indicate connection points of components. Use component leads whenever possible and watch for shorts. L1 consists of a coil of hair-fine magnet wire wound with thousands of turns. This is very difficult to do and it is suggested that the builder locate a high voltage oil burner ignition, neon sign, bug killer or other similar transformer and carefully remove the high voltage secondary winding. The parameters of these transformers are usually around 5000 volts at 10 to 20 mA. These are available through junk yards, etc. Many are rejected to radio and TV interference and yet have an intact secondary winding suitable for this project.

The objective is to obtain one with as many turns as possible, however, most will work quite well. A quick test is to measure the dc resistance of between 5 and 20 k across the coil. A relay coil of high impedance such as 5 to 10 k may also work. The coil must have connecting leads and these should be secured with RTV, wax, or similar material to prevent breaking that may render the coil useless. A soft iron core of 2 to 3 feet is inserted through the coil and serves to offer a lower reluctance to the magnetic flux lines. It is these lines that as they change in the core induce a voltage due to

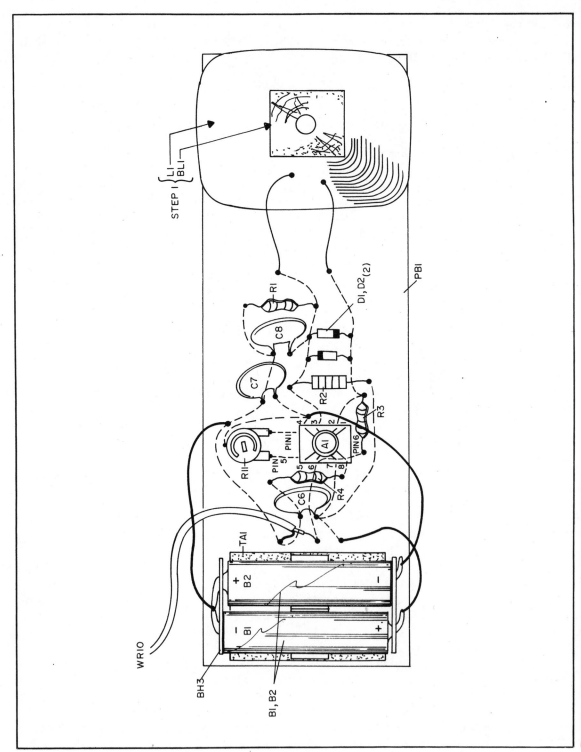

Fig. 16-2. Assembly board layout of sensing head section.

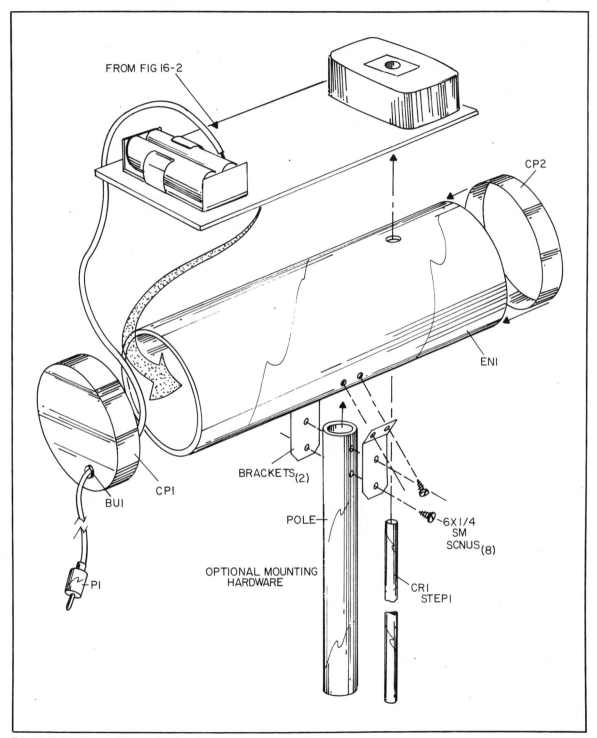

FROM FIG 16-2

CP2

ENI

BRACKETS(2)

CPI

BUI

PI

POLE

OPTIONAL MOUNTING HARDWARE

6X1/4 SM SCNUS(8)

CRI STEPI

Fig. 16-3. Final assembly sensing head.

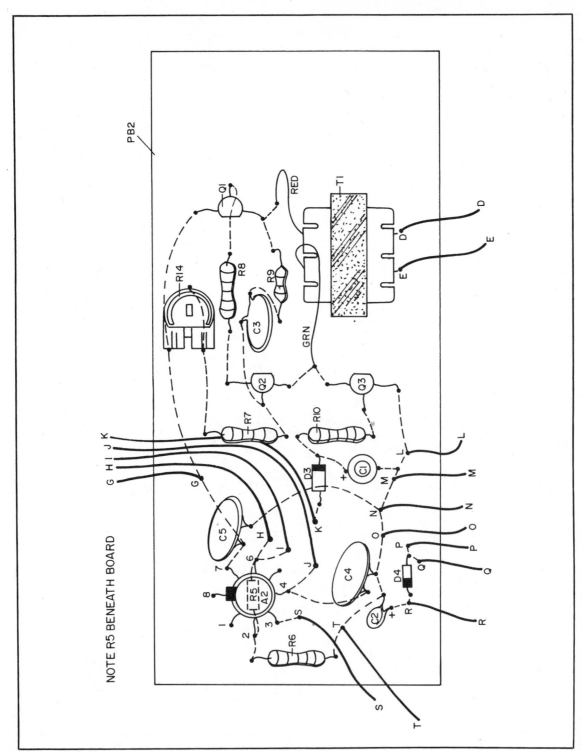

Fig. 16-4. Assembly board layout of control box section.

magnetic induction. It is this voltage that is amplified and detected by the system. A steady magnetic field will not cause this effect and hence produces no signal.

Attach this coil to PB1 perfboard as shown, secure with RTV. You will note wooden block BL1 through center of coil. This sturdies the iron core CR1. The core must fit snugly or it may require gluing or other means for securing. This core should be soft iron or mumetal.

1. Fabricate EN1 from an 8 × 3½″ OD PVC (Fig. 16-3). Note that a small ⅜″ access hole for adjustment of R11 is drilled when all is in place. Fabricate brackets as required for mounting of sense head.

2. Attach battery holder BH3 to PB1 using two-sided tape or RTV (Fig. 16-2).

3. Assemble and solder electronic parts as shown in Fig. 16-2. Be careful of A1 as not to overheat when soldering. A1 may be in an 8-pin can or flatpack DIP configuration. Figure 16-2 shows the can carefully placed in an integrated circuit socket. Attach WR10 through CP1 via busing BU1 (Fig. 16-3).

4. Carefully check for wire and solder errors and install fresh batteries. Connect a voltmeter at the output cable and adjust R11 for zero reading. This should be done outside and away from power lines, etc. If zero cannot be obtained it may be necessary to play around with the value of R2. You will note that R2 for our prototype was approximately 5 megohms.

5. With meter still connected observe a reading when a piece of iron is brought near the sensing coil L1 and moved about. If the piece of iron contains any magnetism it will greatly enhance the reading.

6. Install the assembly into EN1 with caps CP1 and CP2 as shown. Recheck by connecting meter across P1. Zero adjust R11 through access hole if needed. The sense head should now be working correctly.

CONSTRUCTION OF CONTROL BOX

7. Fabricate box (Figs 16-4 and 16-5) for CA1 as shown with holes for meter M1. SPEAKER SP1, R12, R13, and J1, J2. Assemble components as shown in Fig. 16-5. Note battery holders secured via two-sided tape TA1.

8. Assemble PB2 as shown in Fig. 16-4. Observe proper position of A2, and polarity of diodes, transistors and tantalum capacitors. Note that Q1 is a pnp PN2907 transistor.

9. Attach interconnecting leads G through T (see Figs. 16-1B, 16-4, and 16-5). Twist leads S & T together. Note ground lugs of J1 and J2.

10. Connect leads from board assembly to components in box.

11. Position board but do not secure at this point. Turn S1 "off" and insert batteries into BH1 and BH2. Snap on one side of clip and measure 0.5 mA current flow with meter between unclipped snap. This current flow may be eliminated by separating the ground returns of the batteries and switching separately using the contacts on R12 as the second switch.

12. Turn on S1 and measure 2mA (−6V) BH1 and 0.4 mA (+6 V) BH2. Reconnect battery clips.

13. Turn R12 full cw and note meter remaining at zero.

14. Turn R13 full cw and note alarm remaining silent. Lightly short out J1 sense head jack with finger and note meter jumping and alarm starting up. You may wish to check the current reading of BH1 and BH2 (1 mA and 30 mA respectively) with alarm sounding.

15. Connect sense head to J1 and note alarm sounding and meter probably pinning. Turn down R12 and R13 respectively. (This is the result of power lines, motors and other magnetic disturbances.)

16. Preset R14 on board at midrange and adjust R13 until alarm sounds. Readjust R14 for a clean sounding tone if necessary.

17. To appreciate the sensitivity of this unit requires operation in a magnetically quiet location where all controls can be advanced full. Under these conditions of low magnetic background, the unit can detect automobiles, airplanes and other large ferrous objects as they pass by and many experiments involving detection of man-made devices and natural magnetic disturbances are possible with

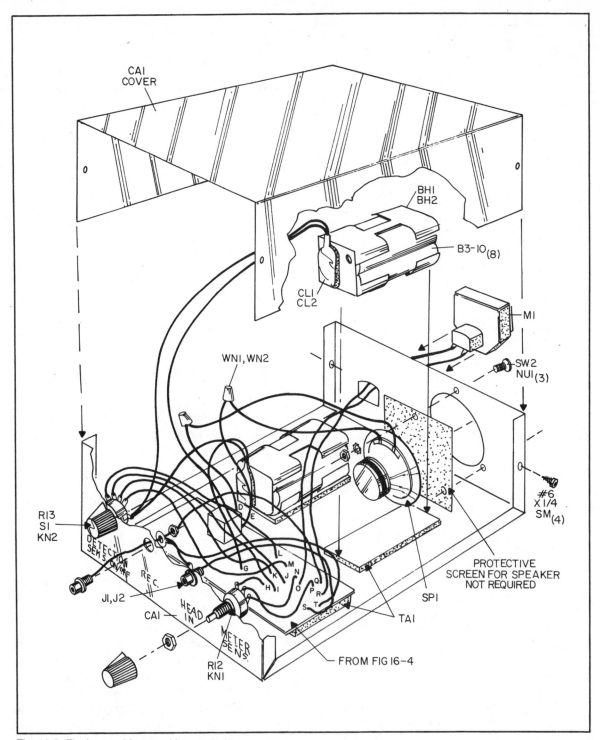

Fig. 16-5. Final assembly control box and wiring.

this device. Also, simply moving a piece of iron near the sensing head should produce an indication.

18. Jack J2 may be used to drive an XY recorder when observing sunspots or other time related disturbances.

Please note that no means was made to filter out 60 Hz disturbances as it appears that many uses of this device involve the detection of power lines and other related equipment. In extremely magnetically quiet environments, it may be advantageous to adjust R11 just to the threshold of meter activation. This fine adjustment eliminates the small dead sensitivity area at this point.

APPLICATIONS

The uses of this device are obviously only appreciated by those who can understand its sensitivity to a minute change in magnetic flux. The unit can be made to detect the intrusion or motion of large steel bodies as they distort the earth's magnetic field. Many applications and experiments are therefore possible with the device. H1Z headphones HS-1 can be plugged into J2 and the experimenter can actually listen in and hear the field changing. One verifying test that is easily done to determine the systems ability to detect the earth's magnetic field is the following:

Rotate the sensing head with short quick jerks in a clockwise direction. Note meter jumping off scale. As rotation is continued, a point will be found where the meter will respond less. The nulling point should occur with the metal core CR1 pointed in a north/south direction. As rotation is continued, it will be found that the short quick jerk must occur in a *counterclockwise* direction to obtain positive meter movement. This is similar to an electrical generator using the earth as its field magnet and the sensing head core as the moving armature.

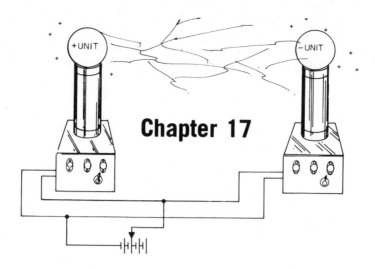

Light-beam Communicators (LBT1)

Your Light-beam Communicator Transceiver project clearly demonstrates the ability of a light beam to be utilized for the transmission of voice communications. The clarity and quality of audio reproduction is crystal clear. Sensitivity and modulation power are more than ample. Range of the unit is line-of-sight for 1000 meters or better. (This is the maximum range we tested these devices, even though they no doubt would have worked much further.) Alignment is easily accomplished by sighting along the barrel of the unit. Short range communications (several hundred meters) is accomplished by hand sighting to each unit. Long range setups are more conveniently obtained using camera tripods.

Units are built in a pistol type configuration with all power and optics being self-contained. A rear panel contains all necessary controls and jacks for convenient operating. The unit may also be used for actually listening to other light sources such as TV pictures, scopes, fluorescent, and many other IR and visible radiation sources.

The unit is suggested to be built using a visible-red emitter for ease and convenience in nighttime alignment. For serious longer range and low noise performance it is suggested to use a filter and the invisible IR diode specified.

CIRCUIT DESCRIPTION

Your Light Beam Transmitter/Receiver consists of a photo transistor receiver that picks up modulated light that is fed through a high gain amplifier and then to headsets or a loud speaker (Fig. 17-1 and Table 17-1). When in the transmit mode the amplifier becomes a sensitive mike preamp that drives a current amplifier modulating a light-emitting diode serving as the transmitter. The receiver section consists of a photo transistor (Q4) positioned at the focal point of lens (LE2) inside of enclosure (EN2). (A separate enclosure, lens, and light-emitting diode for transmitting enhances the flexibility and performance of the device, however it adds to the cost. Duplicating these components

Table 17-1. LBT1 - Light-beam Communications Transceiver

R1,4,8	(3)	390 k ¼ watt resistor
R2	(1)	5.6 M ¼ watt resistor
R3,5,6	(3)	6.8 k ¼ watt resistor
R9,13	(2)	100 k ¼ watt resistor
R10	(1)	2.2 k ¼ watt resistor
R11	(1)	47 k ¼ watt resistor
R12	(1)	1 k ¼ watt resistor
R14	(1)	27 ohm ½ watt resistor
R15	(1)	2 k trimpot vert
R7/S1	(1)	5 k pot and switch
C1	(1)	.05 μF/50 V disc cap
C2,3,5, 6,7,8	(6)	1 μF 50 volts electrolytic cap
C4	(1)	.01 μF/50 V disc cap
C9	(1)	100 μF@25 V El
A1	(1)	CA3018 amp array in can
Q1	(1)	Npn plastic PN2222
Q2	(1)	Pnp plastic PN2907
Q3	(1)	Npn power tab D40D5
LA1	(1)	Red emitter HLMP3750 or equiv. or IR version XC880 or equiv.
Q4	(1)	Photo transistor L1463
T1	(1)	Small transformer 600 ohm/1500 ohm matching
PC1	(1)	Reworked printed circuit board or use perforated board
S2	(1)	Rotary switch 3P ST
M1	(1)	Small 1″ crystal mike cartridge
J1,2	(2)	RCA phono jacks
PL1,2	(2)	RCA phono plugs
WR10	(10″)	Shielded mike cable
CL1,2,3	(3)	Battery snap clips
KN1,2	(2)	Small knob
WR3	(24″)	#24 red hook-up wire
WR4	(24″)	#24 black hook-up wire
WR15	(6″)	#24 buss wire
LE1,2	(2)	54 × 89 lens
BU1,2	(2)	⅜″ plastic bushing
CA1,2	(2)	3 ½″ plastic caps
CA3,4	(2)	2 ⅜″ plastic caps
CA5	(1)	1 ⅝″ plastic cap
SW1	(2)	6 32 × ¼″ screws
SW3	(2)	6 32 × 1″ screw
NU1	(4)	6 32 kep nuts
EN2	(1)	2 ⅜″ OD × 6 ½ PVC sked 40
EN1	(1)	3 ½″ OD × 8 PVC sked 40
DO1	(1)	Wood dowel 2 × 1 ½″
RP1	(1)	3 ¼ × 3 ¼ × 2 gauge fab Fig. 17-2
MP1	(1)	2 ⅞ × 5 ½ × 22 gauge fab Fig. 17-2
TA1	(3″)	Double sided tape 1″
EN3	(1)	6″ × 2″ OD PVC sked 40
HS1	(1)	8 ohm headsets and 8/1000 ohm transformer and PL2
B1,2	(2)	9 V transistor batteries (not incl'd. in kit)
B3,4,5,6	(4)	1.5 V AA batteries (not incl'd. in kit)
FTR1	(1)	Small plastic IR filter (Use with FPE104 only)

Complete kit above available through Information Unlimited, Inc. P.O. Box 716, Amherst, N.H. 03031. Write above or call 1-603-673-4730 for pricing and delivery. Specify Red or IR emitter.

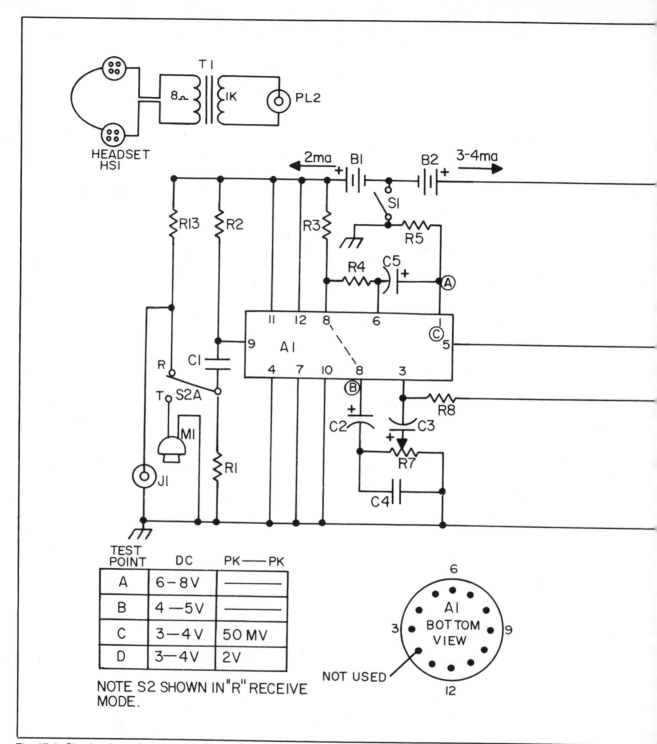

Fig. 17-1. Circuit schematic.

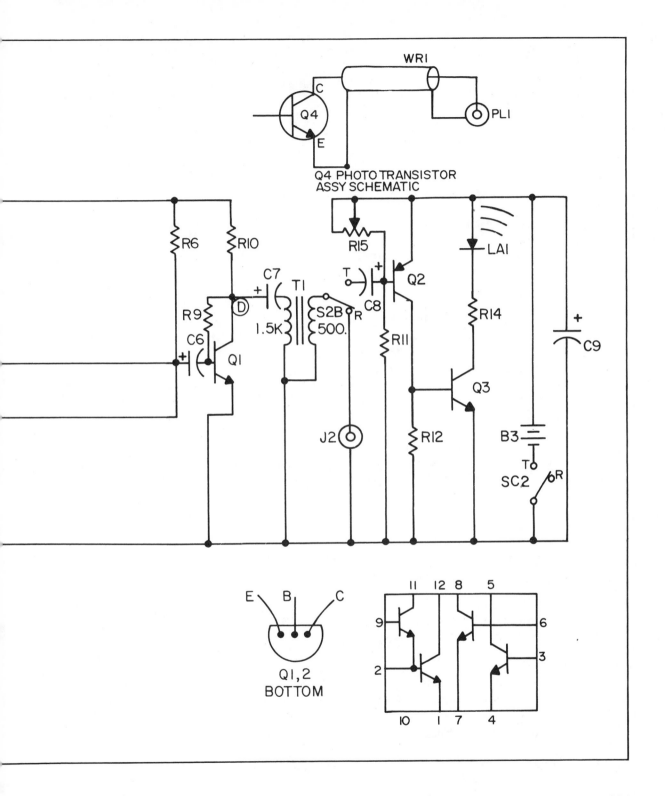

WRI

C

Q4

E

Q4 PHOTO TRANSISTOR
ASSY SCHEMATIC

PLI

R6

RIO

C7

T1

R15

Q2

LAI

R9

D

+

1.5K

S2B
500.

R

T

C8

R14

+

C6

+

Q1

R11

C9

J2

Q3

R12

B3

T

SC2

R

E   B   C

Q1,2
BOTTOM

11 12 8   5

9   6

2   3

10   1   7   4

for both functions could be done, however, at reduced performance.)

Q4 is mechanically secured to a sliding position dowel (DO1) that is adjusted to its proper distance from LE2 and secured with a screw. The signal from Q4 is fed into J1 via a shielded cable (WR10) to keep hum and other electrical pickup to a minimum. Switch (S2A) now selects J1 in the rcvr mode and feeds the signal to the amplifier via C1. The signal is now matched and amplified via the amplifier array (A1). A gain control (R7/S1) controls the sensitivity of the amplifier and also serves as an on/off switch for the receiver section. The output of A1 is now further amplified by Q1 and impedance matched via transformer (T1). S2B now connects T1 to J1 for feeding high-impedance headsets or an external speaker.

The transmitter section consists of a narrow beam visible-red or optional IR light-emitting diode (LA1) located at the focal point of lens (LE1) inside of enclosure (EN1). EN1 also encloses the electronics and controls for the system. Batteries are located in enclosure (EN3) that also serves as a hand grip for the device. A mike (M1) is located on the rear panel (RP1) and is fed into the amplifier (A1) through C1 via mode select switch (S2A). The amplifier now becomes a preamp for the mike. The output of the preamp is further amplified by Q1 and impedance matched by T1. The output of T1 is fed to Q2 via S2B. Q2 is dc coupled to Q3 whose quiescent state is selected via R15 determining the dc current through LA1. The modulation signal is ac coupled to Q2 via C8.

Power for Q2, Q3, and LA1, consisting of the transmitter section, is through a separate battery (B3) and is controlled by S2C being only used during transmit. This enables the device to be used in the rcvr mode with very little current draw.

CONSTRUCTION STEPS

Start construction by making the following fabricated parts (kits will include these items already fabricated). See Fig. 17-2.

1. EN1 main enclosure is an 8-inch piece of 3½″ OD schedule 40 PVC tubing. Cut 1⅞″ hole for EN3 handle at desired slant using hole saw. This hole is approximately 3″ from intended rear end of this piece. Use caution. Drill two ⅛″ holes approximately 2½″ and 6½″ respectively from rear end. These are for securing receiver enclosure EN2 (Fig. 17-3).

2. EN2 receiver enclosure is a 6½″ piece of 2⅜″ OD schedule 40 PVC tubing. Drill ¾″ hole as shown approximately 3″ from rear end. This hole is for optical alignment. Note two mating holes for securing to EN1. These holes are also in the top of this piece for access with screwdriver. The bore axis of these tubes must be parallel. It may be convenient to slot the rear hole to allow slight side movement for final alignment. The large ¾″ hole can be covered with plug, tape, etc., when not needed (Fig. 17-3).

3. EN3 handle and battery enclosure is a 6″ piece of 2″ OD schedule 40 PVC tubing. It is secured via a good tight fit along with some PVC cement. Do not over insert and only glue in place after everything else is assembled. The above PVC tubing can be obtained in most hardware or building supply outlets. However, you will probably have to pay for full-length sections of all the necessary sizes unless you are lucky enough to find some scraps.

4. RP1 rear plate is fabbed from a 3¼″ × 3¼″ square piece of #24 galvanized sheet metal or #22 aluminum. MP1 is fabbed from a 5½″ × 2⅞″ same material. See Fig. 17-2 which may be used for a template by proportionalizing the dimensions.

5. DO1 centering dowel is a 2″ length of 1½″ OD (or thereabouts) for smooth sliding fit into EN2. Q4 is mounted for optical centering via small pin holes in wood for securing via its leads. WR10 cable is fed to Q4 via a slightly off-center feed hole in dowel. Connection is made by soldering to exposed leads of Q4 using the shortest leads possible and then securing with RTV or equivalent. Leads to Q4 should be left long enough to allow touchup repositioning to true optical axis for final alignment (Fig. 17-3).

6. CA1 3½″ plastic cap with 1⅝″ hole in center. Use an Exacto knife or small snips. Be neat as this can ruin the appearance of the device. Note four small pieces of double sided tape TA1 for securing lens LE1 to cap. Place tape so as not to

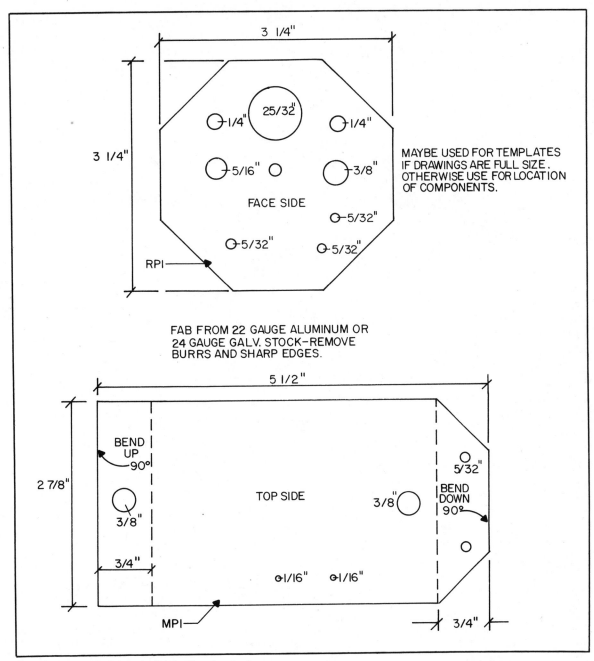

FACE SIDE

3 1/4"

3 1/4"

25/32"

1/4"

1/4"

5/16"

3/8"

5/32"

5/32"

5/32"

RPI

MAYBE USED FOR TEMPLATES IF DRAWINGS ARE FULL SIZE. OTHERWISE USE FOR LOCATION OF COMPONENTS.

FAB FROM 22 GAUGE ALUMINUM OR 24 GAUGE GALV. STOCK–REMOVE BURRS AND SHARP EDGES.

TOP SIDE

5 1/2"

2 7/8"

BEND UP 90°

3/8"

3/4"

1/16"

1/16"

MPI

5/32"

3/8"

BEND DOWN 90°

3/4"

Fig. 17-2. Layout of chassis (fabrication drawing).

contact ridge of EN1 with sticky part when assembling or it will be difficult to remove for checking, etc.

7. CA2 3½" plastic cap remove end with exception of ¼" lip to hold subassembly as per Fig. 17-3 into EN1 via sandwiching of RP1. Use sharp knife using wall of EN1 as a guide.

8. CA3 2⅜" plastic cap. Remove end with

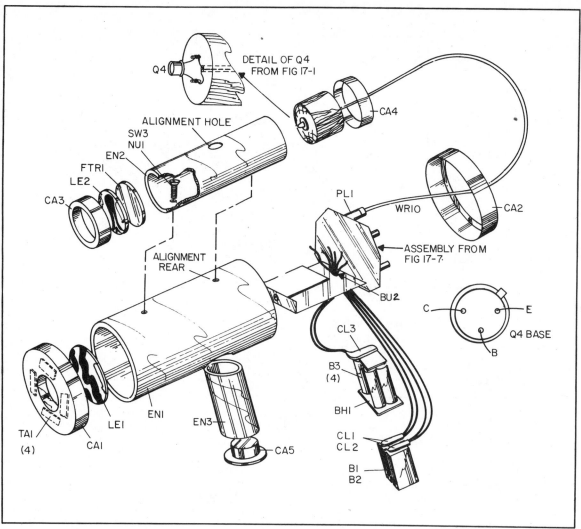

Fig. 17-3. Final assembly.

exception of ⅜″ lip to retain lens LE2 and optional filter FTR1 against end of EN2. Use wall of EN2 as a guide. Fabricate CA4 2⅜″ plastic cap. Place small hole for cable WR10. Hole should create friction hold to prevent DO1 from sliding once set. Secure with RTV when complete and finally aligned.

Note: It is advisable that the following assembly be built on a special PC board unless the builder is experienced in wiring and soldering using perforated board.

9. Assembly board may be perfboard 3″ × 1½″ with 0.1″ grid. Follow layout as shown as close

as possible as detailed in Fig. 17-4. If special PC board is used the following rework must be done as shown in Fig 17-5. This is the removal of foil where noted (one place) and added jumps connection points (4 places). Note polarity of electrolytic capacitors, position of A1, Q1, Q2, Q3, and LA1. Assemble as shown using Fig. 17-4 or Fig. 17-6.

10. Assemble RP1 as shown Fig. 17-7. Note leads are shown long and untwisted for clarity. When assembled, leads from three adjacent pins on R7/S1 must be twisted and as short as possible. Connect C4 as shown. Switch S2 is prevented from

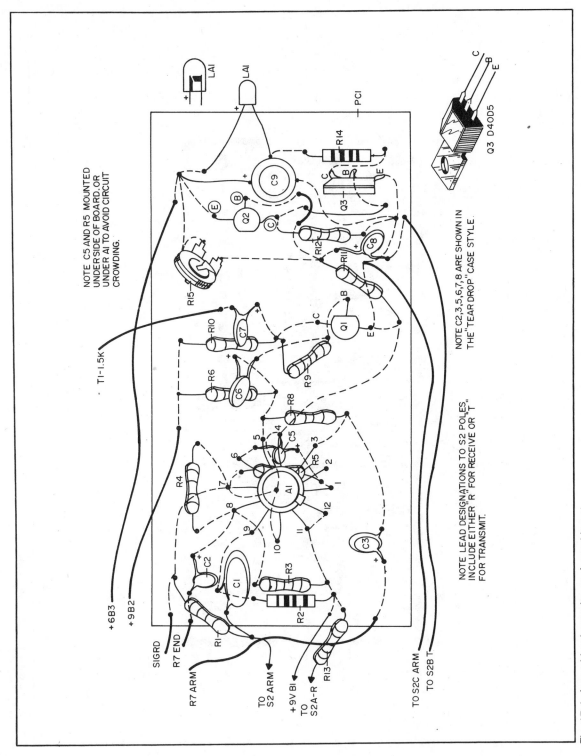

Fig. 17-4. Assembly when using perforated board.

255

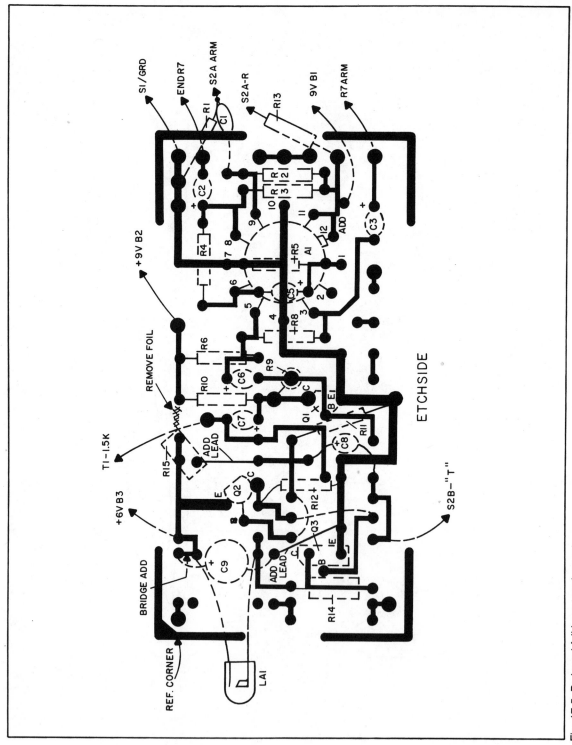

Fig. 17-5. Pc board foil layout and rework.

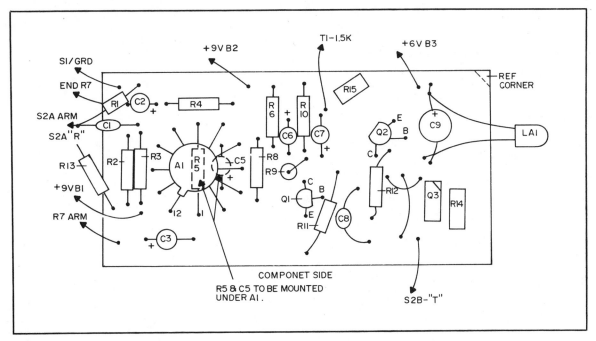

Fig. 17-6. Pc board component layout.

rotating via tab hole in RP1. Twist S2C leads together, twist S2B leads together. Keep all leads short and routed close to the metal chassis. Note switch S2 is in rcvr receive mode as shown in schematic (Fig. 17-1). Attach M1 microphone using RTV. Position and wire as shown. Keep leads short. Assembly jacks J1 and J2. Note ground lug under J2. Attach RP1 to MP1 via screws SW1 (6-32 × ¼″).

11. Connect wires from RP1 assembly to board noting identification of leads shown in drawings. Position as shown in Fig. 17-7. (Do not adhere to tape at this time.) Carefully position LA1 into BU1 as shown. Attach battery clips CL1, 2 and 3 to respective points as shown. It may be advantageous at first to allow board freedom to be moved for total access during preliminary troubleshooting and testing. Leads may be further shortened after proper operating has been verified.

12. Attach T1 and MP1 bending tabs in the small holes. Solder one lead from primary to secondary and sandwich between core of T1 and plate. This makes the ground contact for the transformer. You may want to solder these wires directly to the plate for a positive contact. Use a heavy iron for

this. Note ungrounded 500 ohm lead going to S2B and ungrounded 1.5 k ohm (winding marked P) going to C7 on board.

13. Connect PL1 phone plug to WR10 cable from Q4 section.

PRELIMINARY TESTING

It is assumed that the assembled unit to this point has been wired correctly, with no shorts, and good solder connections. You will note that the complete working unit is conveniently built on a single removable assembly. This assembly should have the three battery clips, CL1, CL2, and CL3.

Receiver Section

14. Turn S2 and R7/S1 full ccw (off positions).

15. Connect one terminal of a fresh 9-volt battery to CL1 and connect a milliameter between the used contact of the battery and the clip (Fig. 17-7). Turn on R7/S1 and note current reading of approximately 2 mA. Fully connect battery and designate B1.

16. Repeat above using a second battery connected to CL2. Turn on R7/S1 and note current

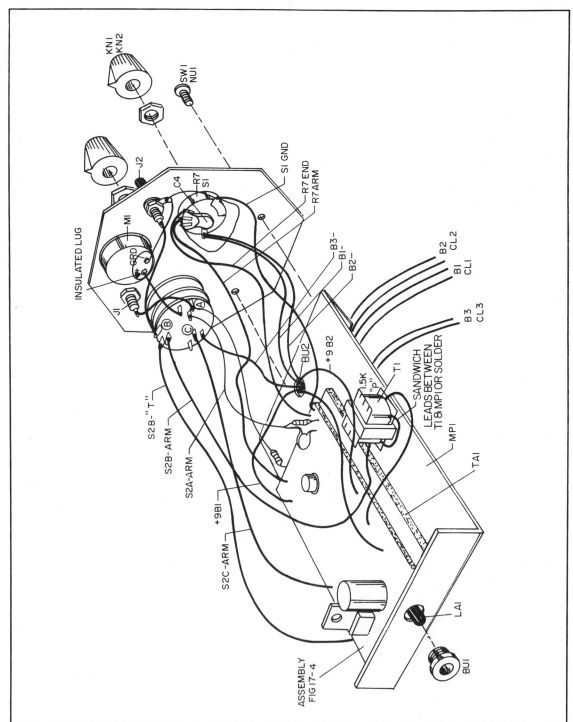

Fig. 17-7. Final wiring.

reading of 3 to 4 mA. Fully connect battery and designate R2.

17. Plug in a high impedance set of headphones into J2. (Note Fig. 17-1 showing standard 8-ohm headsets with spliced in matching transformer stepping up to 1000 ohms is suggested as high impedance headsets are scarce and uncomfortable to wear.) Plug PL1 from Q4 receiving phototransistor into J1.

18. Turn on R7/S1 and slowly turn up gain until a loud 60-cycle hum is heard. This is the normal lighting frequency being picked up by Q4 and at normal ambient lighting conditions will completely block the amplifier. Reduce the gain and attempt to point Q4 at various objects indicating different levels of signal depending on reflection characteristics of surfaces, etc. You will note that the circuit is relatively prone to power-line hum pick up. It is assumed that testing will be done in normal electrical lighting for this step. If not, you may not obtain the 60-Hz hum.

19. If everything above checks out you can proceed to the transmitter section below. If not you must troubleshoot the faulty circuit. It may be convenient to use the test points shown in the Fig. 17-1 schematic and thoroughly familiarize yourself with the "circuit description" given in the beginning of the plans.

TRANSMITTER TEST

20. With all switches fully ccw connect CL3 to 6-volt B3 as done with B1 and B2. Connect a 100 mA meter in series and turn S2 cw to transmit position. Adjust R15 to read about 20 mA and note LA1 lighting to about ½ maximum brilliance. Turn R7/S1 and note LA1 changing brilliance with sounds. Whistle and note current meter jumping to nearly 50 mA. Note LA1 increases in brilliance with sound indicating upward modulation. The device seems to work alright with the LA1 downward modulating but we recommend upward modulating. Certain diodes may require less quiescent current than 20 mA for good upward modulation. You are now ready to final assembly the unit as shown in Fig. 17-3 and can proceed with the optical align-

ment. Note that LA1 should automatically center itself inside of EN1.

OPTICAL ALIGNMENT

In order to obtain maximum performance or range of your LBT1 communicator, it is necessary to properly optically align both the transmitter and receiver. You will note that the receiver tube (EN2) is secured to EN1 via screws. The rear hole is slotted to allow a side-to-side movement of the receiver enclosure in respect to the transmitter enclosure. The up and down position usually self adjusts simply by the abutting of the two enclosures. Remember that both receiver and transmitter sections must be optically aligned to view the same area for maximum two-way communications. The method we demonstrate here is not necessarily the only way to align these devices and is only suggested as a possible means. The builder may have his own ideas and methods to accomplish the above.

The following steps we used at our lab and found to be relatively easy in accomplishing acceptable preliminary alignment. Precise alignment may require the use of a helium-neon laser.

21. Remove transmitter lens/cover and place some thin paper over the open end. Adjust LA1 output to the center of paper (this is the bore sight of enclosure). Secure and replace lens and cover.

22. Secure communicator in vise or other similar holding attachment.

23. Locate reflective surface about 20 feet from the device.

24. With transmitter properly aligned and secured, adjust mirror for reflection of output light occurring in receiver lens. This is adjusted by sighting mirror reflection along "sight line" at surface intersection of the two enclosures. See Figs. 17-8 and 17-9.

With the device and mirror aligned as above, carefully adjust the position of the receiver enclosure so that the focused received beam is centralized (bore sighted) inside the tube. This is best accomplished by placing a thin strip of paper through the "alignment hole" over the phototransis-

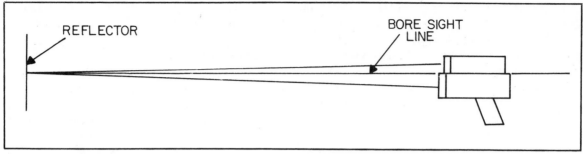

Fig. 17-8. Initial aligning.

tor and adjust the dowel to the focal length of the lens. This should place the focused received light directly on the lens of the phototransistor. Further touch up can be done by careful positioning of the phototransistor with needle nose pliers. Secure dowel, enclosures, etc., to eliminate movement or misalignment. A helium-neon laser aligned parallel to the transmitter output will produce an easily seen reflection for the above, otherwise, use of total darkness will be necessary.

25. Repeat with the second unit. You should now be able to hand sight units along "sight lines" for medium range use. Good reliable long range use should be done with a camera tripod, etc. Nighttime use using the visible-red transmitting diodes is easily accomplished at ranges up to 1000 meters or so by noting the reflection of the transmitting light in the receiving lens as noted at the transmitter station. Daytime operation is best with filter and IR transmitter. Securing of optical components via permanent means should only be done when it is assured of optimum optical alignment.

TO OPERATE THESE UNITS

26. For both transmitter and receiver to be in the "off" mode S2 must be at "R" position and S1/R7 to "off."

27. To use receiver only, plug in headsets to J2 jack and turn on S1/R7 and adjust to desired level (usually no more than ⅛ turn). Point unit at a normal 60-Hz lamp, TV, or other light source and note hum.

28. To use in "transmit" mode, all that is necessary is to place S2 from the "R" to the "T" mode. The modulation level is present by S1/R7 when used in the receiver mode.

One way to test is to look into the transmitter section and note LED flickering with audio signals. S1/R7 can then be readjusted if necessary by this

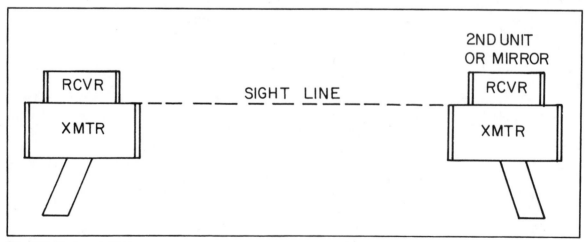

Fig. 17-9. System sighting.

260

indication. Note that LED only has to change ever so slightly for sufficient modulation.

You will note there is a trimpot (R15) on the PC board. This adjusts the quiescent current through this LED and should be just set to where LED is emitting with no audio signal. This saves batteries and prevents downward modulation. This probably should be reset as batteries weaken. Also note that the units pick up 60-Hz hum. The visible red LED (supplied in these units) obviously operates best in darness. For normal daylight operating, the IR LED and filter must be used.

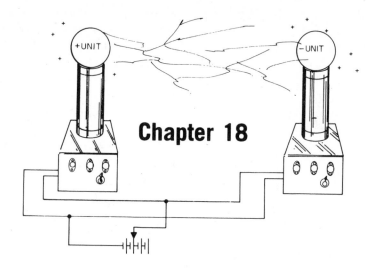

Chapter 18

Solid-state Tesla Coil (TCL3)

The following project shows how to construct a high-frequency, high-voltage device capable of causing a fluorescent lamp to light (without wires), the corona effect (St. Elmo's Fire), and the wireless transmissions of radio-frequency energy. The device as a safety feature may be powered from a 12-volt battery source as well as 120 Vac. It is completely solid-state, simple to construct, and is to be considered as an introduction to the larger generating devices such as the 250 kV Tesla Coil and other high-powered devices.

CIRCUIT THEORY

As can be seen by the schematic (Fig. 18-1 and Table 18-1) the device is a high-voltage inverter circuit taking advantage of the high step up in voltage and low capacitance resonant secondary properties inherent in TV flyback transformers. This high-voltage, high-frequency output energy can light lamps without wires and perform many otherwise impossible feats. Power is obtained either by

an external battery or a conventional transformer full-bridge rectifier circuit.

T2 TRANSFORMER REWORK

1. Two new windings ("*primary*" and "*feedback*" windings) are added to the flyback transformer T1 (Fig. 18-2) that connects to the switching transistors Q1, 2. These windings are hand-wound on the bottom leg of the ferrite core where the original two-turn filament winding was located. In its place, wind a ten-turn, center-tapped winding (designated P1 and P2) using approximately 30 inches of #18 or heavier insulated hook-up wire. This is easily accomplished by winding five turns at one end of the core and bringing out a loop and twisting for a center-tap lead (PCT) before adding the second five turns. The complete ten-turn winding should then be held in place with a turn or two of electrical tape with the two ends (P1 and P2) and the center-tap (PCT) loop all protruding. Connection can be made to the center-tap loop when the insulation has been carefully removed. If it becomes

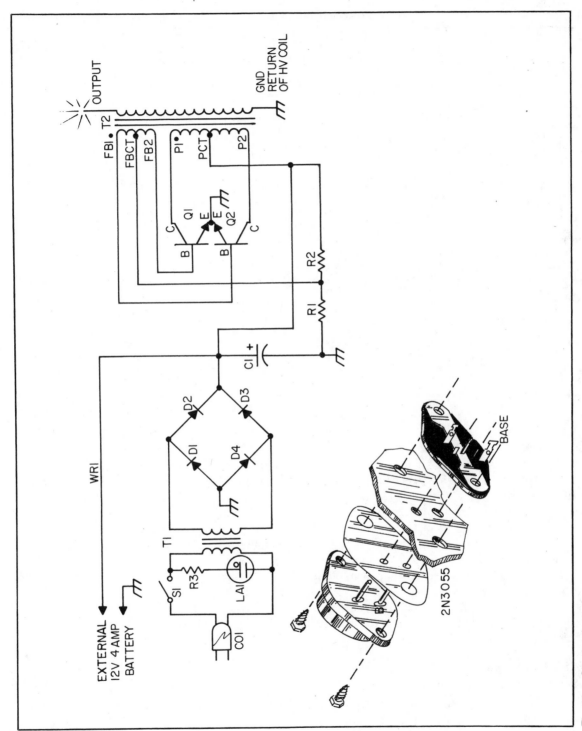

Fig. 18-1. Circuit schematic.

Table 18-1. TCL3 Universal Solid-State Tesla Coil

R1	(1)	27 ohm ½ watt resistor
R2	(1)	270 ohm ¼ watt resistor
R3	(1)	100 k ¼ watt resistor
C1	(1)	8000 μF/16 V electrolytic capacitor
D1,2,3,4	(4)	3 amp 50 volt rectifier
T1	(1)	12 V 3 amp transformer
T2	(1)	TV flyback 20-25 kV
Q1,2	(2)	2N3055 npn TO3 transistor
MK1,2	(2)	TO3 mounting kit
LA1	(1)	Neon lamp with leads
CO1	(1)	Molded power cord
S1	(1)	Toggle switch 3 amps
EN1	(1)	7 × 5 × 3″ al box
HS1	(1)	Dual TO3 heatsink
BU1	(1)	Cord clamp bushing
BU2	(1)	⅜″ plastic bushing
BU3	(1)	½″ plastic bushing
WR1	(3′)	#18 stranded hook-up wire
WR3	(2′)	#24 stranded hook-up wire
TE1	(1)	7 terminal strip
EN2	(1)	3 ½ × 4″ sked 40 PVC tubing
CA1	(1)	3½″ plastic cap
LAB1	(1)	Danger HV label
TER1	(1)	Large smooth metal door knob for output terminal

Complete kit available through Information Unlimited, Inc., P.O. Box 716, Amherst, N.H. 03031. Write above or call 1-603-673-4730 for pricing and delivery.

necessary to cut the center loop, be sure that the two ends are scraped and joined to form a mechanical as well as an electrical connection to the winding.

2. The second winding (feedback) should be wound directly on top of the first, but it should only have a total of four turns—two each side of the center tap. Wind two turns of #22 hook-up wire, pull and twist a center-tap loop (FBCT) and wind the other two turns. Tape this winding in place on top of the first. Do not let the center-tap loops of the two windings touch each other, however, they should not be more than one-quarter of an inch apart. (Note Fig. 18-2 that shows larger separation for the sake of clarity.)

GENERAL CONSTRUCTION

3. Fabricate the metal case as shown from a 7 × 5 × 3″ aluminum minibox. Drill holes for the indicator lamp, feed-through bushing for power, switch on the front panel, three holes for T2, shield EN2, four holes for rubber feet, and two holes for the heatsink (Figs. 18-3 and 18-4).

4. Mount Q1 and Q2 to heatsink using mounting hardware as shown. Be sure that the TO3 mounting cases are insulated from the heatsink. Apply heatsink compound for best results.

5. Solder R1 and R2 as shown. Solder FB1 and FB2 to bases (B) of Q1 and Q2 respectively. Use pliers as a heatsink whenever soldering directly to a transistor.

6. Solder lead from FBCT to TE1 at junction of R1 and R2.

7. Attach +12 volt power lead to PCT connect Q1 and Q2 collectors to P1 and P2. Attach −12 Vdc to emitter buss jump of Q1 and Q2 as shown. This point is chassis ground, connect ground return of T2 to chassis ground.

8. Final wire T1, C1, LA1, R3, and TE1. Watch out for primary 120 V wiring. Make sure there are no ac grounds. Cover leads of A1 with RTV or equivalent.

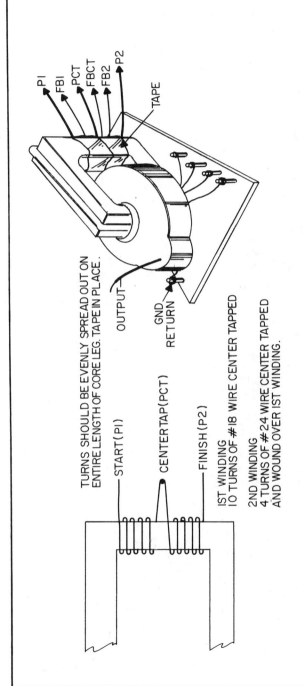

TURNS SHOULD BE EVENLY SPREAD OUT ON
ENTIRE LENGTH OF CORE LEG. TAPE IN PLACE.

START (P1)   OUTPUT

CENTER TAP (PCT)   GND
                    RETURN

FINISH (P2)

IST WINDING
IO TURNS OF #18 WIRE CENTER TAPPED

2ND WINDING
4 TURNS OF #24 WIRE CENTER TAPPED
AND WOUND OVER IST WINDING.

WINDING I

TWO NEW WINDINGS ("PRIMARY" AND "FEEDBACK" WINDINGS) ARE ADDED TO THE FLYBACK TRANSFORMER (T2) THAT
CONNECTS TO THE DRIVER TRANSISTORS. THESE WINDINGS ARE HAND WOUND ON THE BOTTOM LEG OF THE FERRITE
CORE WHERE THE ORIGINAL TWO TURN FILAMENT WINDING WAS LOCATED. REMOVE AND DISCARD THE ORIGINAL
FILAMENT WINDING. IN ITS PLACE, WIND FIRST A TEN-TURN, CENTER-TAPPED WINDING (DESIGNATED P1-P2) USING
APPROXIMATELY 30 INCHES OF #18 OR LARGER INSULATED HOOKUP WIRE. THIS IS EASILY ACCOMPLISHED BY WINDING
FIVE TURNS AT ONE END OF THE CORE AND THEN TWISTING A LOOP IN THE FREE END BEFORE ADDING THE SECOND
FIVE TURNS. THE COMPLETE TEN TURN WINDING SHOULD THEN BE HELD IN PLACE WITH A TURN OR TWO OF
ELECTRICAL TAPE WITH THE TWO ENDS (P1 & P2) AND THE CENTER TAP (PCT) LOOP ALL PROTRUDING. CONNECTION
CAN BE MADE TO THE CENTER TAP LOOP WHEN THE INSULATION HAS BEEN CAREFULLY REMOVED. IF IT BECOMES
NECESSARY TO CUT THE CENTER TAP LOOP, BE SURE THAT THE TWO ENDS ARE SCRAPED AND JOINED TO FORM A MECHANICAL
AS WELL AS AN ELECTRICAL CENTER TAP CONNECTION TO THE WINDING.

WINDING II

THE SECOND WINDING (FEEDBACK) SHOULD BE WOUND DIRECTLY ON TOP OF THE FIRST, BUT IT SHOULD ONLY HAVE
A TOTAL OF FOUR TURNS—TWO EACH SIDE OF THE CENTER TAP. WIND TWO TURNS OF #22 HOOK-UP WIRE, PULL
AND TWIST A CENTER TAP LOOP (FBCT) AND WIND THE OTHER TWO TURNS. TAPE THIS WINDING IN PLACE ON
TOP OF THE FIRST. DO NOT LET THE CENTER TAP LOOPS OF THE TWO WINDINGS TOUCH EACH OTHER

Fig. 18-2. T2 rework winding instructions.

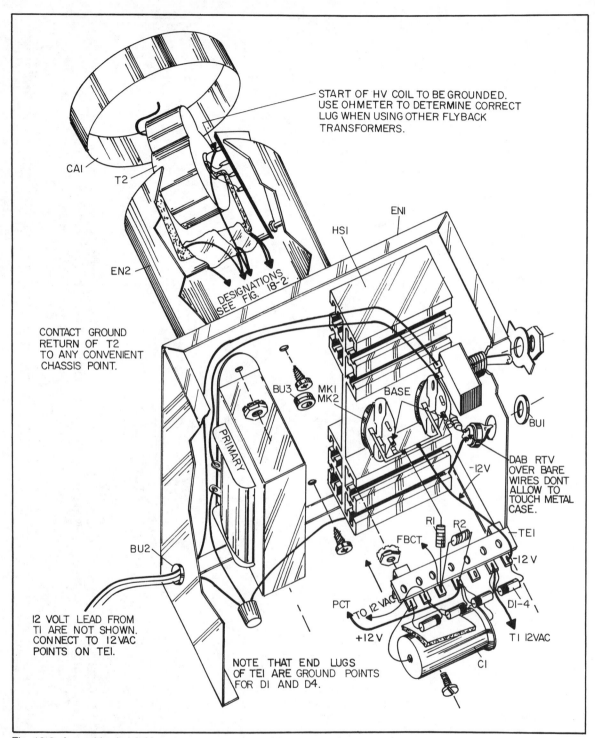

START OF HV COIL TO BE GROUNDED.
USE OHMETER TO DETERMINE CORRECT
LUG WHEN USING OTHER FLYBACK
TRANSFORMERS.

CAI

T2

ENI

HSI

EN2

DESIGNATIONS
SEE FIG. 18-2.

CONTACT GROUND
RETURN OF T2
TO ANY CONVENIENT
CHASSIS POINT.

BU3 MKI
      MK2

BASE

PRIMARY

BUI

DAB RTV
OVER BARE
WIRES DONT
ALLOW TO
TOUCH METAL
CASE.

-12V

RI   R2

TEI

-12 V

FBCT

BU2

PCT

TO I2VAC

DI-4

+12 V

TI I2VAC

12 VOLT LEAD FROM
TI ARE NOT SHOWN.
CONNECT TO I2VAC
POINTS ON TEI.

CI

NOTE THAT END LUGS
OF TEI ARE GROUND POINTS
FOR DI AND D4.

Fig. 18-3. Assembly sketch blowup.

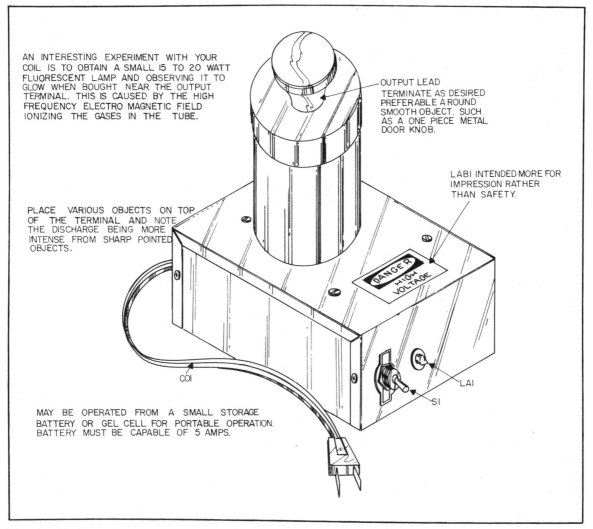

AN INTERESTING EXPERIMENT WITH YOUR COIL IS TO OBTAIN A SMALL 15 TO 20 WATT FLUORESCENT LAMP AND OBSERVING IT TO GLOW WHEN BOUGHT NEAR THE OUTPUT TERMINAL. THIS IS CAUSED BY THE HIGH FREQUENCY ELECTRO MAGNETIC FIELD IONIZING THE GASES IN THE TUBE.

OUTPUT LEAD TERMINATE AS DESIRED PREFERABLE A ROUND SMOOTH OBJECT. SUCH AS A ONE PIECE METAL DOOR KNOB.

LAB1 INTENDED MORE FOR IMPRESSION RATHER THAN SAFETY.

PLACE VARIOUS OBJECTS ON TOP OF THE TERMINAL AND NOTE THE DISCHARGE BEING MORE INTENSE FROM SHARP POINTED OBJECTS.

DANGER HIGH VOLTAGE

COI

LAI

SI

MAY BE OPERATED FROM A SMALL STORAGE BATTERY OR GEL CELL FOR PORTABLE OPERATION. BATTERY MUST BE CAPABLE OF 5 AMPS.

Fig. 18-4. Final assembly.

9. Tape and insulate all bare leads where possible shorting can occur. Check for wiring and shorts.

10. *The following should be done with a dc meter indicating current flow.* Apply power and immediately note that high voltage lead of T1 secondary (usually white lead with cap) emits a corona discharge (purplish sparks) when brought near ground. This should arc to about one inch. If the unit does not work, reverse wires FB1 and FB2, check wire and soldering for accuracy, base and emitter leads of Q1 and Q2 reversed, improper mounting of transistors (without insulating mounting kits). Q1 and Q2 must be completely electrically insulated from the heatsink.

11. Once the unit is functioning properly, check the transistors for heating. You should be able to touch them with your finger after several minutes of operation. If not, a larger heatsink may be necessary for prolonged operation.

Section IV
A Briefing on
Wireless Transmitting Projects

The wireless device projects shown are relatively powerful in that they can be received by an appropriate sensitive communications receiver up to a distance of several miles. A standard FM portable in the fifty to one-hundred dollar range usually can receive these devices for well over an honest mile. Cheap ten to twenty-five dollar radios will not provide good performance but can be used in initially setting up the transmitters and checking out at close range.

There are two legal considerations that the builder must be aware of before constructing these devices. First, the Federal Communications Commission (FCC) takes a dim view of any FM wireless device that can transmit for more than fifty feet. Assembled devices marketed as such are illegal and can create a legal hassle for the manufacturer. However, they do take a lenient view for those who assemble kits for their own personal use and will not interfere providing they receive no complaints. Secondly, it is against the law to use any of these devices to listen to or bug anyone without their permission. Obtain public law pertaining to this subject #90-351.

OBTAINING MAXIMUM PERFORMANCE

Proper use of these devices for obtaining maximum effect and range requires tuning them off of an existing radio station frequency as they will be quickly overridden within several hundred feet by these high-powered transmitters. The units described here are designed to operate in the 88 to 110 MHz range favoring the higher end.

RECEIVERS

A suitable receiver is a must if optimum performance is required. Most FM radios in the fifty to one-hundred dollar range will provide moderate performance up to distances of a mile or more depending, of course, on antenna and terrain. A good receiver must have good front-end sensitivity to dig out these weak signals from the noise level and also must have good selectivity to pick these signals out from between the higher-powered broadcast stations unless, of course, this system is operating above the upper FM band.

The upper frequency limit of most modern FM radios can be extended to provide clear operation without interference from broadcast stations. This is accomplished by tuning in the last station on the dial and adjusting the "osc" trimmer screw and walking the station down one to two MHz. This trimmer is usually found on the rear of the plastic case housing the main tuning capacitor and is the only one that will change the frequency or dial setting. The "ant" trimmer is usually found adjacent to the osc and should be "tweaked" for maximum noise level at *the high end of the dial with the antenna extended*. The FMR20 offered by Information Unlimited, Inc. is factory adjusted as described.

The receiver should have provision for being both portable and ac powered, with an output jack for headphone or signal activating equipment, tape recorder, etc. A signal-strength meter is a positive asset when using these devices for homing or tracking functions. A telescoping antenna may also be used to adjust the signal received to below the AVC threshold and use the orientation of the radio to obtain a bearing on the transmitter source using the audible output as a signal-strength indicator. It should also be noted that signal fading with position will occur out at ½ mile or more. Changing your position by moving several feet can improve reception by a great amount and will allow ranges out to several miles.

Cheap ten to twenty-five dollar radios will not provide good performance but can be used in initially setting up the transmitters and checking out at close range. High quality expensive commercial communication radios obviously will provide the best performance but cost hundreds of dollars.

ANTENNA SELECTION

It is very important that the transmitting frequency of your wireless transmitter be in the clear (not on or near a local FM broadcast station in order to obtain any decent range). A good approach is to slightly retune an FM radio so that the limit of the high end is 110 MHz rather than 108. This allows easy setting of the transmitter frequency without having to tune in between the powerful broadcast stations in your local area.

All of the transmitting devices described have a potential range of several miles when all systems are optimized. This means a *good* receiver, *properly set* transmitter with a *good* antenna, and *decent* terrain. This is not always possible and the user must therefore settle for what is plausible for his situation.

OPERATION AND INSTALLATION

There are several approaches to using this equipment and we shall discuss these in steps.

1. Transmitting range up to 300′ with frequency set at 109 MHz or well away from broadcast stations. Antenna or transmitter need only be 6″ in length and device can be placed anywhere within reason.

2. Transmitter range up to 2000′ with same frequency conditions. Antenna now should be about 30″ or ¼ wavelength of operating frequency and strung out away from conducting objects. Vertical along the wall, etc., works well.

3. Transmitting range over 2000′, same frequency conditions now may require a reflective ground to obtain more range. This can be accomplished by placing the unit on a metal surface, such as the roof of an automobile and running the antenna perpendicular achieving a *ground plane effect*. This may also be simulated by running a second wire (mirror antennas) equal in length to the actual antenna and directly opposite being electrically connected to the ground or common point of the device. This places the unit at the actual feed point of a dipole antenna. This method is capable of ranges in excess of two miles.

4. Further extending of range now requires the radiation portion of the antenna system to be placed as high and in the clear as possible. This now requires the use of a transmission line of coaxial cable connected between the antenna feed point and the actual transmitter. Range now can be many miles depending on terrain and receiving conditions. The use of field strength meters and SWR

indicators are required for further optimizing.

POINTS OF INTEREST

1. *Connecting an antenna may be technically against the rules of the FCC.*

2. The tap on the tank coil may have to be adjusted to compensate for a tuned antenna taking too much power from the oscillator section creating instability and poor performance. This can be remedied by placing the tap closer to the cold end or away from the collector (consequently less antenna loading). The other extreme is that the oscillator may be capable of delivering more power (consequently moving the tap in the other direction).

3. When approaching a range of 1000 feet or more signal fading may occur. This will manifest itself as a rapid flutter in signal strength when driving in an automobile. The user will see that by relocating the receiver only several feet can make the difference between clear noiseless copy to no copy at all. This fading should not limit the usable range of these devices as simple relocating can allow readable signals out to several times the distance where it is first noticed.

4. One of the things to watch for when using a device of this type is proper tuning. The adjustment capacitor is quite sensitive and requires only a slight movement to change the frequency. *Always use a tuning wand.* It is very easy when one is not familiar with this unit to tune to an erroneous signal. This phenomenon is likely to occur when the unit is close to the monitoring receiver. An erroneous signal will be weak, distorted, and unstable (often it is mistaken for the main signal and the unit blamed for poor performance). It is often desirable when the actual signal is located to mark the position of the adjustable capacitor and making note of the frequency spot on the receiver. The main signal will be strong, stable and undistorted if modulated. Several experiments in tuning the unit should be done before attempting to use it for the desired application.

5. Whenever possible, the unit should be used around 108 to 110 MHz which is the lower part of the aircraft band and upper FM broadcast band. When the approximate desired spot is found, final touch-up tuning should be done at the receiver end for clarity, etc. In most areas these frequencies are clear and allow uninterrupted use in contrast to lower in the FM radio band where a slight change in frequency from a clear spot results in the unit being drowned out by a strong broadcast station.

6. *Do not go above 108 MHz if near an airport or air traffic lane, as you may be detected by aircraft and reported plus you may cause them interference.*

ADDITIONAL ANTENNA INFORMATION

The selection of an antenna for these devices results in a toss-up between frequency stability and range (power output). In order to obtain maximum range, the antenna must be receptive to take power from the oscillator. This condition results when the physical length approaches a quarter wave of the frequency that in length (meters) equals $=$

$$\frac{\text{speed of light in meters}}{(4)\ \text{freq in MHz}}$$. However, if too much

power is taken from the oscillator, oscillation will cease or be erratic due to insufficient feedback voltage.

If optimum range is desired, the HA-2 Long Range Antenna offered by Information Unlimited, Inc. can be used, placing it as high as possible and in a clear area. Obviously, this antenna will have to be cut to the following value:

$$\text{Each element length in inches} = \frac{2952}{\text{MHz}}$$

The following procedure is for obtaining maximum performance for your antenna system.

1. Select the approximate operating frequency and cut your antenna elements according to the above formula.

2. Locate the antenna as described above.

3. Connect a milliammeter (0-10 mA) in series with a battery lead. This is accomplished by unsnapping one end of the 9-volt battery clip and connecting the meter in series with the two exposed contacts. (Observe meter polarity.)

4. Tune your receiver to the approximate frequency you desire.

5. Adjust the unit to receiver frequency (Step

4) by adjusting C2. This is indicated by a muting or squealing sound.

6. Note current on meter to read between 5 to 7 mA.

7. Connect antenna and retune C2, if necessary, or retune receiver for slight frequency deviation when an antenna is connected.

8. If current reading drops down when antenna is connected, unsolder and adjust tap on L1 towards C1 at ¼ of a turn at a time until current again reads 5-7 mA as in Step 6. Further adjustment should be done using a field strength meter such as the model #FSM1 offered by Information Unlimited, Inc.

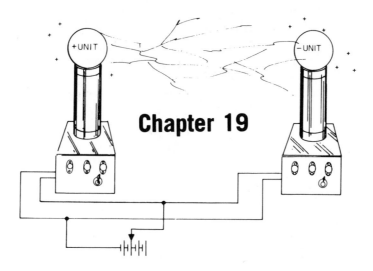

Chapter 19

Infrared Viewer

An infrared viewer is a sophisticated electro-optical instrument designed for direct viewing of infrared. It permits the user to view directly what can be photographed with infrared film. You will be able to view other projects discussed in other chapters of this book. Also, you will be able to directly view the phenomenon which are discussed in the Kodak publication No. M-28, *Applied Infrared Photography*.

Among its numerous applications are seeing-in-the-dark, discerning infrared emissions, and revealing invisible markings. Some typical applications are listed in Fig. 19-1. Typical specifications are given in Fig. 19-2.

The infrared viewer consists of a built-in invisible infrared source, a "T" mount lens for viewing the infrared, a type 6032 image converter tube for transforming the infrared image to a visible picture, and an eyepiece for viewing the visible picture. A high-voltage power supply is incorporated for powering the 6032 tube. Switches are included for turning the viewer on and for controlling the infrared source. Only commonly available

tools are required for assembly. A 6-volt battery is required for power.

This infrared viewer project was designed and written by IR Scientific Inc., the leader in consumer infrared viewing equipment. Infrared viewer projects, kits, and finished viewers are available for the complete range of possible uses from hobby to sophisticated scientific and industrial applications. More information and parts are available from:

IR Scientific Inc.
Box 110
Carlisle, MA 01741

Complete viewers and kits are available from:

Information Unlimited, Inc.
P.O. Box 716
Amherst N.H. 03031

This chapter is supplied to assist you in every way in completing the instrument with the least

Photography—	Observe IR subjects, take IR photographs, use as a darkroom eye
Optics—	Use as IR optics research tool, make IR spectral analysis, observe IR lasers and LEDs
Energy—	Locate hot spots, study convection paths, adjust flames and arcs
Corrosion—	Locate centers of rust, monitor etching processes, study electrolysis effects
Ecology—	Locate oil pollution, appraise growing conditions, study grazing habits
Geology & Mineralogy—	Detect geological hazards, examine rock properties, observe mineral fluorescence
Aerial Overflights—	Make IR aerial surveys, assess crop/forest production, use as aid for crop dusting
Marine—	Locate marine salvage, use as oceanographic research eye, examine aquarium specimens
Crime & Legal—	Make law enforcement observations, work on IR security systems, use as a crime lab tool
Behavioral—	Study habits, assess audience response, monitor status during sleep
Medical—	Use for physiological-psychological testing, make dental and eye examinations, use as medical lab tool
Biology—	Make nocturnal observations, study plant/insect/animal physiology, use for microscopic study
Archaeology & Paleontology—	Decipher aged writings, classify data from pattern character, differentiate fossil remains
Art & Antique—	Detect art flaws, authenticate antiques, use as restoration tool
Printing & Textile	Use for offset color separations, locate flaws in fabrics, inspect leather goods

Fig. 19-1. Typical applications.

possible chance for error. We recommend that you take some time to read the entire chapter through before any work is started. This will enable you to proceed with the parts accumulation and fabrication. It will also allow you to work much faster once construction is started. We also recommend use of

this information once the project is complete. It will furnish you with reference regarding use of the instrument in different applications and its maintenance.

CIRCUIT AND TUBE DESCRIPTION

The heart of the infrared viewer is the RCA or Farnsworth type 6032 image-converter tube. The infrared source, when required, illuminates a scene with invisible infrared light. The lens and eyepiece can be considered as a telescope for viewing the scene. It is the type 6032 image converter which changes the infrared image to a visible picture. The front lens focuses the reflected infrared image, which has been illuminated by the invisible infrared source, on the front of the image-converter tube. The tube converts the infrared to visible light and the eyepiece is used to view the visible scene on the rear of the tube.

Figure 19-3 is a data sheet describing and giving specification data for the image converter tube. Figure 19-4 is the circuit diagram of the electrical circuit for controlling the instrument and providing the high voltage necessary to power the image converter. Electrical current from an external 6-volt battery passes through the on/off switch to the on/off switch and lamp, and simultaneously to the high-voltage power supply.

Electrical current passes through the flyback transformer's 28-turn primary coil to power the 2N3055 transistor. Output from the transistor is magnetically fed back to the transistor. This produces a 6-volt peak-to-peak oscillation of approxi-

Image Converter—	Type 6032
Spectral Response—	500 to 1200 nanometers (800 max.)
Resolution—	50 lines/mm
Lens—	Standardized "T" mount, user definable telephoto and 10× eyepiece
Infrared Source—	900 nanometer, 4500 maximum candlepower
Case Size—	Approximately 7¾″ × 6½″ × 2¾″
Power Requirement—	6 Vdc (4 "D" cells or equivalent)

Fig. 19-2. Specifications.

6032
IMAGE-CONVERTER TUBE

Electrostatic Focus
Good Resolution Capability

4-17/32" Max. Length
2-1/8" Max. Diameter

RCA-6032 is a three-electrode tube of the image-converter type which, in combination with suitable optical systems, permits the viewing of a scene with infrared radiation. It utilizes a semitransparent photocathode at the large end on which the scene to be viewed is imaged by means of an optical objective. The image on the photocathode is focused on the fluorescent screen at the small end of the tube by electron-optical methods to form a reduced image which can be viewed with an optical magnifier. The objective may consist of a Schmidt optical system or a conventional objective lens. The inverted image produced by the optical system on the photocathode is reinverted by the 6032 to give an observed image which is erect.

Features of the 6032 are its good response to radiant energy in the infrared region up to about 12000 angstroms, a faceplate having high optical quality, very low background illumination, minimum resolution of 18 line-pairs per millimeter at the center of the photocathode, and high ratio of light output to infrared energy input.

DATA

General:
```
Spectral Response. . . . . . . . . . . . . . . . . S-1
Wavelength of Maximum Response . .  8000 ± 1000 angstroms
Photocathode, Semitransparent:
  Shape. . . . . . . . . . . . . . . . . . . . . Circular
  Minimum Window Area. . . . . . . . . . . 1 sq. in.
  Minimum Window Diameter. . . . . . . . . . 1-1/8 in.
  Minimum Quality-Circle Diameter within window    1 in.
Phosphor . . . . . . . . . . . . . . . . . . . . . . P20
  Fluorescence . . . . . . . . . . . . . Yellow-Green
  Phosphorescence. . . . . . . . . . . . . Yellow-Green
  Persistence. . . . . . . . . . . . . . Medium-Short
```

Fluorescent Screen:
```
  Shape. . . . . . . . . . . . . . . . . . . . Circular
  Minimum Diameter . . . . . . . . . . . . . . 5/8 in.
Focusing Method. . . . . . . . . . . . . Electrostatic
Overall Length . . . . . . . . . . . . . 4-15/32" ± 1/16"
Maximum Diameter . . . . . . . .•. . . . . 2-3/32" ± 1/32"
Terminals. . . . . . . . . . . See Dimensional Outline
Mounting Position. . . . . . . . . . . . . . . . . Any
Weight (Approx.) . . . . . . . . . . . . . . . . 3.6 oz
```

Maximum Ratings, Absolute Values:
```
GRID-No.2* VOLTAGE (DC or Peak AC)□      20000 max.  volts
GRID-No.1 VOLTAGE□ . . . . . . . . .      2700 max.  volts
AVERAGE PHOTOCATHODE CURRENT
  (Continuous Operation) . . .           1.0 max.   µamp
AMBIENT TEMPERATURE. . . . . . . .        75 max.    °C
```

Characteristics:

Grid-No.2* Voltage . .	16000	20000	volts
Grid-No.1 (Focusing Electrode) Voltage-- 10.75% to 13.25% of grid-No.2 voltage .	1720 to 2120	2150 to 2650	volts
Max. Grid-No.1 Current	0.4	0.5	µamp
Paraxial Magnification Factor♦	0.5	0.5	
Sensitivity:			
Radiant, at 8000 angstroms. . . .	0.0038	0.0038	µamp/µwatt
Infrared●	5	5	µamp/lumen
Minimum Conversion Index†●	8	10	
Minimum Resolution (In central area of photocathode)▲● . .	18	18	line-pairs per mm

* Grid No.2 serves the dual function of high-voltage electrode for accelerating the electron beam and of collector through which the electrons leave the tube after their energy has been transformed within the tube.

□ Referred to photocathode.

● Under the following conditions: 2870°K tungsten light source; light flux of 0.1 lumen incident on Corning No.2540 infrared Filter (Melt 1613, 2.61 mm thick or equivalent); irradiated area of photocathode is 3/4 inch in diameter.

† Ratio of light flux from fluorescent screen to the product of the light flux incident on the infrared filter multiplied by the filter factor.

▲ The resolution, both horizontally and vertically, in a 0.3-inch diameter circle centered on the photocathode is determined with a pattern consisting of alternate black and white lines of equal width. Any two adjacent lines are designated as a "line-pair".

♦ Magnification is defined as the ratio of the distance from the tube axis of an image point on the fluorescent screen to the distance from the tube axis of an object point on the photocathode. Paraxial magnification is the magnification observed along the tube axis. .

Fig. 19-3. Image converter tube data sheet.

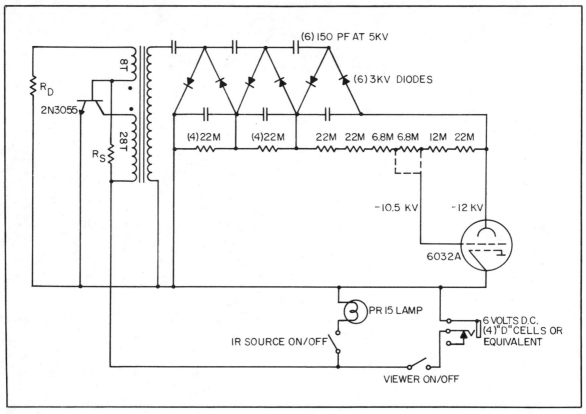

Fig. 19-4. Circuit diagram.

mately 5 kHz. Magnitude of the oscillation is controlled by the value of drive resistor R_d. Oscillation is initiated by start resistor R_s.

The 5 kHz transistor output is also magnetically coupled to the secondary of the flyback transformer. This steps up the voltage to approximately 2000 volts ac. The voltage is then rectified to dc and multiplied by six times to obtain the minus 12,000 volts dc acceleration potential required by the image converter. Rectification and multiplication is accomplished by the network made of six 150 picofarad ceramic capacitors and six 3 kV multiplier diodes. A resistor divider network is placed across the multiplier for stability and to provide the minus 10,500 volt focus potential for the image converter.

Caution

The acceleration and focusing potential for the 6032 image-converter tube are 12,000 volts dc and 10,500 volts dc respectively. These potentials

should be considered hazardous. For safety reasons the output impedance of the dc power supply is extremely high. If accidental contact is made with this high voltage, current will be limited to nonlethal amounts, but the shock can be very unpleasant.

Never turn on or operate in any manner with the covers removed. No person with a heart pacer or any type of heart problem should operate this instrument.

PARTS FABRICATION AND ACCUMULATION

The first step in the construction of the infrared viewer is the accumulation and/or fabrication of parts. A thorough familiarization with all drawings and pictures will help in the task of making parts. You may already have some of these parts on hand or you may wish to obtain parts from various suppliers. They are also available in part pack-

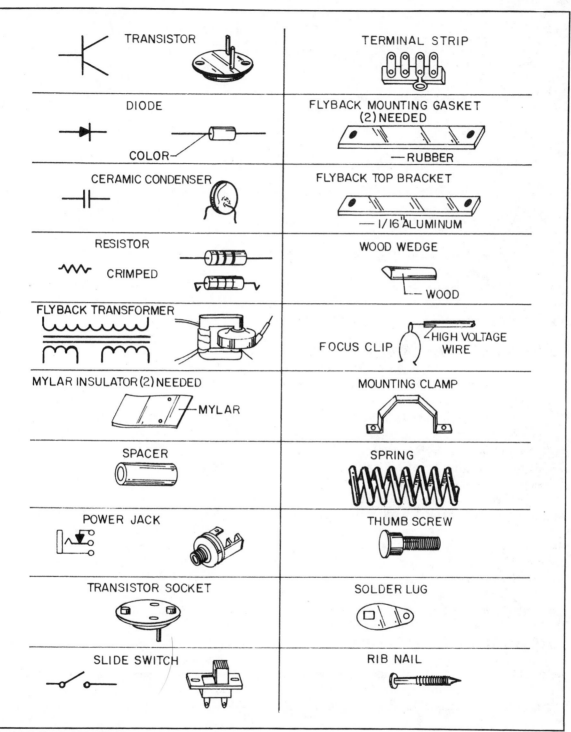

TRANSISTOR   TERMINAL STRIP

DIODE   FLYBACK MOUNTING GASKET (2) NEEDED

COLOR   —RUBBER

CERAMIC CONDENSER   FLYBACK TOP BRACKET

—1/16" ALUMINUM

RESISTOR   WOOD WEDGE

CRIMPED   —WOOD

FLYBACK TRANSFORMER

FOCUS CLIP   HIGH VOLTAGE WIRE

MYLAR INSULATOR (2) NEEDED   MOUNTING CLAMP

—MYLAR

SPACER   SPRING

POWER JACK   THUMB SCREW

TRANSISTOR SOCKET   SOLDER LUG

SLIDE SWITCH   RIB NAIL

Fig. 19-5. Symbol identifications.

277

ages from IR Scientific Inc. Box 110, Carlisle, MA 01741.

Following is a list of numbers of individual parts needed per viewer, description by name and reference to a specific figure.

(1) Type 6032 Image-Converter Tube (Fig. 19-3).

(1) 2N3055 Transistor (Fig. 19-5).

(6) 3 kV High-Voltage Multiplier Leg Diode (Fig. 19-5). Specifications are: fast recovery, 4 kV PIV, 100 mA Fwd. Current, 6-10 chip stack.

(6) 150 pF 5 kV Ceramic Capacitors (Fig. 19-5).

(1) Drive & (1) Start Resistor (Fig. 19-5). See text for explanation of values.

(3) 6.8 Megohm ½ Watt Resistor (Fig. 19-5).

(11) 22 Megohm ½ Watt Resistor (Fig. 19-5).

(1) Flyback Transformer (Fig. 19-5). This is a television type flyback. The high-voltage coil must have a minimum diameter of 2″ to 2½″ × ¾″ thick and the ferrite core must have a minimum cross-section diameter of 7/16″ to ½″. Mounting holes must be added to the left case side for the flyback before the case is assembled, unless a case package is purchased.

(1) Flyback Secondary Coil (Fig. 19-5). This coil is made by first obtaining a wood dowel 1/16″ larger in diameter than the ferrite core. Cut two small grooves into the dowel 90° apart. Cut a piece of manila paper slightly smaller in width than the inside-to-inside dimension of the ferrite core and long enough to wrap around the dowel once. Tape it closed. Centered on the manila tube, wrap 28 turns, in either direction, of #24 to #32 magnet wire. Leaving 4″ pigtails, finish each end by punching a small hole, at the end of the wrap, through the manila and into the groove of the dowel using a sharp pencil. Push through the end of the wire. Over the wire winding, wrap around once with fabric tape (Johnson & Johnson Band-Aid tape). Over this and in like manner, wrap 8 turns of magnet wire, finishing the 4″ pigtail ends, as before, into holes punched into the grooves. Be sure that you wrap the 8 turns in the same direction as the base wrap and that this is centered also. Over this, wrap again with fabric tape. Over the fabric tape, wrap again with clear tape. Slide the whole assembly off of the dowel.

(1) Multiplier and (1) Divider Board (Fig. 19-6). These are made from 1/16″ phenolic or clean etched fiberglass. Layout is according to the drawings in Fig. 19-6. Use Teflon-based posts, swage-based posts or insulated stand-offs to mount parts. A printed circuit board will not work because of the shorts which would occur, due to the high voltages generated.

(2 Different) Mylar Insulators (Fig. 19-7). These should be made from any semi-rigid, high dielectric strength, thin plastic such as mylar, celluloid, etc. Lay out according to dimensions in Fig. 19-7.

(2) #6 × ½″ Long Fiber or Plastic Spacer (Fig. 19-5).

(1) Power Jack (Fig. 19-5).

(1) Transistor Socket (Fig. 19-5). Standard TO-3 socket.

(2) On/Off Slide Switches spst (Fig. 19-5). Be sure the switch tab is long enough to protrude through the ¼″ case back.

(1) 4-Lug Terminal Strip (Fig. 19-5). Has one grounded lug.

(2) Flyback Rubber Mounting Gaskets (Fig. 19-5). Make from tire inner tube.

(1) Flyback Top Bracket (Fig. 19-5). Make from 1/16″ aluminum.

(1) High-Voltage Wood Wedge (Fig. 19-5). Made from any piece of wood. Used to pin high-voltage coil on flyback core.

(1) Image-Converter Focus Clip (Horse Shoe Shaped) (Figs. 19-5 and 19-8). Made from "jumbo" paper clip and bent to shape.

(1) Front and (1) Rear Tube Mounting Clamp (Figs. 19-5 and 19-8). Made from .025″ aluminum ¼″ wide and bent to shape per Fig. 19-8.

(1) Infrared Source Housing (Fig. 19-9). Made from aluminum can which food spreads and mushrooms come in. Dimensions: 2½″ × 2½″.

(1) Rubber Grommet (Fig. 19-9).

(1) Reflector Assembly (Fig. 19-9) (includes Lamp and Cap). Made from flashlight reflector available at Radio Shack. Cut to fit inside of infrared source housing and glue on glass or acrylic disk.

(1) Infrared Source Gasket (Fig. 19-9). Made

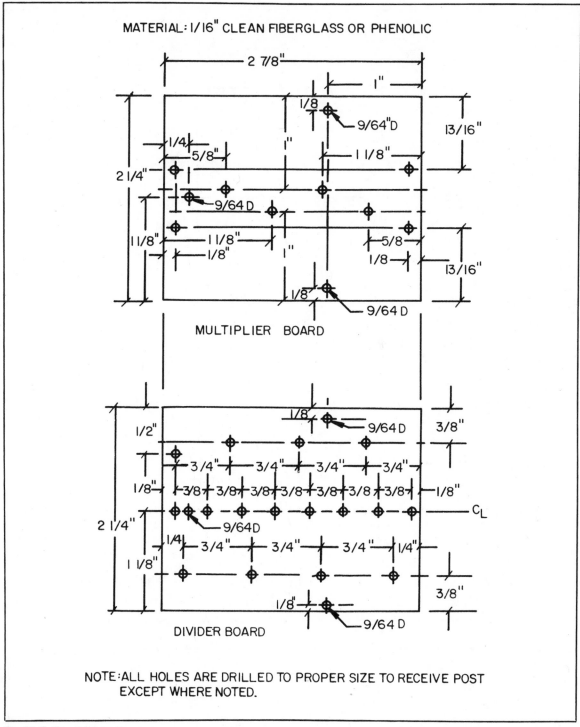

Fig. 19-6. Board fabrication.

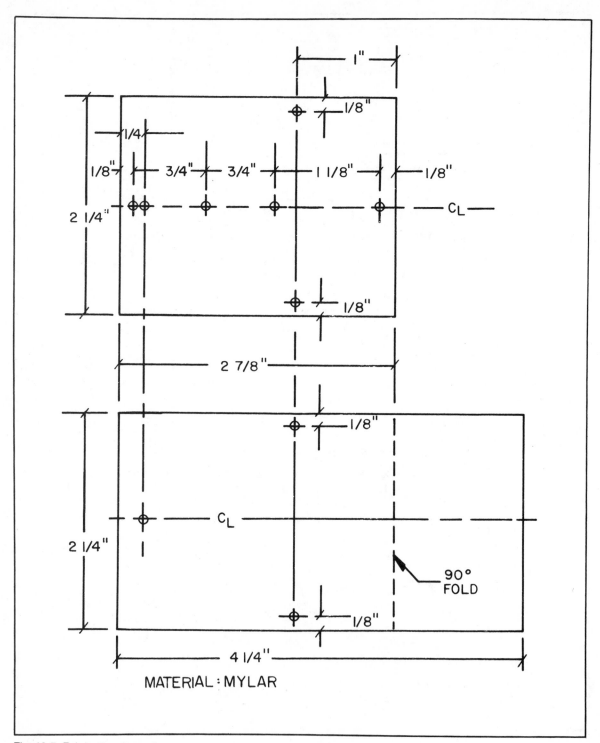

Fig. 19-7. Fabrication dimensions.

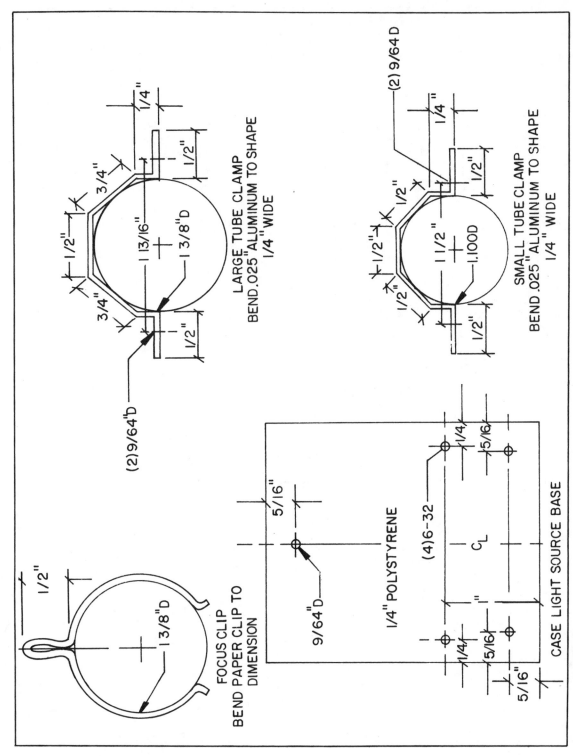

LARGE TUBE CLAMP
BEND .025" ALUMINUM TO SHAPE
1/4" WIDE

1/4"
1/2"
3/4"
1 13/16"
1 3/8" D
1/2"
3/4"
1/2"
(2) 9/64" D

SMALL TUBE CLAMP
BEND .025" ALUMINUM TO SHAPE
1/4" WIDE

(2) 9/64 D
1/4"
1/2"
1 1/2"
1.100 D
1/2"
1/2"
1/2"

FOCUS CLIP
BEND PAPER CLIP TO
DIMENSION

1/2"
1 3/8" D

5/16"
9/64" D
1/4" POLYSTYRENE
(4) 6-32
C_L
1/4"
5/16"
5/16"
1/4"
5/16"
5/16"

CASE LIGHT SOURCE BASE

Fig. 19-8. Fabrication drawings.

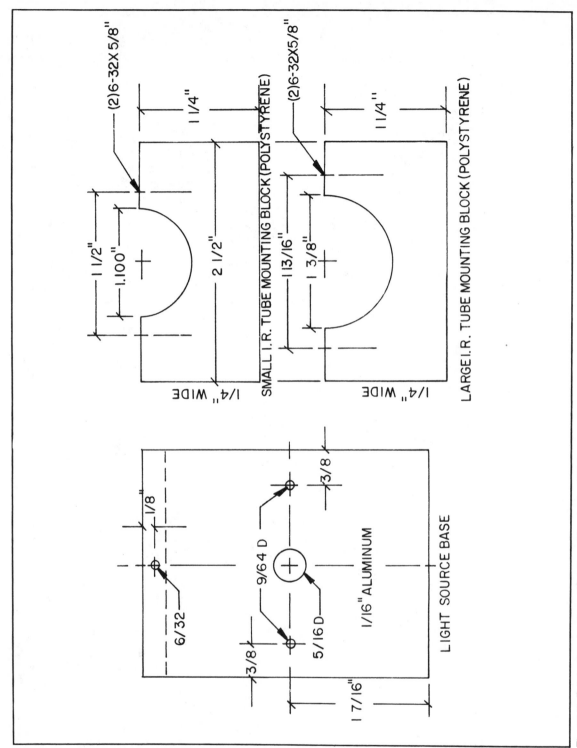

Fig. 19-8. Fabrication drawings. (Continued from page 281.)

Fig. 19-9. Light source and filter.

from rubber tube (non cord) which has been cut to length, slit down the middle and the ends glued together. Length should be long enough to fit circumference of infrared source housing and contain reflector assembly.

(1) Infrared Filter (Fig. 19-9). This is a disk made from Kodak Wratten Filter. It is available through Kodak Photography Labs. Use Kodak #87B filter. This will allow the maximum amount of transmission while still being invisible to the eye.

(1) Infrared Source Base (Figs. 19-8 and 19-9). Made from 1/16″ aluminum according to drawing in Fig. 19-8.

(2) 3/16″ Outside Diameter × ¾″ Spring (Fig. 19-5).

(1) Objective Lens and/or Mount. In all of the pictures of this project, it is presented with a built-in 95 mm F1.8 lens. This manner of lens mount may be used or a "T" adaptor, which is available from a camera store may be utilized. A "T" mount is a standardized photographic thread mount. It is then possible to use a wide variety of "T" mount lenses available at any camera store. This also facilitates interchangeability. If a "T" mount is used, the inner thread portion of the mount is all that will be mounted. In any case, the two lens mounting blocks (2½″ × 2½″ polystyrene) will have to be machined to accept the built-in lens or lens mount. Care should be taken to make sure that the lens focuses properly on the front of the image tube. The dimensions of the case may have to be changed to accept the proper focus.

(1) Eyepiece. This is a telescope, jeweler's or microscope eyepiece of 7× to 10× power. Make sure that focus is on the rear screen of the image tube.

(2) Lens Mounting Blocks (See Objective Lens and/or Mount, above).

(5) Corner Braces, (1) Case Right Side, (1) Case Left Side.

(1) Case Infrared Source Back, (1) Case Rear, (1) Image-Converter Front Mounting Block and (1) Image Converter Rear Mounting Block (Figs. 19-8 through 19-12). All made from ¼″ polystyrene according to drawings.

(1) Case Partition (Figs. 19-10 through 19-12) made from ⅛″ polystyrene according to drawings.

(1) Top and (1) Bottom Case Covers (no drawing). These are made from ⅛″ polystyrene to cover the top and bottom of the viewer once assembly is complete. Dimensions will have to be determined once modifications to the case (to accept the objective lens) are finished. Mounting screw locations for the covers will also have to be determined.

(1) 6-32 × ⅝″ Thumbscrew (Fig. 19-5). If this part cannot be found, make from a 6-volt lantern battery thumbnut and glue a 6-32 screw shank in place.

(1) Roll Hook-up Wire.

(6″) TV High-Voltage Wire.

Below is a list of the screws, nuts, etc., which will be needed to complete the viewer.

(2) 6-32 × ¾″ Black Nylon Screws
(1) #6 × 3/16″ Panhead Sheet Metal Screw
(2) #6 × ⅜″ Flathead Sheet Metal Screws
(2) #6 × ½″ Panhead Sheet Metal Screws
(2) 6-32 × 5/16″ Flathead Screws
(6) 6-32 × 7/16″ Roundhead Screws
(2) 6-32 × ½″ Panhead Screws
(4) 6-32 × ½″ Flathead Screws
(2) 6-32 × 3″ Panhead Screws
(8) 6-32 × ¼″ Nuts
(4) #6 Starwashers
(4) #6 Splitwashers
(4) #6 Solder Lugs
(2) #4 × ⅝″ Long Rib Nails
(6) #4 × ¾″ Long Rib Nails

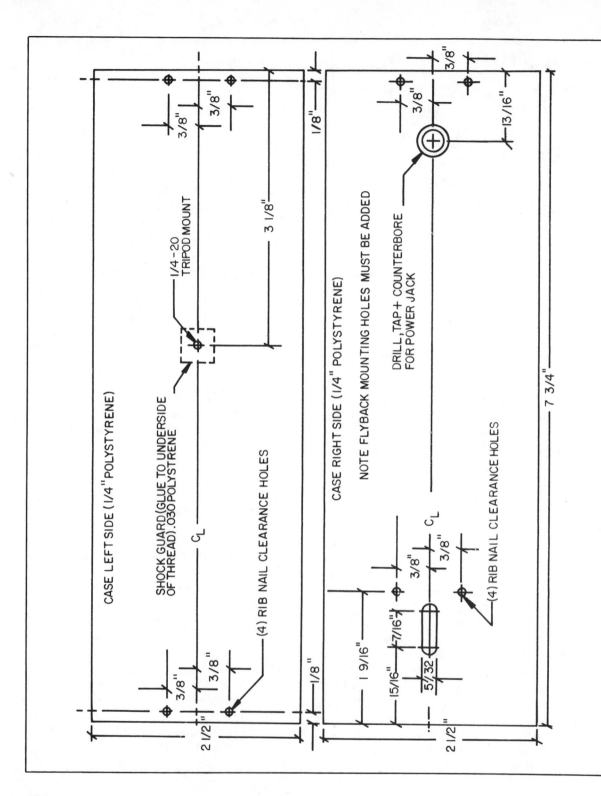

CASE LEFT SIDE (1/4" POLYSTYRENE)

SHOCK GUARD(GLUE TO UNDERSIDE OF THREAD).030 POLYSTYRENE

1/4-20 TRIPOD MOUNT

3/8"
3/8"

3 1/8"

C_L

(4) RIB NAIL CLEARANCE HOLES

3/8"
3/8"

2 1/2"

1/8"

CASE RIGHT SIDE (1/4" POLYSTYRENE)

NOTE FLYBACK MOUNTING HOLES MUST BE ADDED

DRILL, TAP + COUNTERBORE FOR POWER JACK

3/8"
3/8"

13/16"

7 3/4"

C_L

3/8"
3/8"

1 9/16"

15/16"

7/16"

5/32

(4) RIB NAIL CLEARANCE HOLES

2 1/2"

1/8"

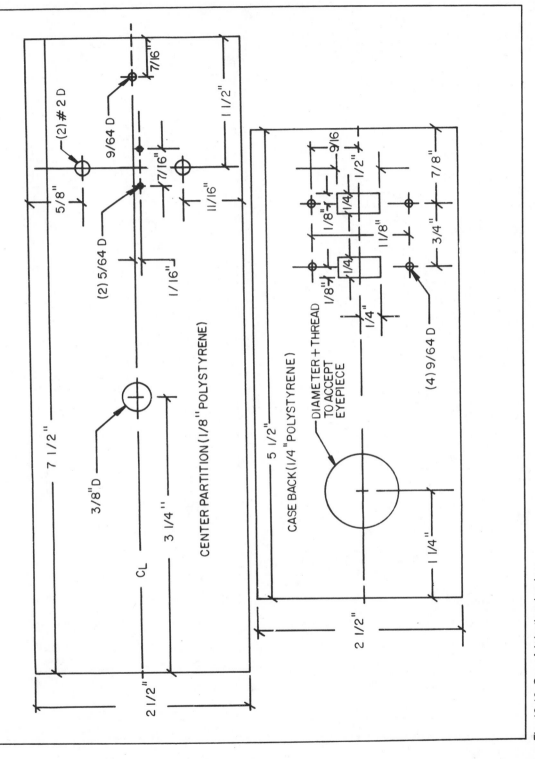

Fig. 19-10. Case fabrication drawings.

285

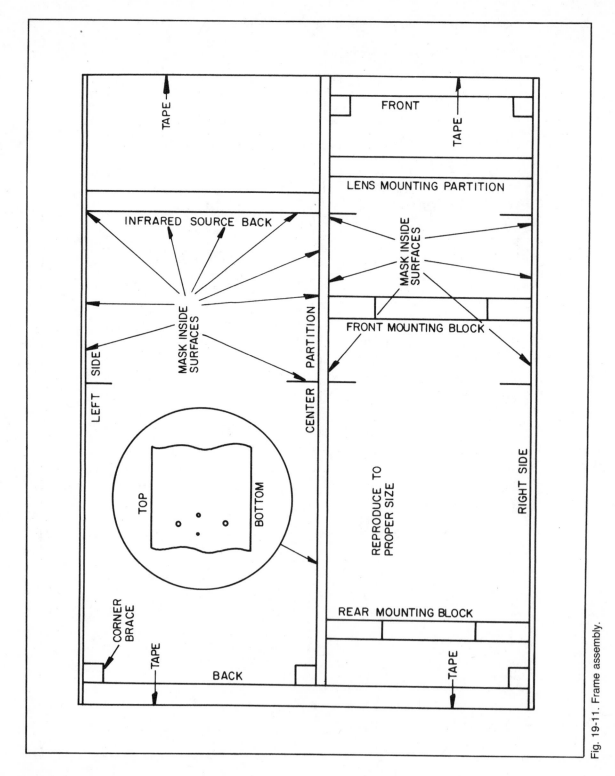

Fig. 19-11. Frame assembly.

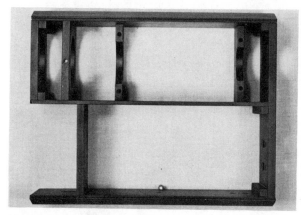

Fig. 19-12. The fabricated frame.

Other miscellaneous items you will need:

(1) Tube Model Airplane Plastic Cement
(1) Can of *low-carbon* Flat Black Spray Paint
Rosin Core (60-40) Solder

ASSEMBLY INSTRUCTIONS

Once the objective lens mounting dimensions have been determined, and the eyepiece and fly-back mounting holes have been located, the viewer assembly may begin. Do not assemble any case parts until all dimensioning has been thoroughly figured out.

CASE ASSEMBLY

Figure 19-11 is a layout plan for the case parts. This may be reproduced or redrawn to actual scale. This is then used as a layout sheet, upon which the case parts may be glued together. It is recommended that wax paper be put in between the case parts and the paper. This will avoid the glue sticking to the paper plan. After the case parts are glued together, and have dried for several hours, lead holes can be drilled for the rib nails and the rib nails installed.

The next step is painting the finished case, and the top and bottom covers. If the carbon content of the flat black spray paint is 1% or higher, the inside plastic surfaces which will be adjacent to the high voltage must be masked prior to painting. This will avoid a serious high-voltage corona problem. The

inside sections to be masked are shown in Fig. 19-11.

MULTIPLIER ASSEMBLY

The multiplier assembly consists of the multiplier board and the divider board. Once put together separately, they are assembled together into the single multiplier assembly pictured in Fig. 19-13.

The first step is to assemble the capacitors onto the multiplier board, as pictured in Fig. 19-13. It is important that the capacitors are located accurately as pictured, and that room is left for later installation of the #6 × ½″ spacers at the two mounting holes located along the sides near the top.

Caution

The acceleration and focusing potentials for the 6032 image-converter tube and 12,000 volts dc and 10,500 volts dc respectively. These potentials should be considered hazardous. For safety reasons, the output impedance of the dc power supply is extremely high. If accidental contact is made with this high voltage, current will be limited to nonlethal amounts, but the shock can be very unpleasant.

Never turn on or operate in any manner with the covers removed. No person with a heart pacer or any type of heart problem should operate this instrument.

Before any focusing can be done, the power supply must be trimmed. This is done to determine the start-up voltage (4.0 volts or lower) and to determine the maximum output voltage (12 kV). R_d and R_s values will be added now. Maximum output voltage should be measured at the output post of the multiplier assembly.

R_d and R_s resistor values are determined keeping in mind that R_s values are usually 9 times the value of R_d, using the nearest standard resistor value. R_d is usually 33 to 150 ohms. R_s is usually 270 to 1200 ohms. So therefore, examples such as these are common: 100 ohms and 820 ohms, 47 ohms and 390 ohms. Using a high-voltage television probe or a resistor network, which can be determined by using Ohm's Law, determine the values now of R_d and R_s, keeping in mind that the ratio of 9× controls

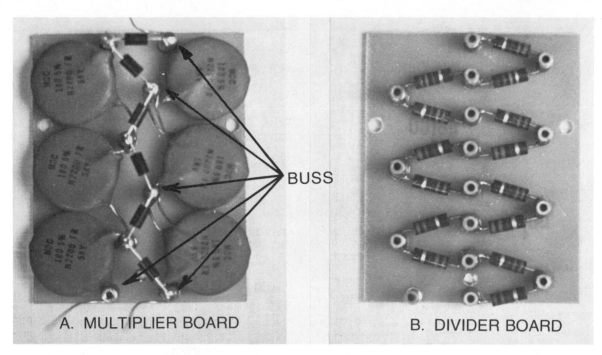

A. MULTIPLIER BOARD

BUSS

B. DIVIDER BOARD

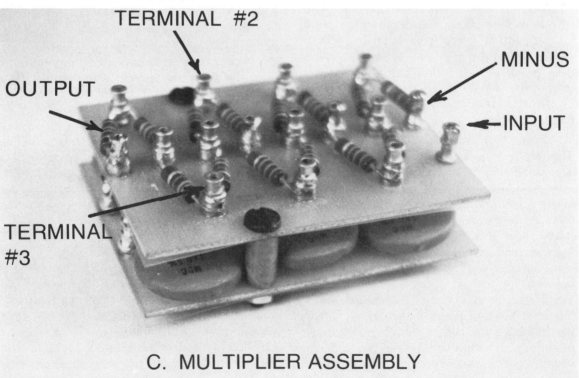

TERMINAL #2

OUTPUT

MINUS

INPUT

TERMINAL #3

C. MULTIPLIER ASSEMBLY

Fig. 19-13. Multiplier assembly.

TOP VIEW

BOTTOM VIEW

Fig. 19-14. The completed case assembly.

the start-up voltage. This ratio is for a 6-volt input voltage. If other input voltages are desired, change the ratio accordingly. Once values are determined, install the two resistors in their proper spots on the terminal strip.

Next, solder 5 buss wires to the five posts marked in Fig. 19-13 and orient properly, as shown, for connection to the proper post in the divider board network. Position all diodes as pictured and install. Note that all diode leads which are colored for polarity indications always go to the left, when the multiplier board is oriented as shown. Be sure all connections are soldered properly, and have no sharp protrusions.

Install all resistors, as marked in Figs. 19-13 and 19-4, onto the divider board and solder. Again take care to make a good solder joint and to avoid protrusions coming out of the solder. Assemble the divider board and multiplier board together with the small mylar insulator and the two #6 × ½″ spacers in between them. Install into the case, as pictured in Fig. 19-14, using the two 6-32 × ¾″ black nylon screws and the large mylar insulator. All buss wires should be fed through their proper posts, cut to length, and solder.

FINAL ASSEMBLY

The infrared source should be prewired, assembled and installed. Install the objective lens and/or mount, tube, eyepiece, switches, power jack and flyback. All wiring can be completed at this point, with the exception of the focus wire, focus connection, R_d and R_s. The adjustment and installation of these are explained in the next section.

FOCUS ADJUSTMENT

At this point all assembly operations have been completed. To operate the instrument, it is necessary to adjust the focus. Both optical focus of the lenses and electronic focus of the image converter and required.

In daytime, but in a darkened room, position the instrument several feet back from a window and orient to look out the window. It is necessary that the scene permits viewing to several hundred feet.

Place a flat piece of white paper against the front face of the tube. Rotate the lens until the view through the window is sharply focused. Adjust so that objects several hundred feet away are sharply focused. This serves to focus the lens to the front of the tube, with the unit turned off. Once focused, remove the white paper, but do not change the setting of the lens.

It is necessary to stop down the lens to approximately F-8.0. This will avoid over-driving the tube. This can be done with the iris in the lens, or by taping a piece of cardboard with a ⅜″ hole centered on it to the front of the lens. By doing this, a nighttime situation will be approximated, which is what the image tube is built for.

Locate a 6.8 megohm and a 12 megohm resistor. With long nose pliers, bend the leads of each resistor so that they will fit into the holes of multiplier assembly terminals #2 and #3. Refer to Fig. 19-13 for locations of terminals #2 and #3. Cut off ¼″ below each bend. These resistors will be used for adjustment of the image converter's electronic focus. Insert the focus lead from the image tube into terminal #3, but do not solder.

Locate the top and bottom covers. Temporarily tape in place with masking tape. With the main switch and the infrared source switch both off, connect a 6-volt power source to the power jack. Four "D" cells or equivalent will supply adequate power.

With the instrument still oriented to look out the window, switch the main switch to the "on" position. A high pitched sound will be heard and the image-converter's screen will light up, as seen through the eyepiece. **Do not operate the instrument with either cover removed.**

Rotate the eyepiece for the best focus of the screen. Concentrate on maximizing the grain (sandy appearance) of the screen, rather than the image. Readjust the lens to obtain the sharpest focus of objects several hundred feet away. Switch the instrument off. Remove the top cover and insert the 12 megohm resistor between multiplier terminals #2 and #3. Replace the covers, turn on and observe if sharpness of image is better or worse. Repeat using 6.8 megohm resistor. Move the focus lead to terminal #3. Repeat the above two steps, trying the 6.8 megohm resistor first. Lastly, try with no resistor. Determine which combination gives the sharpest focus.

Once the best combination has been found, turn power off, remove the covers, remove the cardboard from the front of the lens and solder the optimum combination in place. Now all that has to be done to complete the instrument is to install the top and bottom covers. All assembly is completed. The instrument is ready for operation.

OPERATION

Initially, it is recommended that operation of the instrument be checked in daylight conditions, using the cardboard disk or iris used for focus adjustment. For normal operation in darkened or dimly lighted conditions, this cardboard disk or iris is not needed, and would only hamper the infrared light-gathering ability of the tube.

It should be recognized that the instrument is extremely sensitive to the brightness of the infrared. Focal sharpness of the image is not only dependent on careful adjustment of the lens and eyepiece, but also on control of infrared brightness.

Infrared, either too bright or too dim, will cause the image to "bloom" or "starve". This is similar to the result obtained with a television set if the brightness control is not adjusted properly.

When used for viewing sources emitting infrared, the infrared source is not used. Typical sources are infrared lasers, infrared light-emitting diodes and heat sources over 600° F. When used for viewing in darkened conditions, the infrared source is switched on and used as a light, flashlight or lantern beam would be. Vertical adjustment is preset by the two screws under spring tension at the base of the infrared source. Horizontal adjustment is accomplished by sliding the thumb screw in its slot.

With practice, you will become proficient in the operation of the instrument. You will have the satisfaction of investigating the many interesting applications of infrared with an instrument you have built. A few applications were listed in Fig. 19-1 of this manual. For additional application, Kodak Publication M-28, *Applied Infrared Photography* is recommended.

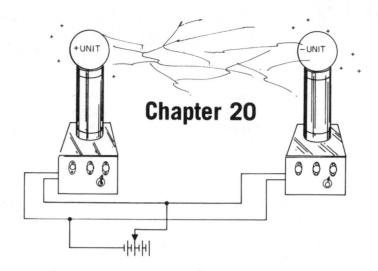

Chapter 20

Miniature Long-range
FM Voice Transmitter (MFTI)

This project describes a device that when properly constructed and adjusted will pick-up sounds in a room at the level of whispers. The device with a proper antenna can clearly transmit to over one mile when used with a good FM receiver. It is obvious that if good antenna systems are on both the unit and the receiver, that ranges far in excess of that stated above are possible. The advantage of this unit is that it can detect extremely low level sounds such as whispers or when properly and centrally located can detect voices and sounds throughout an entire house. It also should be noted that the turning range of the device covers the standard FM band and the lower aircraft band for complete privacy. The unit is intended to be placed in a location and operated without being handled.

CIRCUIT DESCRIPTION

This unit is a super-sensitive, mini-powered FM transmitter (Table 20-1) consisting of an rf oscillator section interfaced with a high-sensitivity wide-passband audio amplifier and capacitance

mike with built-in FET that modulates the base of the rf oscillator transistor (Fig. 20-1). Transistor (Q2) forms a relatively stable rf oscillator whose frequency is determined by the value of L1 and tuning capacitor (C8). The setting of C8 determines the desired operating frequency and is in the standard FM broadcast band with tuned circuit design favoring the high end up to 110 MHz. Capacitor C7 supplies the necessary feedback voltage developed across R11 in the emitter circuit of Q2 sustaining an oscillating condition. Resistors R9 and R10 provide the necessary bias of the base emitter junction for proper operation while capacitor (C10) bypasses any rf to ground fed through to the base circuit. C9 provides an rf return path for the tank circuit of L1 and C8 while blocking the dc supply voltage fed to the collector of Q2.

The audio section utilizes a high sensitivity capacitance mike and built-in FET transistor (field-effect) and will clearly pick-up all low level sounds in the speech audio spectrum. The speech voltage developed across R1 by M1 is capacitively

Table 20-1. MFT1 Miniature Long-Range FM Voice Transmitter Parts List.

R1	(1)	10 k resistor ¼ watt
R2,6,7,8	(4)	1 k resistor ¼ watt
R3	(1)	100 k resistor ¼ watt
R4	(1)	8.2 k resistor ¼ watt
R5,10	(2)	15 k resistor ¼ watt
R9	(1)	3.9 k resistor ¼ watt
R11	(1)	220 ohm
C1	(1)	100m/25 V small electrolytic cap
C2	(1)	470 pF/50 V disc cap
C3,5	(2)	10 μF/25 V electrolytic cap
C4,9,10	(3)	.01/50 V disc cap
C6	(1)	1 μF/50 V electrolytic cap
C7	(1)	5 pF mica zero temp cap
C8	(1)	6 35 pF trimmer cap
Q1, 2	(2)	Select for high B 100 MHz-npn PN2222
M1	(1)	FET cap mike
PB1	(1)	1⅝ × 1⅝ perfboard per Fig. 20-2
CL1	(1)	6″ battery clip
WR1	(1)	6″ #24 wire
B1	(1)	9-volt transistor battery
L1	(1)	Coil 8 turns #16 per text
CA1	(1)	1 ⅞ plastic cap

BLM3 Battery eliminator replaces battery and allows unit to operate on 120 Vac.

Note: Most available FM radios can easily be detuned slightly to shift the dial readings down where 108 is 109. This is accomplished by carefully adjusting the "osc" padding trimmer located on the main tuning capacitor and "walking" a known station down the necessary megahertz or two. An "antenna peaking" trimmer should now be adjusted for maximum signal at the high end. See section on general use and helpful hints using wireless devices, antennas, etc.

Complete kit with optional PC board and battery eliminator available through Information Unlimited, Inc., P.O. Box 716, Amherst, N.H. 03031, write or call 1-603-673-4730 for pricing and delivery.

coupled by C4 to the base of Q1. The base of Q1 is biased by R3 and R5. You will note that the emitter of Q1 is based via R6 and C5. A signal voltage developed across R4 is capacity coupled through C6 to the base of Q2 through R8. Q1 emitter is ac bypassed by C5 while being dc biased by R6 to allow full signal excursion without junction clipping. This amplified signal is developed across Q1 collector resistor R4 and is ac coupled through C6 to the base of Q2 where this amplified speech signal causes FM and AM modulation of the oscillator circuit by slight shifting of the operating point. Also, note R7 and R8. These resistors along with C1 and C2 (if required) decouple the oscillator and audio circuits and are necessary to prevent feedback and other undesirable effects. When properly assembled, the above circuit should produce crystal clear quality when the receiver is properly tuned to the unit.

Note that a shunt capacitor may be connected across the base lead of Q2 to reduce sensitivity. Also, R7, C1, C2 miniature 100 μF electrolytic capacitor and 470 pF disc are not shown in pictorial, for clarities sake and are only required if audio instability in the form of low frequency rumble or motorboating is encountered. If required connect C1 (+) lead to junction of R1, R3 and (−) lead to junction of R5, C5.

The circuit with the components listed operates best in the upper FM band. This is a clear spot without interference from FM radio stations. However, satisfactory performance is obtained above 110 MHz for limited range use as this is the aircraft band and discretion is advisable.

CONSTRUCTION AND PRELIMINARY ALIGNMENT

1. Identify all components (Fig. 20-2).
2. Position perfboard piece so that border holes match those of pictorial showing coil in lower

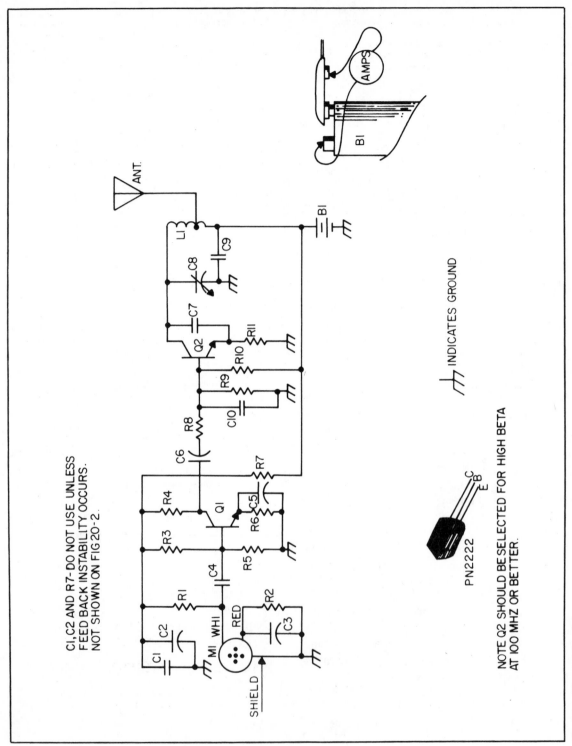

C1, C2 AND R7 - DO NOT USE UNLESS FEED BACK INSTABILITY OCCURS. NOT SHOWN ON FIG 20-2.

INDICATES GROUND

NOTE Q2 SHOULD BE SELECTED FOR HIGH BETA AT 100 MHZ OR BETTER.

PN2222

Fig. 20-1. Circuit schematic.

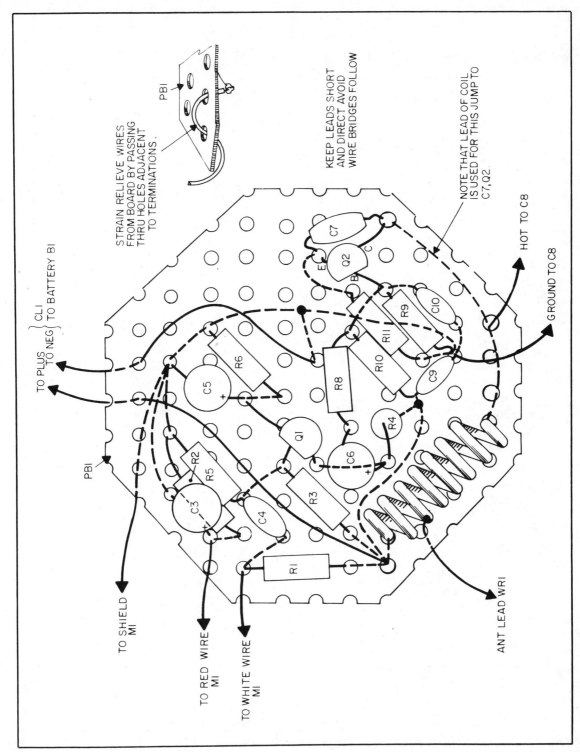

Fig. 20-2. Assembly layout.

295

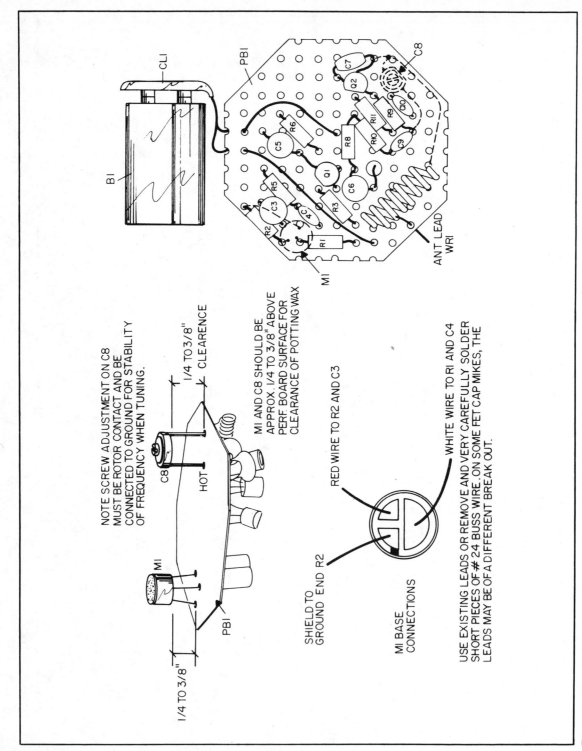

NOTE SCREW ADJUSTMENT ON C8 MUST BE ROTOR CONTACT AND BE CONNECTED TO GROUND FOR STABILITY OF FREQUENCY WHEN TUNING.

M1 AND C8 SHOULD BE APPROX. 1/4 TO 3/8" ABOVE PERF. BOARD SURFACE FOR CLEARANCE OF POTTING WAX

RED WIRE TO R2 AND C3

SHIELD TO GROUND END R2

WHITE WIRE TO R1 AND C4

M1 BASE CONNECTIONS

USE EXISTING LEADS OR REMOVE AND VERY CAREFULLY SOLDER SHORT PIECES OF #24 BUSS WIRE. ON SOME FET CAP MIKES, THE LEADS MAY BE OF A DIFFERENT BREAK OUT.

1/4 TO 3/8"

1/4 TO 3/8" CLEARENCE

Fig. 20-3. Assembly showing M1 and C8.

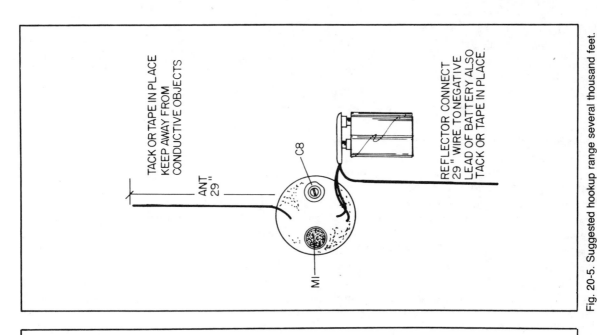

TACK OR TAPE IN PLACE
KEEP AWAY FROM
CONDUCTIVE OBJECTS

ANT
29"

C8

M1

REFLECTOR CONNECT
29" WIRE TO NEGATIVE
LEAD OF BATTERY ALSO
TACK OR TAPE IN PLACE.

Fig. 20-5. Suggested hookup range several thousand feet.

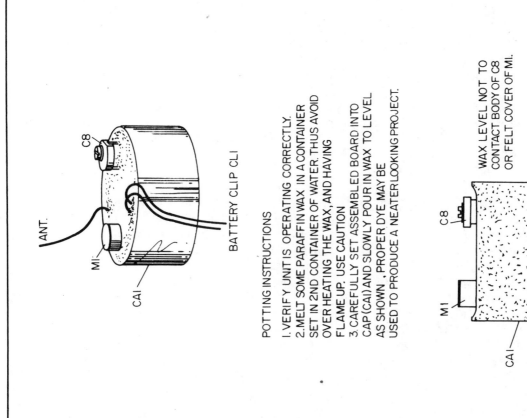

C8

ANT.

M1

CAI

BATTERY CLIP CLI

POTTING INSTRUCTIONS

1. VERIFY UNIT IS OPERATING CORRECTLY.
2. MELT SOME PARAFFIN WAX IN A CONTAINER
SET IN 2ND CONTAINER OF WATER. THUS AVOID
OVER HEATING THE WAX, AND HAVING
FLAME UP. USE CAUTION
3. CAREFULLY SET ASSEMBLED BOARD INTO
CAP (CAI) AND SLOWLY POUR IN WAX TO LEVEL
AS SHOWN . PROPER DYE MAY BE
USED TO PRODUCE A NEATER LOOKING PROJECT.

WAX LEVEL NOT TO
CONTACT BODY OF C8
OR FELT COVER OF MI.

C8

M1

CAI

Fig. 20-4. Final assembly showing wax potting.

left-hand corner (mark for reference, etc.). This step is important if pictorials are to be followed exactly.

3. Form L1 by tightly wrapping 8 turns of #16 buss wire on a #8 wood screw. This produces a finished 8 turn coil of approximately .135″ inner diameter of about .625″ in length. Insert in proper holes as shown and dress leads as shown in pictorial.

4. Insert and solder remaining components (observe polarity of noted components) using their respective leads for connection (Fig. 20-2). Keep leads short and direct. Note dashed lines indicating connections on underside of board. Watch for shorts and bad solder joints. Note that layout as shown eliminates connecting wire bridges and potential shorts.

5. Connect CL1 battery clip using holes in perfboard as strain reliefs. Twist the wires (Figs. 20-2 and 20-3).

6. Connect a 6″ piece of wire to 2nd turn on L1. Note that the wire is strain-relieved through hole in perfboard. This is the antenna wire.

7. Tune an FM receiver to a fairly strong station at the high end of the band (108 MHz or higher).

8. Connect and solder M1 mike and C8 tuning capacitor. Note position of C8 and proper connections to M1 (Fig. 20-3).

9. Connect a 9-volt battery to clip. If a multimeter or 1-10 mA meter is available, connect in series with battery lead. This can be done by removing one of the fasteners of the clip and connect the meter to the free contacts (Fig. 20-1). Meter should read 5 to 7 mA. Pick-up a short piece of the bare wire and touch the coil L1 a turn at a time starting from the C1 end. Note that as you progress turn by turn away from C1, that the indicated meter current will drop or change.

10. If device performs as follows with the battery connected, slowly rotate C8 until the station being received by the radio at approximately 108 MHz breaks out in audio feedback or is blocked out. It may be difficult at first to spot the signal as this

adjustment is very touchy. Also, note that several spots in the adjustment *may be* erroneous and will be weak and unstable. Use an insulated tuning wand.

11. Once the desired setting of C8 is found it should be marked with alignment marks with the frequency noted.

12. When operation is verified it may be desired to pot assembly as shown (Figs. 20-4 and 20-5). This allows easier tuning and handling.

OPERATING AND TUNING

One of the things to watch for when using a device of this type is proper tuning. The adjustment capacitor C8 is quite sensitive and requires only a slight movement to change the frequency. Always use a tuning wand. It is very easy when one is not familiar with this unit to tune to an erroneous signal. This phenomenon is likely to occur when the unit is close to the monitoring receiver. An erroneous signal will be weak, distorted, and unstable (often it is mistaken for the main signal and the unit blamed for poor performance). It is often desirable when the actual signal is located to mark the position of the adjustable capacitor C8 and making note of the frequency spot on the receiver. The main signal will be strong, stable, and undistorted (if modulated). Several experiments in tuning the unit should be done before attempting to use it for the desired application.

Also, whenever possible, the unit should be used around 108 to 109 MHz, which is the lower part of the aircraft band and upper FM broadcast band. When the approximate desired spot is found, final touch-up tuning should be done at the receiver end for clarity, etc. In most areas these frequencies are clear and allow uninterrupted use in contrast to lower in the FM band where a slight change in frequency from a clear spot results in the unit being drowned out by a strong broadcast station. *Do not go above 108 MHz if near an airport or air traffic lane* as this is illegal. See section on general use and helpful hints using wireless devices, antenna, etc.

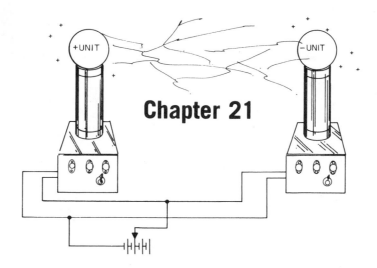

Chapter 21

Miniature Long-range
Telephone Transmitter (VWPM-5)

This project shows how to construct a device allowing the user to monitor his own phone on both incoming and outgoing calls without wires. This system is intended to transmit both sides of a telephone conversation to any FM radio or headset-type receiver and associated earpiece. It allows the user to perform outside activities such as mowing the lawn or working outside the house, etc., while monitoring his phone and clearly hearing both sides of the conversation.

The unique feature of this device is that the unit is only activated when the phone is off of the hook rather than at all times transmitting a continuous signal. This prevents potential interference and unnecessary wearing down of the batteries. It also allows listening to a radio station while monitoring the phone. When the phone is used, the radio station is replaced by the transmission of the unit. The device is assembled on a perfboard and is intended to be housed in an enclosure at the discretion of the builder.

CIRCUIT DESCRIPTION

Transistor (Q1) forms a relatively stable rf oscillator whose frequency is determined by the value of L1 and tuning capacitor (C2). Frequency range is between 88-110 MHz on the standard FM broadcast band favoring the higher end (Fig. 21-1 and Table 21-1). The circuit with the components listed operates best at about 109 MHz. This is clear spot without interference from FM radio stations. However, satisfactory performance is obtained on the remaining FM broadcast band.

Capacitor (C3) supplies the necessary feedback voltage developed across R2 in the emitter circuit of Q1 sustaining an oscillating condition. Resistor (R1) and (R3) provides the necessary bias of the base emitter junction for proper operation while capacitor (C4) bypasses any rf to ground fed through to the base circuit. C1 provides an rf return path for the tank circuit of L1 and C2 while blocking the dc supply voltage fed to the collector of Q1.

You will note that the junction of the base bias

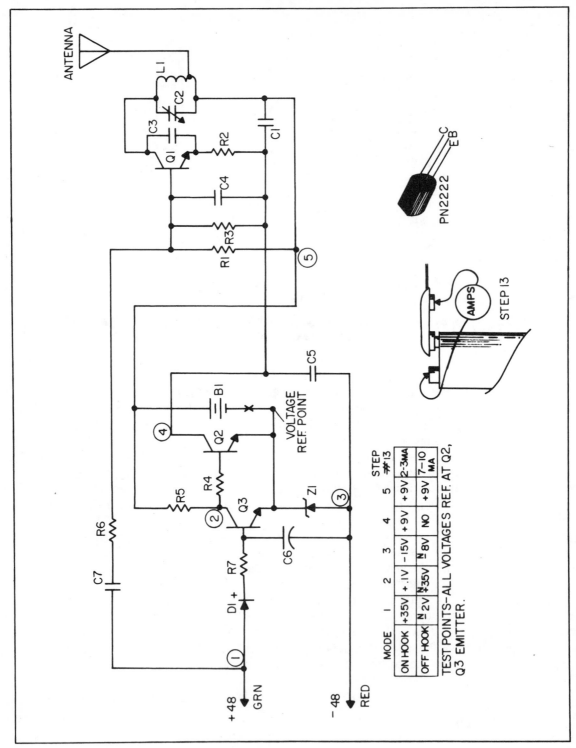

MODE	1	2	3	4	5	STEP #13
ON HOOK	+35V	+.1V	-15V	+9V	+9V	2-3MA
OFF HOOK	≈2V	≈35V	≈8V	NO	+9V	7-10 MA

TEST POINTS– ALL VOLTAGES REF. AT Q2, Q3 EMITTER.

Fig. 21-1. Circuit schematic.

R1	(1)	Resistor 15 k ¼ W
R2	(1)	Resistor 220 ohm ¼ W
R3,5	(1)	Resistor 3.9 k ¼ W
R4	(1)	Resistor 2.2 k ¼ W
R6	(1)	Resistor 150 k ¼ W
R7	(1)	Resistor 100 k ¼ W
C1,4,7	(3)	Cap .01 μF/50 V disc
C2	(1)	Cap 6 35 pF mini trimmer
C3	(1)	Cap 5 pF zero temp
C5	(1)	Cap .1 μF/25 V disc
C6	(1)	Cap 10 μF @ 25 V elec
Q1,2,3	(3)	Semicond. npn plastic PN2222
D1	(1)	Semicond. 50 V lamp rect 1N4002
Z1	(1)	Semicond. 1N5245 15 V zener
L1	(1)	Coil 8 turns #16 tightly wound on a #8 wood screw used for a form
CL2,3	(2)	Connector alligator clips
PB1	(1)	1″ × 2 ½″ perfboard
WR1	(18″)	#24 hook-up wire
CL1	(1)	Battery clip and leads
B1	(1)	9-volt battery (not included in kit)

Note: Most available FM radios can easily be detuned slightly to shift the dial readings down where 108 is 109. This is accomplished by carefully adjusting the "osc" padding trimmer located on the main tuning capacitor and "walking" a known station down the necessary megahertz or two. An "antenna peaking" trimmer should now be adjusted for maximum signal at the high end. See section on general use and helpful hints using wireless devices, antennas, etc.

Complete kit with optional PC board available through Information Unlimited, Inc., P.O. Box 716, Amherst, N.H. 03031, write or call 1-603-673-4730 for pricing and delivery.

You may wish to use the battery eliminator with this project described in Chapter 30.

resistors R1 and R3 is a feed point consisting of capacitor and resistor C7 and R6. Because of the nature of the oscillator frequency being subject to change by varying the base bias condition, a varying ac voltage superimposed at this point causes a corresponding frequency shift (FM) along with an amplitude modulated (AM condition). It is this property that allows this circuit to clearly FM modulate intelligent speech that is detected in any FM receiver when properly tuned.

You will note that before the phone receiver is lifted off of the hook that a dc voltage of approximately 50 volts along with some ac hum is measured across the green and red wires of the phone line. When the receiver is lifted, commencing normal conversation, the dc voltage drops to near zero and an existing ac voltage corresponding to the speech conversation now modulates the oscillator as described above. C7 blocks the dc component of the phone line when an on-hook condition exists and allows little attenuation of the varying ac speech

voltage. R6 is necessary for attenuation of this speech voltage that could cause overmodulation and unnecessary sidebands, etc., if not properly selected.

The 50 volts appearing across the phone lines now is used to turn on Q3 via forward bias through D1 and R7 saturating Q3 below the turn-on point of Q2. You will note off-setting the emitters of Q2 and Q3 by around 15 volts. This offset allows Q3 to turn "off" when phone is used (less than 10 volts on line) turning on Q2 that clamps the negative return of Q1 (rf oscillator) to the battery (B1), initiating operation. The only current used during an "off" mode is the collector current of Q3.

CONSTRUCTION AND PRELIMINARY ALIGNMENT

1. Identify all components (Fig. 21-2).
2. Position perfboard piece so that border holes match those of pictorial showing coil in upper left-hand corner (mark upper left-hand corner for

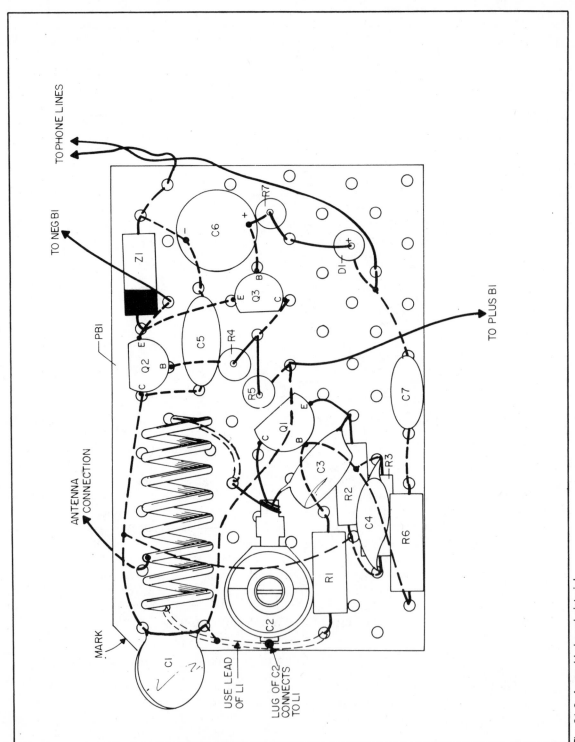

Fig. 21-2. Assembly board pictorial.

302

reference, etc.). This step is important if Fig. 21-2 is to be followed exactly.

3. Enlarge C2 mounting hole "A" to .125 as shown in Fig. 21-2 to accept rotor lug. Clip off the unused solder tab from C2 and insert other used tab in hole "A". Position as shown. Caution: Do not stress or force C2 in any way. Note that some trimmers used for C2 only have 2 pins and fit nicely without above rework to perfboard.

4. Insert coil L1 in proper holes as shown and dress leads as shown by dashed lines. Use these heavy leads for junctioning of other components.

5. Insert and position all components as shown and check location as per Fig. 21-2.

6. Note polarity of zener diode Z1, diode D1, capacitor C6, and position of Q2 and Q3.

7. Solder components as shown using leads for direct connection points. Keep these as direct as possible. Avoid bridging wires without adequate clearance. The layout shown involves only one wire bridge connecting junctions of L1, C1, C2, R1, and red lead of battery clip. See Fig. 21-2 and note dashed lines indicating connections on underside of assembly board.

8. Attach CL1 battery clip leads, 8″ leads that connect to phone lines, and a short piece of wire carefully soldered to the 2nd turn of L1 as shown. (Do not short turns on L1.) Strain-relieve these wires through holes in perfboard as shown before soldering to respective connection points. Connect alligator clips or proper plug to the 8″ wires designated to go to phone line. Note the polarity.

9. The next step is to check your wiring for accuracy, quality of solder joints, and potential shorts. Watch the case of Q1 if it is metal as it must not contact anything since it is the collector terminal and is above ground potential.

10. Tune an FM receiver to a fairly strong station at the high end of the band (108 MHz).

11. Connect a 9-volt battery to CL1. If a millimeter or 0-10 mA meter is available, connect in series with battery lead. See Fig. 21-1. This can be done by removing one of the fasteners of the clip and connecting the meter to the free contacts. Meter should read 5 to 7 mA. Pick up a short piece of the bare wire and touch the coil L1 a turn at a time

starting from the C1 end. Note that as you progress turn by turn away from L1 that the indicated meter current will change.

12. If device performs as follows with the battery connected, slowly rotate C2 until the station being received by the radio at approximately 108 MHz is blocked out or disappears. It may be difficult at first to spot the signal as this adjustment is very touchy. Also, note that several spots in the adjustment *may* be found to produce a signal. Only two are valid, the rest (being erroneous) will be weak and unstable. You may wish to mark C2 with alignment marks when the proper adjustment is found.

13. Connect operating unit to phone lines with phone on hook and note meter reading dropping to below 3 mA and signal disappearing on radio. You may have to reverse the connections to the phone lines. Check the following voltage points indicated on schematic for verification of proper circuit operation (Fig. 21-1). When phone is off of hook you should hear a distinct hum being transmitted to the receiver.

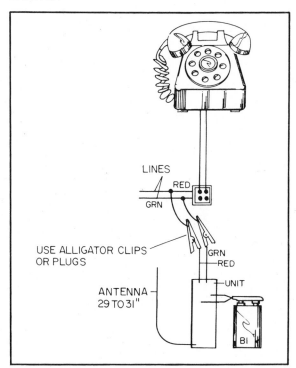

Fig. 21-3. Hookup.

OPERATING—TUNING

One of the things to watch for when using a device of this type is proper tuning. The adjustment capacitor C2 is quite sensitive and requires only a slight movement to change the frequency. *Always use a tuning wand.* It is very easy when one is not familiar with this unit to tune to an erroneous signal. This phenomenon is likely to occur when the unit is close to the monitoring receiver. An erroneous signal will be weak, distorted, and unstable (often it is mistaken for the main signal and the unit blamed for poor performance). It is often desirable when the actual signal is located to mark the position of the adjustable capacitor C2 and making a note of the frequency spot on the receiver. The main signal will be strong, stable, and undistorted (if modulated). Several experiments in tuning the unit should be done before attempting to use it for the desired application.

Whenever possible, the unit should be used around 108 to 109 MHz, which is the lower part of the aircraft band and upper FM broadcast band. When the approximate desired spot is found, final touch-up tuning should be done at the receiver end for clarity, etc. In most areas these frequencies are clear and allow uninterrupted use in contrast to lower in the FM radio band where a slight change in frequency from a clear spot results in the unit being drowned out by a strong broadcast station. *Do not go above 108 MHz if near an airport or air traffic lane.*

Most available FM radios can easily be detuned slightly to shift the dial readings down where 108 is 109. This is accomplished by carefully adjusting the "osc" padding trimmer located on the main tuning capacitor and "walking" a known station down the necessary megahertz or two. An "antenna peaking" trimmer should now be adjusted for max signal at the high end. See section on general use and helpful hints using wireless devices, antennas, etc.

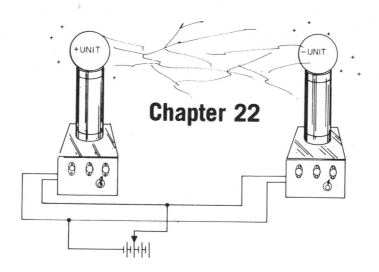

Automatic
Telephone Recording Device (TAT2)

This project turns any standard tape recorder into an automatic telephone recording system that clearly records both sides of a telephone conversation. Ideal for anyone who requires a detailed record of phone conversations, an absolute must for salesmen and order desks. Also, discourages unauthorized use of the telephone. Unit does not consume power from phone line for relay operation as do other similar devices that are apt to interfere with normal phone use. It starts and stops tape recorder everytime the phone is used completely and automatically.

CIRCUIT DESCRIPTION

The device is a dc switch that is normally on via the forward biasing of Q1 via R3 (Fig. 22-1 and Table 22-1). Q1 now clamps Q2 into a forward state by biasing its complement well into a saturated state via R4. The dc switch is turned off via a negative voltage above that of the zener (D1). This voltage is usually about 48 and is the on-hook value of the phone line. This negative voltage overrides

the effect of R3 and keeps the circuit "off". When the phone is off the hook, the 48 volts drops to 10 volts, that is below the zener voltage of D1 and R3 now turns the circuit on. Audio signal is via attenuator resistor R1 and dc isolating capacitors C1, C2.

The device is really nothing more than a high impedance switch that isolates the recording controlled device from the phone line via some relatively simple electronic circuitry. It requires no battery and obtains power for operating via the remote jack that in most recorders is a source of 6 volts. When clamped to ground it initiates recorder operation. The unit nicely interfaces with most portable cassette recorders providing they contain a remote control jack.

CONSTRUCTION STEPS

1. Layout perfboard as shown and assemble using standard wiring and soldering techniques (Figs. 22-2 and 22-3).
2. Attach P1 and P2 to respective points. P1

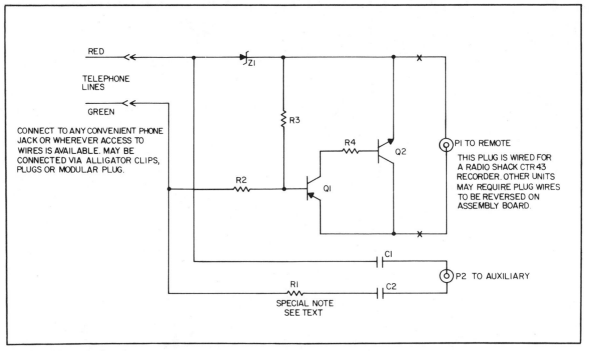

Fig. 22-1. Circuit schematic.

is usually a 2.5 mm plug while P2 is a 3.5 mm respectively for "remote" and "aux" inputs of most recorders.

3. Verify proper wiring and soldering.

4. Obtain a recorder and commence playing a tape.

5. Plug in P1 and P2 and note recorder still playing. Note: some recorders may be oppositely polarized requiring that P1 leads be reversed to properly bias the collector of Q2.

6. Connect leads to phone line as shown with phone on hook and note recorder stopping until

Table 22-1. TAT2 Automatic Telephone Recording Device Parts List.

R3	(1)	220 ohm 1/4 watt resistor
R2	(1)	39 k 1/4 watt resistor
R1	(2)	100 k 1/4 watt resistor
C4	(2)	.01 μF/50 V disc cap
Q6	(1)	PN2222 npn silicon
Q5	(1)	PN2907 pnp silicon
Z1	(1)	15 V zener diode
P7	(1)	Sub-mini plug molded diamond tip connector 2.5 mm
CL1,2	(2)	Alligator clips
P8	(1)	Mini-plug molded diamond tip connector 3.5 mm
CA1	(1)	Plastic cap or suitable enclosure
PB1	(1)	1 1/4 × 1 1/6 perfboard .1 × .1
WR4	(18″)	#24 hook-up black wire
WR3	(18″)	#24 hook-up red wire
CASTO		Potting castolite or paraffin wax

Complete kit with optional PC board available through Information Unlimited, Inc., P.O. Box 716, Amherst, N.H. 03031, write or call 1-603-673-4730 for pricing and delivery.

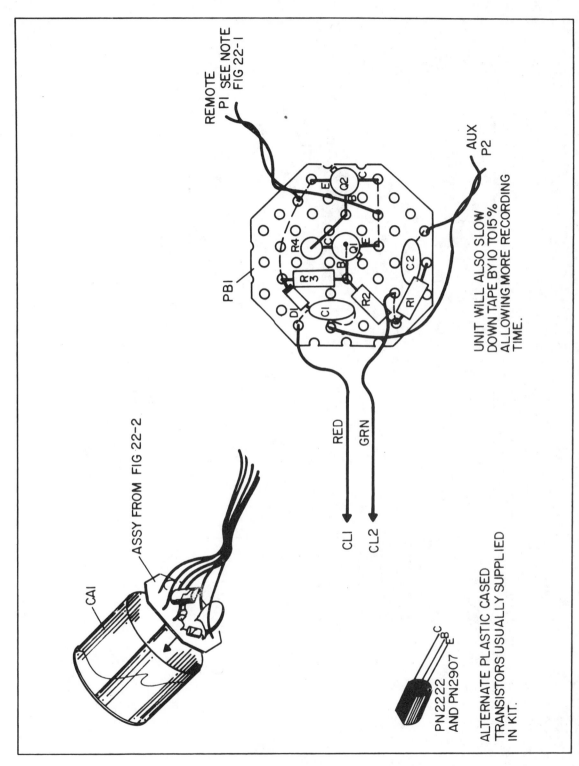

REMOTE
P1 SEE NOTE
FIG 22-1

AUX
P2

UNIT WILL ALSO SLOW
DOWN TAPE BY 10 TO 15%
ALLOWING MORE RECORDING
TIME.

PB1

R4

Q1

R3

C2

R2

R1

D1

C1

RED

GRN

CL1

CL2

ASSY FROM FIG 22-2

CA1

PN 2222
AND PN2907

ALTERNATE PLASTIC CASED
TRANSISTORS USUALLY SUPPLIED
IN KIT.

Fig. 22-2. Assembly layout.

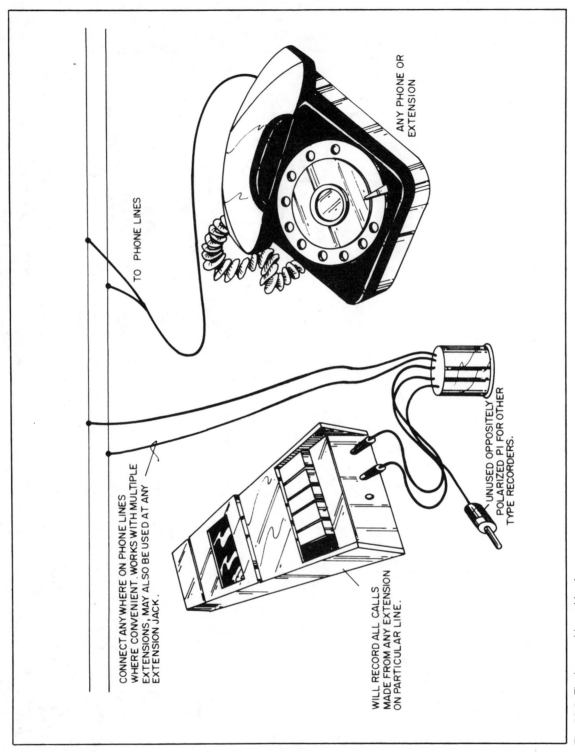

TO PHONE LINES

ANY PHONE OR
EXTENSION

CONNECT ANYWHERE ON PHONE LINES
WHERE CONVENIENT. WORKS WITH MULTIPLE
EXTENSIONS, MAY ALSO BE USED AT ANY
EXTENSION JACK.

WILL RECORD ALL CALLS
MADE FROM ANY EXTENSION
ON PARTICULAR LINE.

UNUSED OPPOSITELY
POLARIZED PI FOR OTHER
TYPE RECORDERS.

Fig. 22-3. Final assembly and hookup.

phone is lifted off of hook.

      7. Set recorder in the record mode and make a phone call. Note recorder starting instantly when phone is off hook and stopping when phone is placed on hook.

      8. Note that P1 should only be inserted in recorder during the record time and that the other recorder functions such as rewind, fast forward, etc., should be done with P1 removed.

      9. Play back previously recorded tape and note absence of excessive hum and distortion. Note both sides of conversation are readable. It may be necessary to decrease the value of R1 to obtain more volume or to increase it if phone is overloading the recorder. Also test other recorder input to determine best combination. The "aux" input of most recorders is the least sensitive while the "mic" is most sensitive.

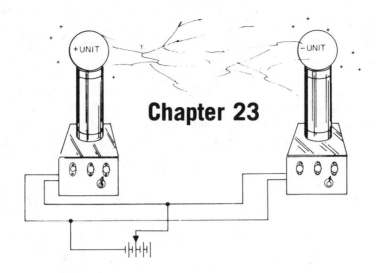

Chapter 23

Snooper Phone Listening Device (SNP-2)

This project shows how to construct a simple electronic device intended as an intrusion detection and listening device for checking home, office, etc., while away or on vacation, etc. It may also be used to trigger other electrical devices. The circuitry does not require a tone encoder. It does not provide for selected activation by a particular caller. This limits the device from being used for illegal purposes and allows the sale of the kit or fully assembled unit without the requirement for special authorization.

Unlike the controversial Infinity Transmitter, our system without the decoder and encoder circuitry once connected to the phone lines will open a mike to anyone who happens to dial the number. This feature obviously negates the use of the device for illegal interception of oral communications. Even though the phone does not ring it would not be long before the potential victim of an illegal surveillance would be aware of the setup. It is therefore quite obvious that the applications are strictly limited to use as a security device checking for unauthorized intrusion or the checking of household appliances by placing a pick-up device in certain favorable locations to overhear these desired sounds and noises. I am sure that anyone with this capability would breathe a sigh of relief to be able to dial his home or office phone while away and hear the familiar sounds of appliances and other systems properly performing their duties and functions.

CIRCUIT OPERATION

Your SNP-2 circuit (Fig. 23-1 and Table 23-1) consists of a high-gain amplifier fed into the telephone lines via transformer (T1). The circuit is initiated by the action of a voltage transient pulse occurring across the phone line at the instant the telephone circuit connection is made. This transient is a voltage change of from 48 Vdc to about 5 Vdc and usually occurs before the ring signal. It is this change that immediately triggers a timer (TM1) whose output pin 3 goes positive turning on transistors, Q2 and Q3. Timer TM1 now remains in this state for a period depending on the values of R17 and C13 (usually about 10 seconds for the values shown). You will note that when Q3 is turned

Table 23-1. SNP-2 Snooper Phone Listening Device Parts List.

R1,4,8	(3)	390 k ¼ watt resistor
R2	(1)	5.6 M ¼ watt resistor
R3,5,6	(3)	6.8 k ¼ watt resistor
R7/S1	(1)	5 k pot/switch
R9,16	(2)	100 k ¼ watt resistor
R10	(1)	2.2 k ¼ watt resistor
R13,18	(2)	1 k ¼ watt resistor
R14	(1)	470 ohm ¼ watt resistor
R15	(1)	10 k ¼ watt resistor
R17	(1)	1 M ¼ watt resistor
C1	(1)	.05 μF/25 V disc cap
C2,3,5,6,7	(5)	1 μF 50 V electrolytic cap or tant (preferably non-polarized)
C4,11,12	(3)	.01 μF/50 V disc cap
C8,10	(2)	100 μF @ 25 V electrolytic
C9	(1)	5 μF @ 150 V electrolytic cap
C13	(1)	10 μF @ 25 V electrolytic cap
TM1	(1)	555 timer dip
A1	(1)	CA3018 amp array in can
Q1,2	(2)	PN2222 npn sil transistor
Q3	(1)	D4OD5 npn pwr tab transistor
D1,2	(2)	50 V 1 amp rect. 1N4002
T1	(1)	1.5 k/500 matching transformer
M1	(1)	Large crystal mike
J1	(1)	Phono jack optional for sense output
WR3	(24")	#24 red and black hook-up wire
WR4	(24")	#24 black hook-up wire
CL3,4	(2)	Alligator clips
CL1,2	(2)	6" battery snap clips
PB1	(1)	1 ¾ × 4 ½" .1 × .1 perfboard
CA1	(1)	5 ¼ × 3 × 2 ⅛ grey enclosure fab (Fig. 23-4)
WR15	(12")	#24 buss wire-use for wiring Fig. 23-2
KN1	(1)	Small plastic knob
BU1	(1)	Small clamp bushing
B1.2	(2)	9 volt transistor battery or 9 V ni-cad battery

Complete kit minus batteries available through Information Unlimited, Inc., P.O. Box 716, Amherst, N.H. 03031, write or call 1-603-673-4730 for pricing and delivery.

A selective activated device using an encoder is available from Information Unlimited, Inc. #INF1K—Infinity Transmitter Kit.

on by timer TM1 that a simulated "off-hook" condition exists by the switching action of Q3 connecting the 500-ohm winding of the transformer directly across the phone lines. Simultaneously Q2 now clamps the ground of A1, amplifier, and Q1, output transistor, to the negative return of B1, B2, therefore, enabling this amplifier section. Note that B2 is always required by supplying quiescent power to TM1 during normal conditions. Systems is off/on controlled by S1.

A crystal mike picks up the sounds that are fed to the first two transistors of the A1 array connected as an emitter follower driving the remaining two transistors as cascaded common emitters. Output of the array now drives Q1 capacitively coupled to the 1500 ohm winding of T1. R7 controls the pick-up sensitivity of the system.

Diode D1 is forward biased at the instant of connection and essentially applies a negative pulse at pin 2 of TM1, initiating the cycle. D2 clamps any high positive pulses. C9 dc-isolates and desensitizes the circuit. The system described will oper-

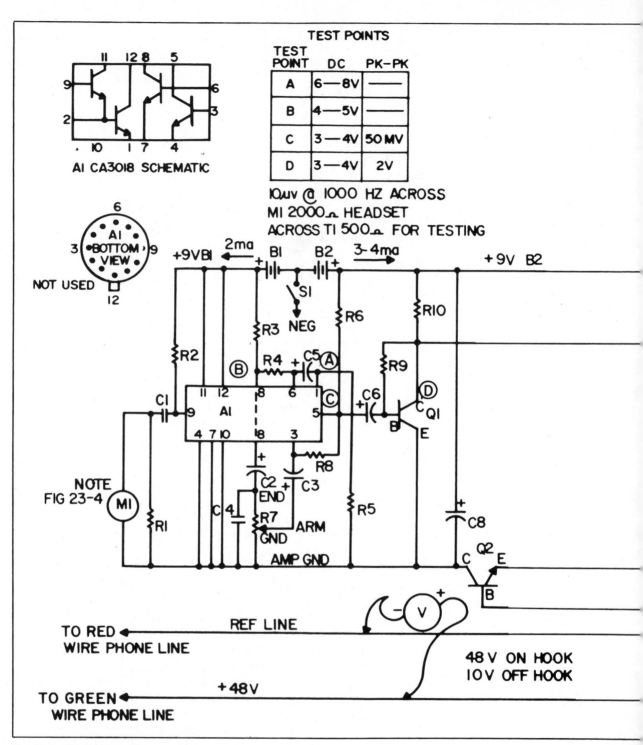

Fig. 23-1. Circuit schematic.

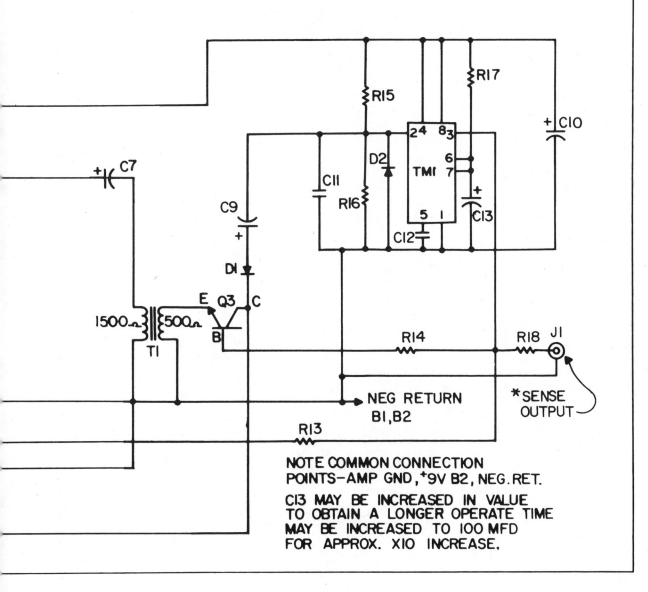

* THE "SENSE" OUTPUT IS INTENDED TO INTERFACE WITH OUR MODEL
SCU-I SENSE CONTROL ACTIVATING SWITCH. THIS ALLOWS YOUR UNIT
TO EXTERNALLY CONTROL OTHER ELECTRICAL DEVICES.

R15

R17

C10

C7

TMI

R16

C11

D2

C9

C13

C12

DI

E Q3 C

B

1500Ω   500Ω

TI

R14

R18   JI

NEG RETURN
BI,B2

*SENSE
OUTPUT

R13

NOTE COMMON CONNECTION
POINTS-AMP GND, +9V B2, NEG.RET.

C13 MAY BE INCREASED IN VALUE
TO OBTAIN A LONGER OPERATE TIME
MAY BE INCREASED TO 100 MFD
FOR APPROX. XIO INCREASE.

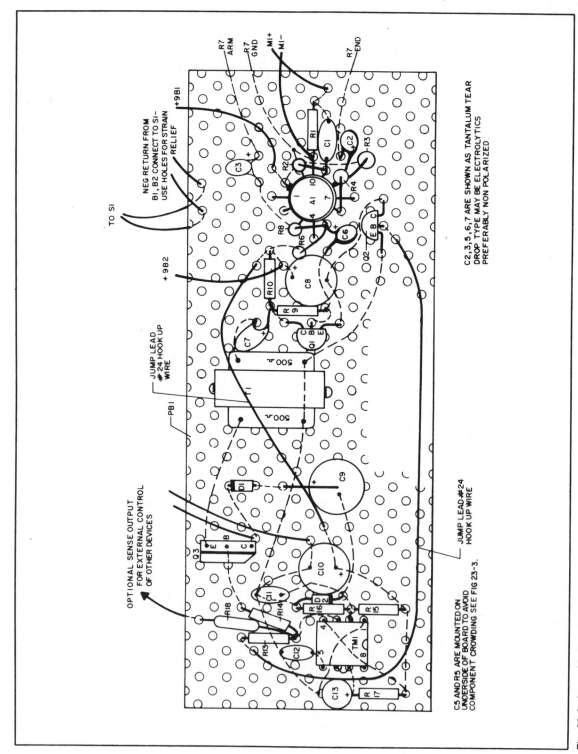

Fig. 23-2. Assembly board.

314

ate when any incoming call is received without the phone ever ringing.

CONSTRUCTION STEPS

1. Identify all components (Fig. 23-2). Layout of this circuit should be followed as shown, especially positioning and lead length of A1, R7 and input components. Use leads of components wherever possible.

2. Fan out the leads of A1 and note position of tab (adjacent to pin 12). Insert as shown with pin 1 and 7 opposite one another and 4 and 10. Note that the lead layout forms a square with pin 1, 4, 7, and 10 being the corners.

3. Connect pins 4, 7 and 10 as shown in Fig. 23-3. Also wire R5 and C5 as shown. Note polarity of C5.

4. Insert and wire components up to and including T1 transformer. Temporarily connect R13 to (+9 V B2) to turn on Q2. Note polarity on all tantalums and electrolytic capacitors if not using nonpolarized.

5. Connect and twist a short pair of wires to board as shown connecting to mike element M1. Note ground lug of mike going to wire that connects to pin 10 of A1. Also connect short leads from underside of board to R7 as shown. Connect C4 across R7 (Fig. 23-3).

6. Connect negative wires of battery clips to S1. Strain-relieve by inserting through holes in perfboard as shown. Connect positive wires (+9) of clips as shown.

7. Check carefully all wiring and soldering to this point. Obtain a pair of 2000 ohm headsets or a scope and connect across 500 ohms winding of T1.

8. Insert one of the contacts of B1 into its respective clip and connect meter in series with other contacts (Fig. 23-3). Measure 1-2 mA with S1 on. Repeat with B2 measuring 3 to 4 mA. Check on/off action of S1.

9. With headsets connected across T1, slowly adjust R7 until sounds are clearly heard. Advance R7, noting no feedback or motorboating (with the exception of perhaps at full gain). Hum may be present if unit is near power lines. Headsets

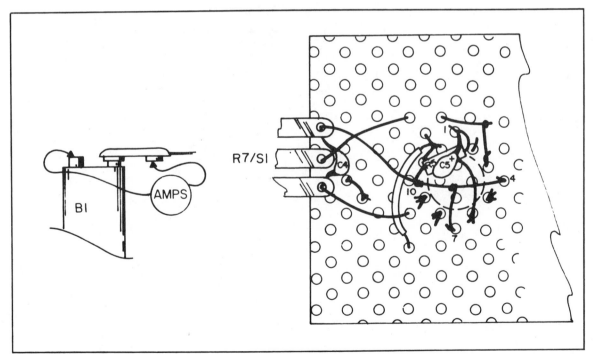

Fig. 23-3. Detail of A1 wiring.

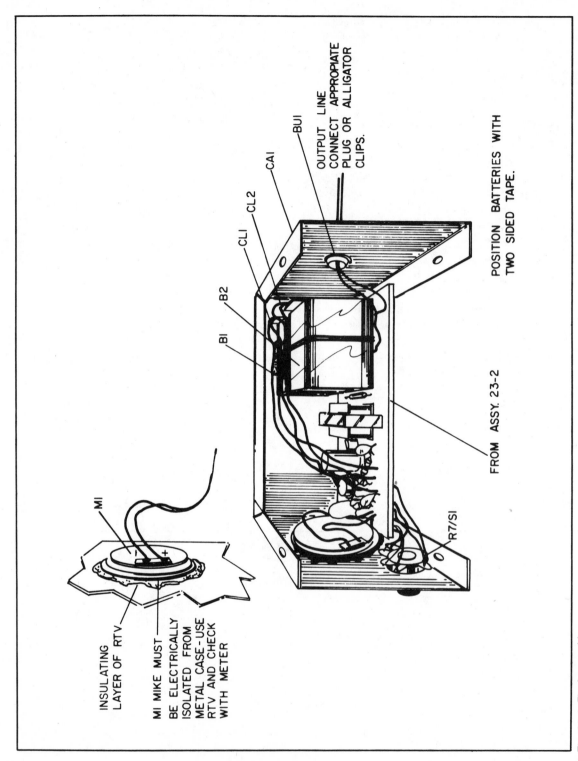

CL1
CL2
CA1
BU1

OUTPUT LINE
CONNECT APPROPRIATE
PLUG OR ALLIGATOR
CLIPS.

B2
B1

M1

INSULATING
LAYER OF RTV

M1 MIKE MUST
BE ELECTRICALLY
ISOLATED FROM
METAL CASE - USE
RTV AND CHECK
WITH METER

FROM ASSY. 23-2

R7/S1

POSITION BATTERIES WITH
TWO SIDED TAPE.

Fig. 23-4. Final assembly.

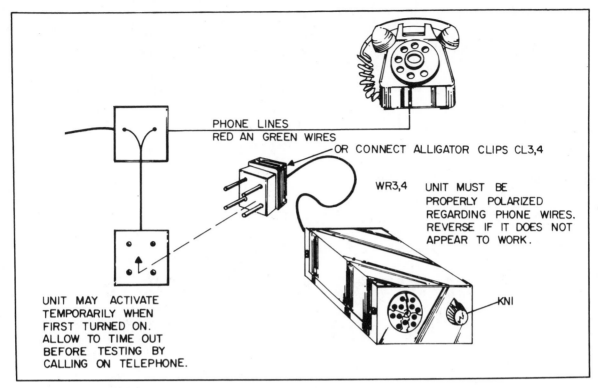

PHONE LINES
RED AN GREEN WIRES

OR CONNECT ALLIGATOR CLIPS CL3,4

WR3,4   UNIT MUST BE
PROPERLY POLARIZED
REGARDING PHONE WIRES.
REVERSE IF IT DOES NOT
APPEAR TO WORK.

KNI

UNIT MAY ACTIVATE
TEMPORARILY WHEN
FIRST TURNED ON.
ALLOW TO TIME OUT
BEFORE TESTING BY
CALLING ON TELEPHONE.

Fig. 23-5. Suggested hookup.

may acoustically leak causing painful squealing and screeching.

10. Observe test points in sketch if necessary for troubleshooting. Output should be crystal clear with more than enough gain. Once this section is functioning properly proceed to the next step.

11. Connect and solder the remaining components to perfboard. Note that extra holes must be drilled for every other pin for Timer TM1. Insert observing position of pin 1 and do not overheat.

12. Connect leads that are to be connected to phone lines. Observe color codes and polarity. Green to collector of Q3, red to negative return. These leads are strain-relieved by passing through perfboard holes as shown in Fig. 23-2.

13. Verify all wiring for errors and quality of solder joints. Turn S1 to "off" and reinsert batteries.

To test this unit on the phone requires a buddy to call or the use of two separate phone lines in the same house. Remember with the unit connected to the phone lines and S1 turned "on" any call made will *not* cause the phone to ring, but instead will turn on the mike and amplifier enabling the caller to hear all sounds in the area until Timer TM1 resets the system. When testing it is a good idea to connect a high resistance voltmeter (20,000 ohm/volt) across the lines as shown in the schematic observing the correct polarity. As soon as the telephone connection is made, this voltage will drop from 48 to below 10 developing a negative pulse at pin 2 of TM1.

The unit is now ready to put into an aluminum case. Fabricate as shown in Fig. 23-4 and secure assembly board with RTV, etc. Strain-relieve wires to phone line with a clamp bushing. Place battery in position with the strips of foam rubber, etc. Connect to telephone lines (Fig. 23-5).

A complete kit of the more sophisticated Infinity Transmitter capable of being selectively activated by a tone encoder allowing normal use of the telephone is available from Information Unlimited, Inc. P.O. Box 716, Amherst, N.H. 03031.

317

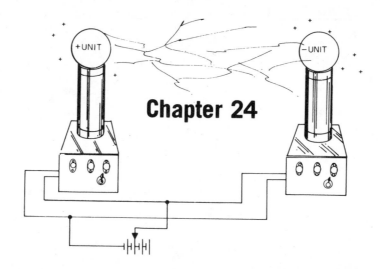

Chapter 24

Long-range Super-sensitive Parabolic Microphone (PWM3)

This project is intended for the selective listening of sounds and voices coming from a particular area or direction. These selected sounds, by virtue of the geometry of the device, are focused to a mike element and fed to a low-noise, high-gain amplifier powering headsets, recorders, transmitters, etc.

Sound, like light, due to phase incoherence has a tendency not to focus at the same point for a given band of frequencies. This is overcome to some extent by the use of a large mike located at the focal point of the unit. Even though the exact focal point is not the same for all frequencies, little difference in performance is gained by changing it for selected frequencies.

The advantages of a system of this type is the ability to capture a large cross section of sonic energy equal to that intercepted by the area of the reflector. This energy is focused onto a smaller area (mike element) where its displacement or amplitude value is increased by the quotient of the captured area divided by the area focused down to. Therefore, an acoustical gain of:

$$\text{Gain dB} = 20 \, \text{Log} \, \frac{\text{Area Parabola}}{\text{Area Mike}}$$

The acoustical gain of the device is about 20 dB. The parameters are based on a narrow band of frequencies and will be less for normal sounds due to interference and their lower inherent frequency.

A disadvantage of a system of this type is the moderate rejection of back and side noise. Hence, the full capabilities of the system (as far as long distance listening) is often masked by the inability to run the amplifier at high gain. However, on a still winter night, the unit can be run at almost full gain with amazing results, hearing voices and sounds from unbelievable distances. The unit does provide a good positive gain over normal hearing in any situation if the amplifier is set not to overload.

Figures 24-1, 24-2, and 24-3 show several applications using other related equipment described on the materials list of these plans. See Table 24-1.

DESCRIPTION OF OPERATION

Sounds are collected and picked up by reflec-

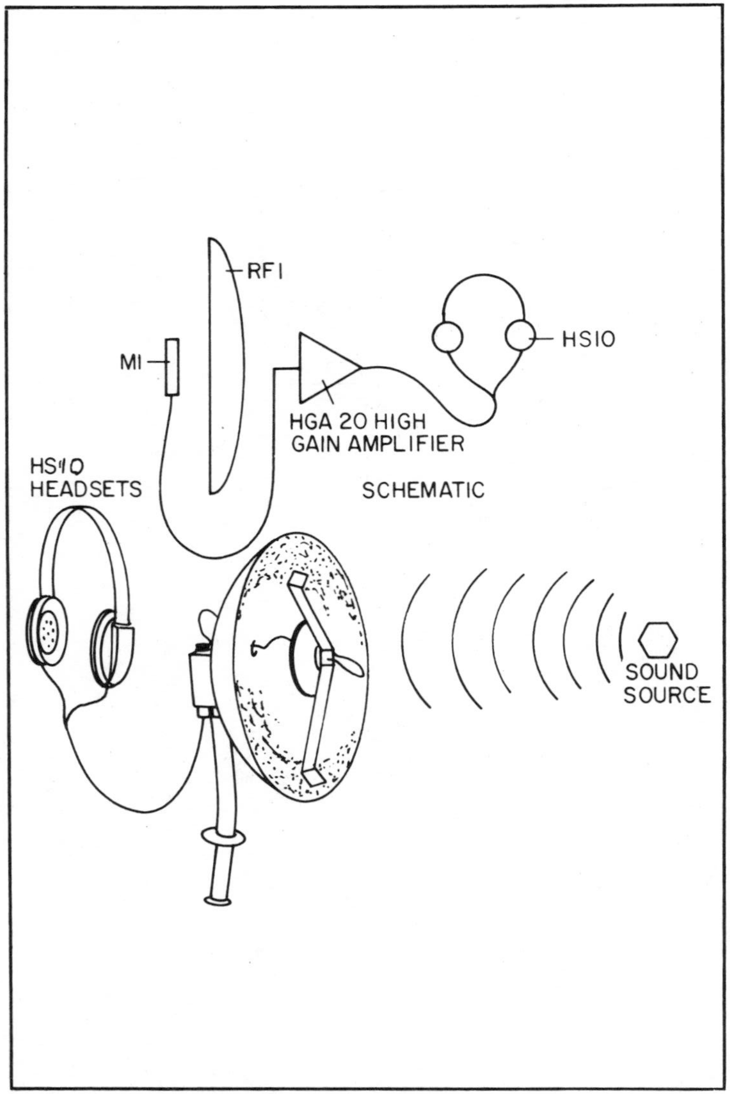

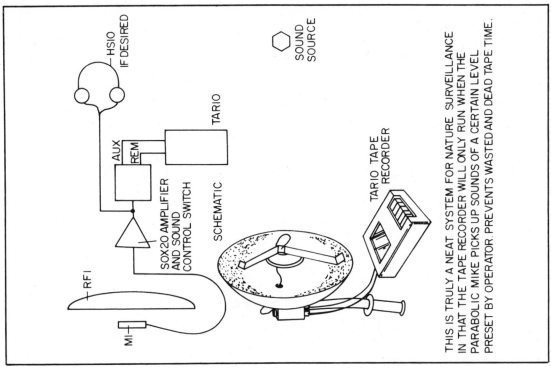

HSIO
IF DESIRED

SOUND
SOURCE

TARIO

AUX
REM

SOX20 AMPLIFIER
AND SOUND
CONTROL SWITCH

SCHEMATIC

RFI

MI

TARIO TAPE
RECORDER

THIS IS TRULY A NEAT SYSTEM FOR NATURE SURVEILLANCE
IN THAT THE TAPE RECORDER WILL ONLY RUN WHEN THE
PARABOLIC MIKE PICKS UP SOUNDS OF A CERTAIN LEVEL
PRESET BY OPERATOR. PREVENTS WASTED AND DEAD TAPE TIME.

Fig. 24-2. Ultrasensitive directional sound activated recorder system.

HSIO

HGA 20 HIGH
GAIN AMPLIFIER

SCHEMATIC

SOUND
SOURCE

RFI

MI

HSIO
HEADSETS

Fig. 24-1. Normal use PM30 parabolic mike picks up sounds
and amplifies them for the headsets.

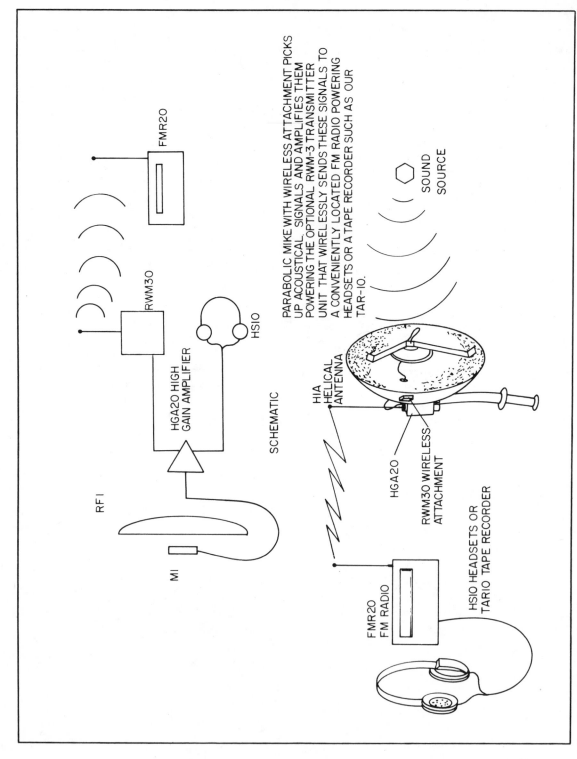

PARABOLIC MIKE WITH WIRELESS ATTACHMENT PICKS UP ACOUSTICAL SIGNALS AND AMPLIFIES THEM POWERING THE OPTIONAL RWM-3 TRANSMITTER UNIT THAT WIRELESSLY SENDS THESE SIGNALS TO A CONVENIENTLY LOCATED FM RADIO POWERING HEADSETS OR A TAPE RECORDER SUCH AS OUR TAR-I0.

SOUND SOURCE

HIA HELICAL ANTENNA

HGA20

RWM30 WIRELESS ATTACHMENT

FMR20 FM RADIO

HSIO HEADSETS OR TARIO TAPE RECORDER

FMR20

RWM30

HSIO

HGA20 HIGH GAIN AMPLIFIER

RFI

MI

SCHEMATIC

Fig. 24-3. Remote listening hookup.

Table 24-1. Long-range Sensitive Parabolic Mike System.

RF1	(1)	24″ clear plastic parabolic reflector
M1	(1)	Large 2″ crystal mike
P1	(1)	RCA phono plug
WR10	(18″)	18″ shielded mike cable
FUN1	(1)	3″ nylon funnel or aluminum funnel
BU1	(1)	⅜″ bushing
FRUB1		Foam rubber
RB1	(1)	Rear bracket Fig. 24-5
FB1	(1)	Front bracket Fig. 24-5
HA1	(1)	Handle 15″ by ⅞″ conduit tubing Fig. 24-5
HG1	(1)	Handgrip for above handle—use bicycle, etc.
SW12	(4)	6 32 × ½″ self tapping type F
SW2/NU1	(2)	6 32 × ½″ screw and nut
SW8	(2)	6 × ¼″ sheet metal
HGA20	(1)	High-gain amplifier assembled and tested
HGA2K	(1)	High-gain amplifier kit—Use for Figs. 24-1 and 24-3
SOX20	(1)	High-gain amplifier and sound switch assembled and tested.
SOX2K	(1)	High-gain amplifier kit—Use for Fig. 24-2
RWM3K	(1)	Remote wireless repeater transmitter kit— Use for Fig. 24-3.

Optional items as described for above systems

   TAR10—Cassette recorder—Intended to interface with any of our existing devices. External jacks for other supporting equipment. Sound operated when used with our SOX2.

   FMR20—Portable/battery or ac AM/FM radio—Peaked up and retuned for operation on high end of band. Receiver interfaces with all our equipment described and provides excellent performance. External jacks for recorders, auxiliary inputs and other functions. Complete with telescoping antenna.
   HS10—Cushioned headphones—excellent quality.

   HA10—Helical antenna—for short range use with RWM3 up to 500 feet.

   Complete kit of above parts is available from Information Unlimited, Inc., P.O. Box 716, Amherst, N.H. 03031. Write or call 1-603-673-4730 for pricing and delivery.

tors (RF1). This collected energy is now focused down to the microphone element (M1). High school mathematics describe the properties of a parabolic shaped device stating that all sound waves that are in parallel will be focused to a common focal point (the microphone element).

This phenomenon is even more pronounced when using radio, or light waves. In order for efficient collection and focusing to take place the reflector should be as large as possible in relation to the wavelength of the energy in interest. A given diameter reflector will be more efficient when receiving the highest frequencies of the sound target.

The reflector described is 24″ in diameter and gives reasonable gain at voice frequencies. However, far better results are obtained when listening to higher frequency sounds that are produced by birds, bats and insects. A system targeted for voice frequencies would work far better if it had a reflector of from 4′ to 10′ in diameter (impractical in most cases).

The microphone now converts the sound pressure energy into electrical waves that are fed into a high-gain amplifier via shielded cable WR10. The electronics shown here is our (SOX2) high-gain amplifier/sound controlled switch. This device is a separate project and is described in detail in the SOX2 project plans. A simpler approach is our model HGA2 amplifier without the option of the sound control. See Figs. 24-1 and 24-2.

The output of the SOX2 can be fed into a tape recorder with its control plug automatically ac-

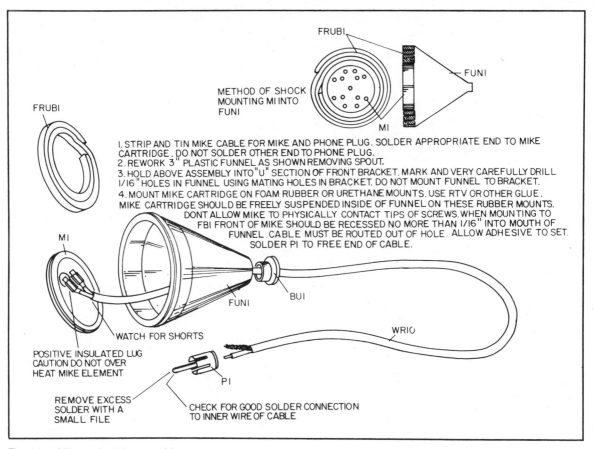

Fig. 24-4. Mike and cable assembly.

tivating the tape motors. Gain control is accomplished via R7/S1 (external control) with a secondary audio-level control inside of the SOX2. This allows setting of the threshold for tape recorder activation with a separate control for the desired levels of sound to be recorded.

Using the (HGA2) High-Gain Amplifier (as in Fig. 24-1) only involves plugging in the headsets (HS1) and simply adjusting the gain for the lowest value that provides comfortable listening. Figure 24-3 shows a secondary low level output of the HGA2 being fed into our RWM3 remote wireless repeater transmitter. This is a positive asset when performing nature studies in mosquito and bug infested areas allowing listening over any remote or auto FM radio within range (usually up to 1 mile or more).

CONSTRUCTION STEPS

1. Select the particular system you desire from Figs. 24-1, 24-2, or 24-3. Obtain or assemble and test the required support electronics necessary per the instructions. Verify proper operation.

2. Assemble mike and cable as shown in Fig. 24-4.

3. Assemble rear bracket RB1, FB1 and HA1 (Figs. 24-5 and 24-6). Note angle on tabs with mounting holes.

4. Mount electronics to HA1 as shown in Fig. 24-7. Use two sheet metal screws (SW12) and position as shown. The HGA2 and SOX2 will require two 3/16″ holes drilled in the bottom of the channel section of the enclosure to mate with the holes in handle HA1. Note that it must be offset as shown to clear the reflector.

5. Assemble RB1 and FB1 assembly together sandwiching the RF1 reflector in between. Make sure center of reflector coincides with the center of the bracket assembly or the device will look lopsided and lack in performance. Use screw and nuts, SW2/NU1.

6. Route mike cable through hole in reflector and plug into appropriate jack in SOX2 or HGA2 adjacent to "adjust knob".

7. Plug headsets into connector at opposite end from mike jack. Note that 8-ohm headsets require a matching transformer. Complete headset/transformer/plug ready to use is available through Information Unlimited, Inc. Order HS10. If SOX20 is used connect TAR10 or equivalent recorder as shown.

8. Rotate "adjust knob" until it clicks "on" and slowly advance until sounds are heard. *Be careful of your ears as this device can cause pain if gain is set too high for anticipated sounds.* Rotate "adj

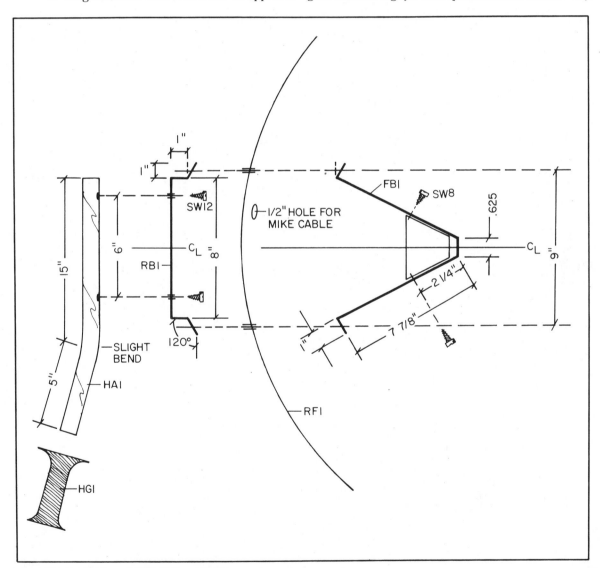

Fig. 24-5. Fabrication of RB1, FB1, and HA1.

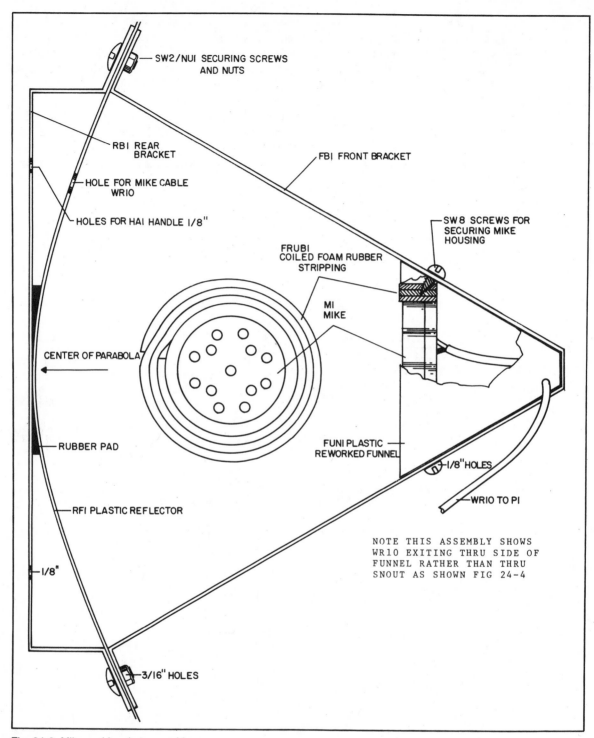

SW2/NUI SECURING SCREWS AND NUTS

RBI REAR BRACKET

FBI FRONT BRACKET

HOLE FOR MIKE CABLE WRIO

HOLES FOR HAI HANDLE 1/8"

SW8 SCREWS FOR SECURING MIKE HOUSING

FRUBI COILED FOAM RUBBER STRIPPING

MI MIKE

CENTER OF PARABOLA

RUBBER PAD

FUNI PLASTIC REWORKED FUNNEL

1/8" HOLES

WRIO TO PI

RFI PLASTIC REFLECTOR

NOTE THIS ASSEMBLY SHOWS WR10 EXITING THRU SIDE OF FUNNEL RATHER THAN THRU SNOUT AS SHOWN FIG 24-4

1/8"

3/16" HOLES

Fig. 24-6. Mike and bracket assembly.

324

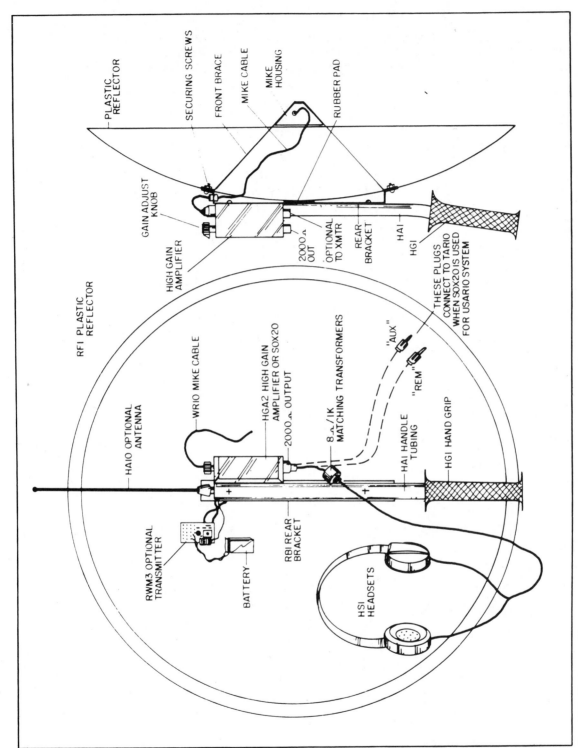

PLASTIC REFLECTOR

SECURING SCREWS

FRONT BRACE

MIKE CABLE

MIKE HOUSING

RUBBER PAD

GAIN ADJUST KNOB

HIGH GAIN AMPLIFIER

2000 Ω OUT

OPTIONAL TO XMTR

REAR BRACKET

HAI

HGI

THESE PLUGS CONNECT TO TARIO WHEN SOX20 IS USED FOR USARIO SYSTEM

RF1 PLASTIC REFLECTOR

HAIO OPTIONAL ANTENNA

WRIO MIKE CABLE

HGA2 HIGH GAIN AMPLIFIER OR SOX20

2000 Ω OUTPUT

8 Ω/1K MATCHING TRANSFORMERS

"AUX"

"REM"

HAI HANDLE TUBING

HGI HAND GRIP

RWM3 OPTIONAL TRANSMITTER

BATTERY

RBI REAR BRACKET

HSI HEADSETS

Fig. 24-7. Final assembly.

325

T knob" just back of point where recorder starts when using SOX20. Also you may wish to experiment with the setting of R20 in the SOX20 to get the proper recording level for the recorder while at the same time obtaining the correct recorder activating level for the desired sounds.

9. To use your PM3, simply point it at the area under observation and advance the gain adjust knob to the point of comfortable listening.

SOME CAUTIONS WHEN USING THIS DEVICE

☐ Don't run the gain up to a point where background noise masks the desired pick-up.

☐ Don't expect the unit to be directional inside a building or enclosure since interfering echoes will mask most of the directional properties.

☐ Don't expect to obtain the maximum range capability of the device when operated in excessively noisy environments. Remember the reflector provides about 20 dB gain front-to-back and slightly more to the sides. If the resultant interfering noise exceeds this value, obviously the usable overall gain even in front of the unit can only be 20 dB below rear and side noise.

☐ Don't use the unit to intercept oral communications without the consent of the parties being listened to. Such surveillance is illegal.

☐ Don't use the unit in high winds as it will not work and can be blown right out of your hands.

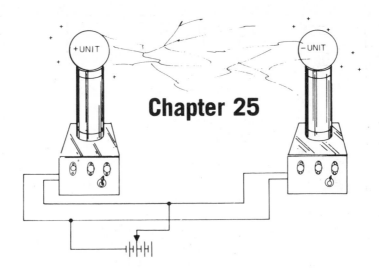

Invisible Beam
Property Guard Alarm System (ISP2)

This project shows how to construct a device for guarding against unauthorized intrusion of guarded areas. It detects people, animals, vehicles, etc., and is designed for driveways, walkways, paths, or wherever potential traffic or unauthorized trespassing may take place. It can be made to detect either exiting or entering traffic. The device when properly constructed can guard a space of up to several hundred feet in width. It is not bothered by lights, sunlight, snowflakes, leaves, or other potentially false triggers. It is virtually fail-safe in that any fault results in the alarm being activated.

When received the signal is fed to a special selective receiving circuit whose output remains in a normally low state until interrupted. The signal is fed to the alarm box (usually placed by the bedside or wherever convenient). The alarm box can also house the batteries that operate the receiver section. This method prevents defeating the system by removing the power. It should be noted that the alarm box will always sound even without the proper functioning of the support circuitry such as the receiver and transmitter, thus indicating a fault or attempted disabling of the system.

TRANSMITTER SECTION

The transmitter section is shown built on a piece of perfboard and mounted inside of a tubular enclosure. See Figs. 25-1 through 25-4. (The builder may have other ideas about enclosing.) Note position of the light-emitting diode (LA1) in respect to the lens (LE1). The circuit is nothing more than a low-duty-cycle relaxation oscillator designed to minimize battery drain yet provide sufficient signal for reliable operation over considerable distance. The batteries used in the system shown are two "C" cells in a series arrangement for 3 volts. The pulser will run for several months between battery changes. Suggested batteries are Radio Shack #23-581.

1. Fabricate PB1, RP1, MP1, EN1, FU1, and CA2 as shown and described in the parts list (Tables 25-1 through 25-4).

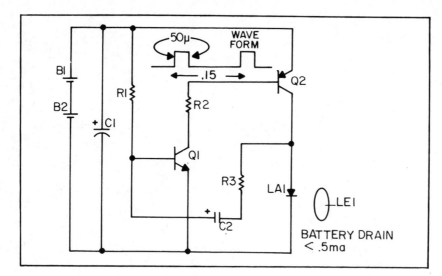

Fig. 25-1. Schematic XMTR.

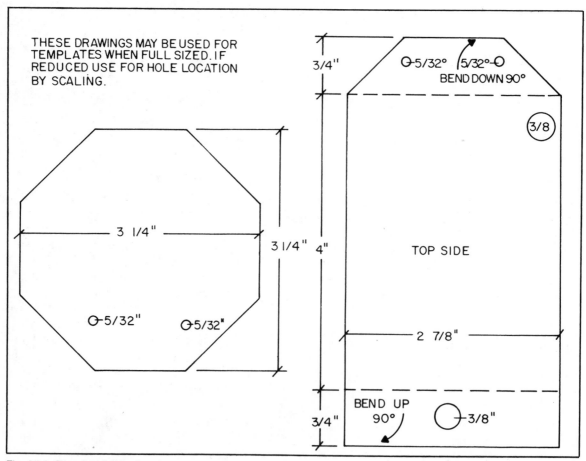

THESE DRAWINGS MAY BE USED FOR TEMPLATES WHEN FULL SIZED. IF REDUCED USE FOR HOLE LOCATION BY SCALING.

Fig. 25-2. Template fabrication drawing.

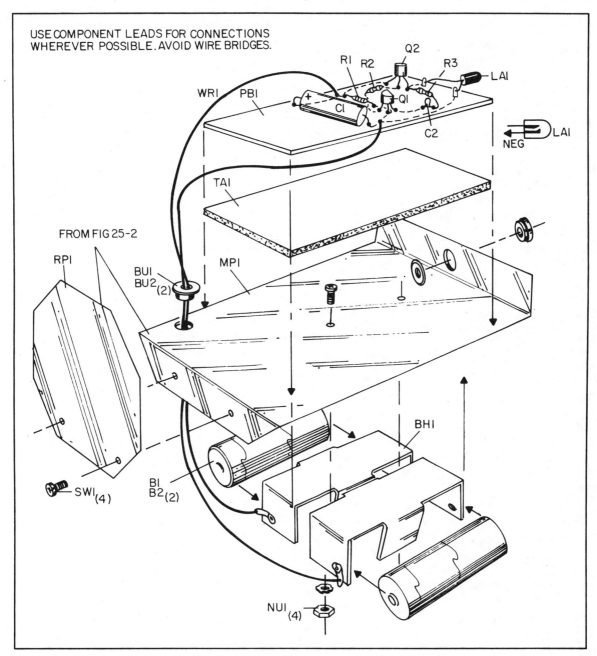

USE COMPONENT LEADS FOR CONNECTIONS WHEREVER POSSIBLE. AVOID WIRE BRIDGES.

Fig. 25-3. Transmitter wiring blowup.

2. Assemble board using standard wire and soldering techniques. Note polarity of LA1, C2, and C1. Wires to battery pack are shown passing through bushing (BU2) on mounting plate (MP1).

Note centering position of LA1 in respect to lens (LE1). This must be reasonable close for ease in final alignment. It should be mentioned at this time that temporary use of a visible-light-emitting diode

329

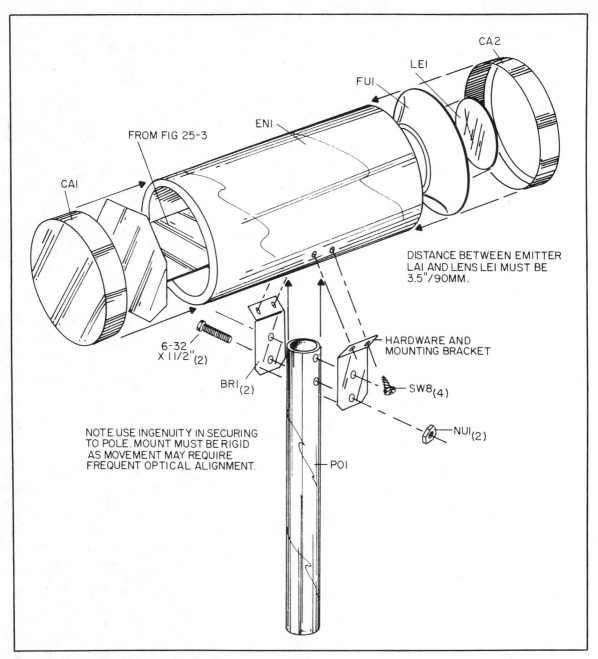

CA2

LEI

FUI

ENI

FROM FIG 25-3

CAI

DISTANCE BETWEEN EMITTER
LAI AND LENS LEI MUST BE
3.5"/90MM.

6-32
X 1 1/2"(2)

HARDWARE AND
MOUNTING BRACKET

BRI(2)

SW8(4)

NOTE USE INGENUITY IN SECURING
TO POLE. MOUNT MUST BE RIGID
AS MOVEMENT MAY REQUIRE
FREQUENT OPTICAL ALIGNMENT.

NUI(2)

POI

Fig. 25-4. Transmitter final assembly drawing.

such as the HLMP 3750 may make initial alignment easier due to being able to see the output in darkness.

    3. Connect battery as shown and note LA1 flashing at about 10 pps. (Just about at the flicker fusion rate.) Note waveform diagram. When the receiver section is properly working you can direct the output to the phototransistor and note the

Table 25-1. ISP2 Transmitter Section Parts List.

R1	(1)	390 k ¼ watt resistor
R2	(1)	27 ohm ¼ watt resistor
R3	(1)	10 ohm ¼ watt resistor
C1	(1)	470 μF/16 V electrolytic cap
C2	(1)	47 μF/35 V tant cap
Q1	(1)	PN2222 silicon npn transistor
Q2	(1)	PN2907 silicon pnp transistor
LA1	(1)	XC880 240 beam IR emitter-use visible red
Visible version		HLMP 3750 for initial alignment if required
BH1,2	(2)	Dual "C" cell battery holders only one shown Fig. 25-3
B1,2	(2)	1.5 volt "C" cells—Radio Shack #23-581
LE1	(1)	Lens 54 × 89 mm shown Fig. 25-4.
BU1,2	(2)	⅜" plastic bushings
RP1	(1)	Fab as per Fig. 25-2
MP1	(1)	Fab as per Fig. 25-2
PB1	(1)	3" × 1 ¾" perfboard
EN1	(1)	8" × 3 ½" OD PVC tube fab per Fig. 25-4
FU1	(1)	3 ½" plastic funnel—fab to retain lens LE1, Fig. 25-4
CA1	(1)	3 ½" plastic cap
CA2		3 ½" plastic cap—remove end as shown with sharp knife using walls of EN1 as a guide when cutting
SW1	(2)	6 32 × ¼" PH screws
NU1	(2)	6 32 nuts
TA1		Piece of two sided tape for securing board to mounting plate
WR1	(12")	#24 hook-up wire

Table 25-2. ISP2 Receiver Section Parts List.

R4	(1)	100 ¼ watt resistor
R5,12	(2)	10K ¼ watt resistor
R6	(1)	500 k trimpot
R7	(1)	470 k ¼ watt resistor
R8,9	(2)	6.8 k ¼ watt resistor
R10	(1)	15 k ¼ watt resistor
R11,13	(2)	1 k ¼ watt resistor
C3,5	(3)	.01 μF/25 V disc cap
C4	(1)	.1 μF/25 V disc cap
C6	(1)	10 μF/25 V electrolytic cap
C7	(1)	470 μF/16 V electrolytic cap
Q3,4,5	(3)	PN2222 npn silicon transistors
Q6		Photo transistor L14G3
BU3,4	(2)	⅜" plastic bushings
PB2	(1)	2 ¾ × 2" perfboard
RP2	(1)	Fab as per Fig. 25-2 similar to RP1
MP2	(1)	Fab as per Fig. 25-2 similar to MP1
CA3	(1)	3 ½" plastic cap
CA4	(1)	3 ½" plastic cap—remove end as shown with sharp knife using walls of EN2 as a guide when cutting
LE2	(1)	54 × 89 mm lens
FIL1	(1)	IR filter plastic 3" diameter or equivalent
FU2	(1)	3 ½" plastic funnel—fab to retain lens LE1—Fig. 25-8
EN2	(1)	8 × 3 ½" OD PVC tube fab Fig. 25-8
TA1	(1)	Two-sided tape for mounting board to plate
WR11		3 conductor cable for connecting receiver to system
SW1	(2)	6 32 × ¼" PH screws
NU1	(2)	6 32 kep nuts

"sense" level immediately dropping 1 to 2 volts. Interrupting the beam with a piece of cardboard will immediately cause the sense level to go back to 4 to 5 volts. LE1 is easily mounted in the reworked plastic funnel. The lens is carefully adhered in place as shown using RTV or some other glue. The assembly board when placed as shown should optically align LA1 with LE1. Batteris are simply changed by removing cap (CA1) and sliding out assembly.

RECEIVER SECTION

The receiver section is also built in a tubular enclosure similar to the transmitter section (see Figs. 25-2, 25-5, 25-7, and 25-8). The mounting of the completed enclosure is left to the builder to suit his own need. (See Figs. 25-13 through 25-15 for possible methods.) A phototransistor (Q6) is mounted so that it can be adjusted to the optical axis

Table 25-3. ISP2 Alarm Section.		
R14	(1)	39 k ¼ watt resistor
R15	(1)	180 k ¼ watt resistor
R16	(1)	22 k ¼ watt resistor
R17,23	(2)	10 k ¼ watt resistor
R1 8	(1)	100 k ¼ watt resistor
R19,24	(2)	1 k ¼ watt resistor
R20	(1)	100 k trimpot resistor
R21	(1)	100 ohm ¼ watt resistor
R22	(1)	4.7 k ¼ watt resistor
D1-4	(4)	Signal diode 1N914
C10	(1)	.1 µF/25 V disc cap
C8,11	(2)	.01 µF/50 V disc cap
C9	(1)	100 µF @ 25 V cap
C12	(1)	1000 µF/16 V
Q7,8,10	(3)	Semiconductor npn plastic PN2222
Q9	(1)	Semiconductor npn plastic PN2907
I1	(1)	555 timer dip 8 pin
T1	(1)	Transformer 1 k/8 ohm matching
PB3	(1)	2 ½ × 4 ½ .1 × .1 perfboard
S1,2	(2)	Medium toggle dpst
S3	(1)	Mini pushbutton
J1	(1)	Phono jack
BUS	(1)	Plastic bushing ⅜″
SK1	(1)	8 ohm 2 ¾″ speaker
SCR1	(1)	3″ × 3″ screen mesh
CA1	(1)	5 ¼ × 3 × 2 ⅛ al box

Table 25-4. Power Supply Section (Parts numbers start new sequence).		
R1,2	(2)	1 ohm ½ watt resistor
R3	(3)	27 ohm ½ watt resistor
C1	(1)	470 µF/16 V capacitor
CR1,2	(2)	Semiconductor 50 V 1 amp rect
Z1	(1)	Semiconductor 6.2 V zener
T1	(1)	Transformer 12 V 100 mA power
CO1	(1)	Line cord
BU1	(1)	Cord clamp bushing
WN1,2	(2)	Connector small wire nuts
WR11	(3′)	3 cond #22 wire
CA1	(1)	5 ¼ × 3 × 2 ⅛ grey enclosure
TA1	(2″)	two sided tape
PB1	(1)	2 ¼ × 4 ½″ .1 × .1 perfboard
BU2	(1)	Nylon bushing

of the lens (LE2) and be near its focal point. You will note the lens is mounted similarly as the transmitter section with the addition of IR filter (FIL1). The circuit operates in the following way: Lens (LE2) collects the infrared energy pulses from the transmitter and focuses them onto the phototransistor (Q6). Filter (FIL-1) attentuates ambient background light that would otherwise saturate Q6. These pulses are ac-coupled to a discriminator circuit consisting of Q5 and Q4. As long as Q5 receives

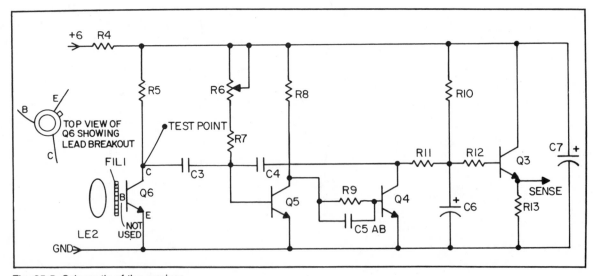

Fig. 25-5. Schematic of the receiver.

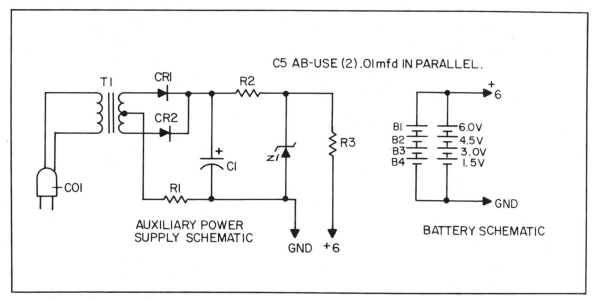

Fig. 25-6. Schematic of the power supply.

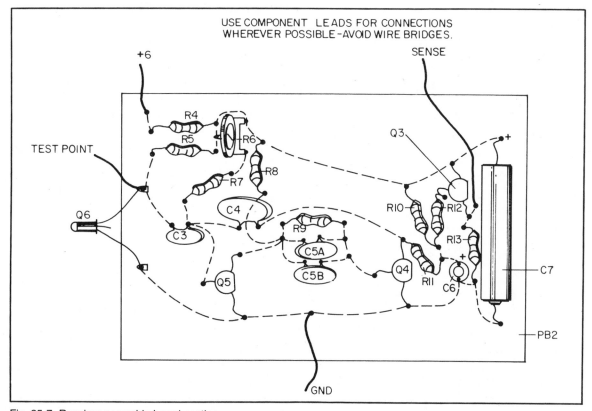

Fig. 25-7. Receiver assembly board section.

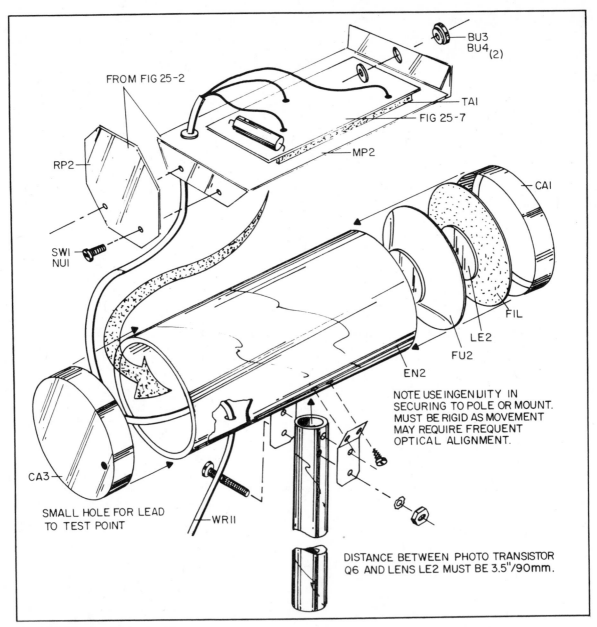

FROM FIG 25-2

BU3
BU4 (2)

TAI

FIG 25-7

MP2

RP2

CAI

SWI
NUI

FIL

LE2

FU2

EN2

NOTE USE INGENUITY IN
SECURING TO POLE OR MOUNT.
MUST BE RIGID AS MOVEMENT
MAY REQUIRE FREQUENT
OPTICAL ALIGNMENT.

CA3

SMALL HOLE FOR LEAD
TO TEST POINT

WRII

DISTANCE BETWEEN PHOTO TRANSISTOR
Q6 AND LENS LE2 MUST BE 3.5"/90mm.

Fig. 25-8. Receiver final assembly.

negative pulses from Q6 it remains "off" with Q4 being "on", holding the junction of R10 and R11 (and Q3 buffer) at a low level. C6 eliminates false triggering due to snowflakes, leaves, etc., and can be varied for faster or slower response depending on use and environment.

When the optical path is broken, Q5 now sees a positive bias turning it "on" and Q4 "off" thus causing Q3 to conduct and place a high level on the "sense" line. R6 selects the threshold of triggering for Q5 and Q4 and is adjusted during final adjustment.

1. Fabricate RP2, MP2, PB2, FU2, and CA4 as shown and described on parts list.

2. Insert solder lugs for wiring Q6 or use leads. Note that Q6 must be on axis with lens (LE2), similar to transmitter sections. Attach to these lugs using full length of Q6 leads for proper positioning without straining.

3. Proceed to assembly board as shown using standard wiring and soldering techniques. Check wiring and solder.

4. Fabricate enclosure as shown and ¼" hole for WR11 as shown near rear. Note holes for screws and bracket for mounting.

5. Secure lens as shown to reworked funnel using a thin layer of RTV or equivalent. Place lens as shown and let set until secured. Check for centering and make sure lens is clean. Attach filter as shown using cutout CA4.

6. Attach board to MP2 plate via two sided tape. Terminate the three-conductor cable to +6, "grd" and "sense" points and strain-relieve with a knot and BU4. A connector should be used for convenience on the end of the cable. Snake out "test point" lead through CA3.

7. Place a flashlight about 30 feet away and shine it into lens (LE2) along optical axis. Position Q6 to intercept focused beam (use a piece of white paper as an indicator). This roughly positions Q6 on the optical axis of LE2. If the fabrication is done with Q6 at optical center, alignment should be automatic.

8. Apply 6 volts as shown and measure 4 to 5 volts on "sense" line.

9. Position operating transmitter next to lens along optical axis of receiver and position until "sense" drops to 1-2 volts. Placing unit where floor abuts at wall provides a good verification of alignment. This verifies that the system is performing correctly. Units should be separated by at least 6 feet for this test.

ALARM SECTION

The alarm section (Figs. 25-9 through 25-11) is built on a piece of perfboard and housed in a small aluminum minibox along with the speaker, on/off switch and alarm duration switch. A bushing at the rear of the box feeds the necessary three-conductor cable to its terminating points on the perfboard.

The circuit performs as follows: A normal low (sense signal) exists with the system receiver intercepting the optical pulses. Diode D1-D4 improves noise immunity and provides better isolation between "sense" line and Q7. C11 eliminates transients that might falsely trigger the system such as fluorescent lights, etc. When the optical path is broken, the "sense" line goes high (4 to 5 volts) and turns Q7 on and initiates I1 timer sounding the alarm. The time "on" period is determined by R17, R18, and C9 and is selectable between approximately 2 seconds and 10 seconds via S1. I1 during its triggered period causes Q10 to conduct, thus turning on the tone oscillator consisting of Q8 and Q9. R20 adjusts the frequency of the tone. S3 is a test to verify system operation.

1. Cut perfboard as shown and drill extra holes for I1 and T1.

2. Mount components as shown and solder. Use component leads for connection where convenient. Use standard wire and solder techniques. Note dashed lines indicate connecting wires beneath board.

3. Fabricate case for speaker using a piece of window screen for grill. Fabricate holes for S1, S2, S3, and bushing BU5. Attach using glue or screws and washers.

4. Connect wires from assembly board to speaker (SK1), switch S1, S2, and terminate three-conductor cable assembly (WR11) to board as shown (+6, sense, and ground). Strain-relieve via a knot. Note that connectors could be used for convenience, as WR11 will probably be quite long.

5. Verify correct wiring and soldering procedures.

6. Connect 6 volts to unit and place S1 in "short" position and S2 on. Depress S3 and note speaker sounding for about 2 seconds. Repeat with S1 in "long" position and note alarm sounding for 10 seconds before turning off.

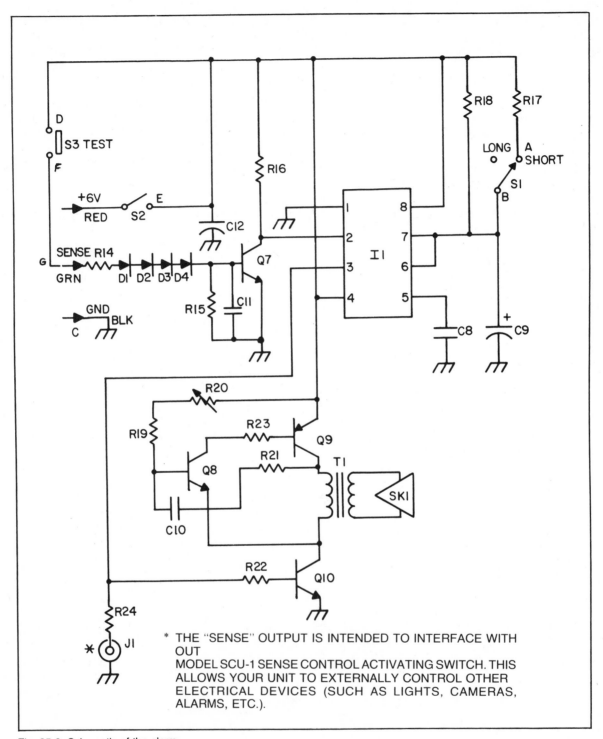

* THE "SENSE" OUTPUT IS INTENDED TO INTERFACE WITH OUT
MODEL SCU-1 SENSE CONTROL ACTIVATING SWITCH. THIS
ALLOWS YOUR UNIT TO EXTERNALLY CONTROL OTHER
ELECTRICAL DEVICES (SUCH AS LIGHTS, CAMERAS,
ALARMS, ETC.).

Fig. 25-9. Schematic of the alarm.

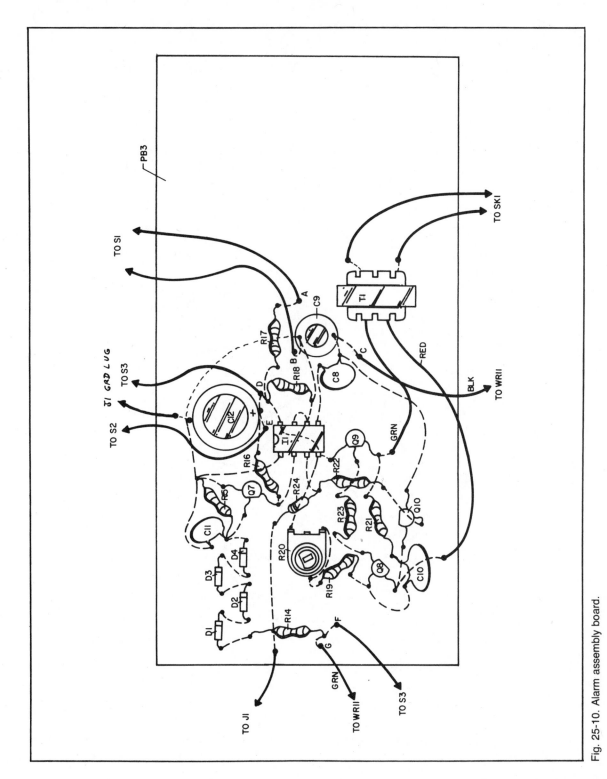

Fig. 25-10. Alarm assembly board.

337

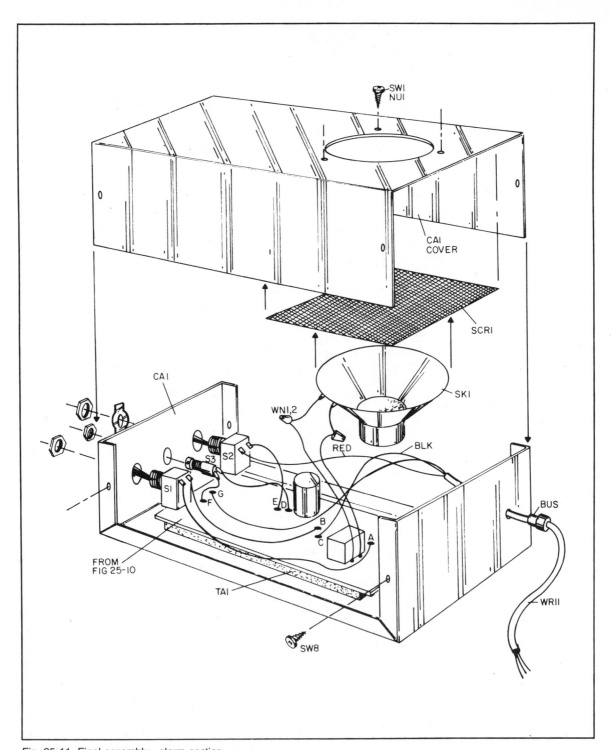

Fig. 25-11. Final assembly—alarm section.

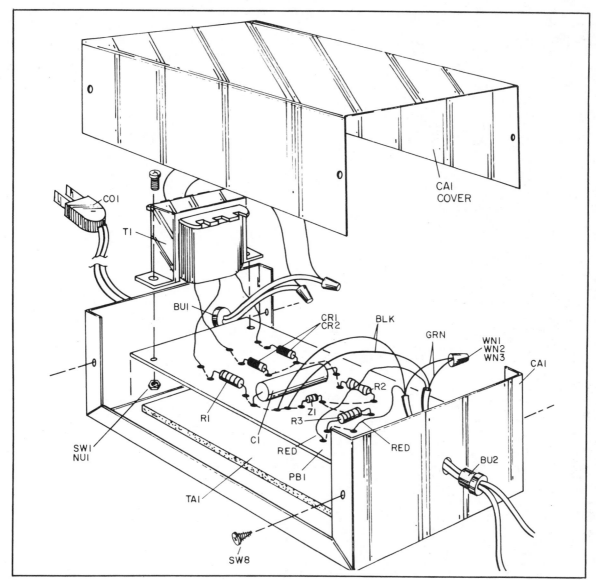

Fig. 25-12. Auxiliary power supply final.

7. Unit is completed and tested.

POWER SUPPLY AND BATTERY SECTION

Construct as shown Figs. 25-6 and 25-12. Use standard wiring and soldering techniques. This system is designed to keep the batteries charged at all times so in case of power failure alarm will still be functionable.

SYSTEM SETUP AND ALIGNMENT

It is assumed at this point that the builder has used his own ingenuity in housing and mounting of the transmitter and receiver section (Fig. 25-13). One different method is to enclose the transmitter section inside of a driveway lamp housing such as those available through hardware stores. This is possible using a short focal length lens. (See op-

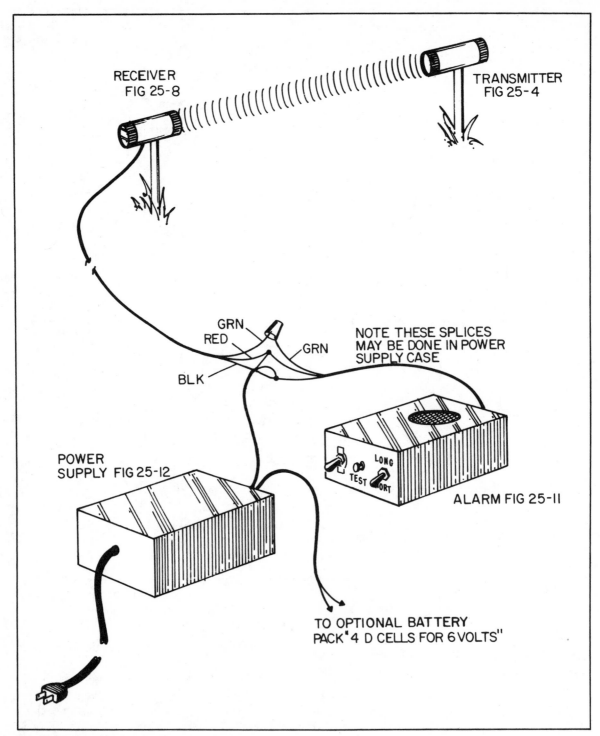

RECEIVER
FIG 25-8

TRANSMITTER
FIG 25-4

GRN

RED

GRN

BLK

NOTE THESE SPLICES
MAY BE DONE IN POWER
SUPPLY CASE

POWER
SUPPLY FIG 25-12

ALARM FIG 25-11

LONG

TEST

SHORT

TO OPTIONAL BATTERY
PACK "4 D CELLS FOR 6 VOLTS"

Fig. 25-13. Completed system layout.

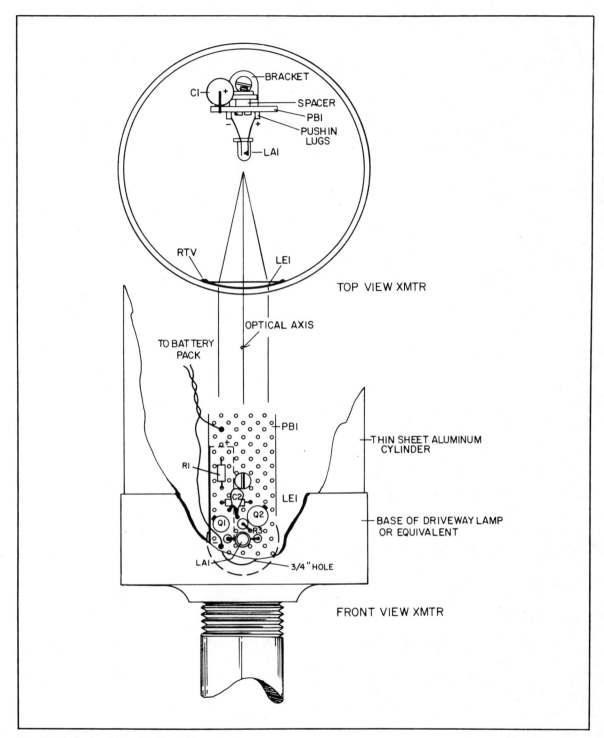

Fig. 25-14. Optional housing ideas for concealment.

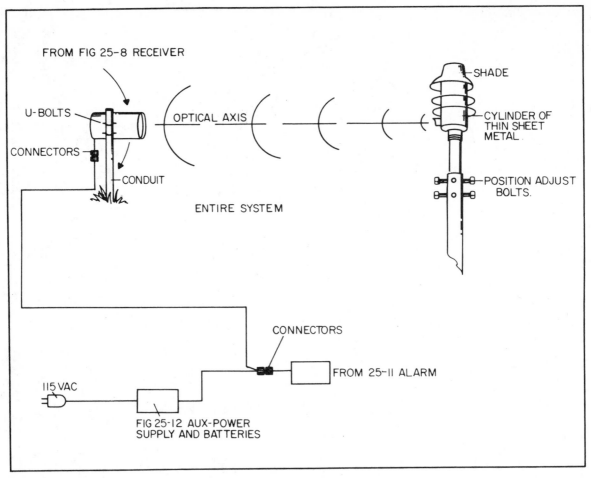

FROM FIG 25-8 RECEIVER

U-BOLTS

CONNECTORS

OPTICAL AXIS

CONDUIT

SHADE

CYLINDER OF THIN SHEET METAL

POSITION ADJUST BOLTS.

ENTIRE SYSTEM

CONNECTORS

FROM 25-11 ALARM

115 VAC

FIG 25-12 AUX-POWER SUPPLY AND BATTERIES

Fig. 25-15. System using optional housing ideas for concealment.

tional transmitter construction Figs. 25-14 and 25-15.) Also, the low drain of this section allows about six to eight weeks using four "C" cells connected two cells in series-parallel (allowing the section to operate without any connection wires). Using this approach also camouflages this section allowing placement in plain view as a regular driveway light. The receiver section can be conventionally mounted via "U" bolts, screws, etc., attaching it to a metal pole secured in the ground. This method allows vertical adjustment via the "U" bolts and horizontal adjustment by rotation. The receiver can be hidden in a tree, cinder blocks, side of hill, etc.

Note the attention to selection of the proper

height of the beam from the ground will avoid small animals such as raccoons, cats, etc., from sounding this alarm. These parameters must be selected by the builder.

1. Place transmitter and receiver in position. Allow vertical and horizontal movement of both.

2. This step should be performed in darkness or low ambient light. Obtain a small pen-type flashlight. Remove rear cover (CA1) of transmitter section and slide out transmitter assembly. Place flashlight so that lamp is approximately positioned where LA1 of transmitter would be. Use modeling clay to temporarily position, etc.

Remove receiver assembly via rear cover (CA3). Remove IR filter (FIL1). Obtain an empty toilet paper roll and cover one end with white paper. Insert paper covered end of roll into receiver enclosure through rear. You should obtain a focused spot of light from the flashlight in the transmitter section on the paper cover of the roll. Adjust both receiver and transmitter enclosure for maximum brightness of this spot. Repeat placing flashlight in receiver and vice-versa. This is usually not necessary if protected distance is less than 25 feet as simple mechanical sighting along the enclosure usually suffices for a rough initial position. The builder may have to improve on the alignment using his own ingenuity if other methods of housing are used.

3. Final adjustment is accomplished by connecting a vtvm, oscilloscope, or other high resistance voltmeter to the "Test Point" lead on schematic and select low range ac volts. With luck, a small voltage resulting from the optical coupling will be read on the meter. Place hand in path and note if meter drops at all. If it does, simply proceed to make slight mechanical adjustment for maximum reading. Proper positioning of the phototransistor and LED inside the housing should be okay. Proceed by mechanically positioning receiver and transmitter for maximum reading.

SPECIAL NOTE ON OPERATION

This device is extremely reliable and effective against intrusion, etc. There are, however, several items to be observed. Lightning can raise havoc with this system causing triggering nearby lightning discharges. Also, after a severe storm, the system should be verified for proper operation. Severe, heavy rain can sometimes trigger this alarm. High humidity conditions upon condensation of water on lenses such as early morning may cause triggering. Use silicone or other lens cleaner to cause moisture to bead up. Lenses should be cleaned with windex as often as the environment requires. Also, protective roofs over the lenses may help keep them from getting wet. The unit performs reliably in temperatures from over $+100°$ F to $-25°$ F.

All the above problems along with other faults will create an "on" alarm condition indicating a problem, therefore, eliminating guesswork as to verifying alarm operation. You will note the power supply schematic along with emergency batteries providing alarm operation in the absence of power while constant trickle-charging of the batteries keeps them in top operating condition.

The receiver and transmitter may be located up to several hundred meters from the alarm box and power supply. A three conductor PVC jacketed cable can be used and should be buried at least several inches in the ground. Cable should be kept away from power lines or other sources of electrical noise.

Extended control of spotlights, electrification devices, cameras, traps, other alarms, or anything that can be activated electrically may be retrofitted to this system via our sense and control unit #SCU1. Please note that all parts for this project are available from Information Unlimited, Inc. P.O. Box 716, Amhurst, NH 03031.

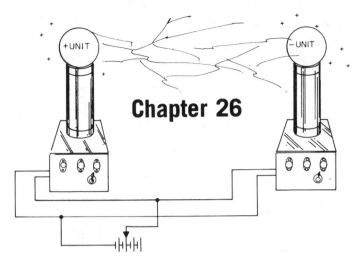

Chapter 26

Phaser Burning
Protection Device/Portable
3000 Volt/4 mA Dc Supply (PSW3)

The following project describes how to construct a device that will produce up to 4000 volts of continuous direct current at 3-5 mA that is capable of burning damp paper or living tissue. It is intended for producing severe electrical shocking power to ward off attacks by dogs or other vicious animals. It is not to be confused with the inductive discharge type animal prods and shock sticks that are available on the market today, but utilizes a microelectronic oscillator and power multiplier stage. The unit is about the size of a small flashlight and is completely self-contained with a rechargeable ni-cad battery. It also may be used as a portable power supply for many projects requiring these output parameters. Charge level is indicated by a small neon lamp. Contact from the device is made via two electrodes protruding out of the end. These electrodes can be sharpened for penetrating thick fur or animal hide or made as convenient connection points to other circuits. The circuit is activated via a simple momentary normally "off" push-button switch conveniently placed on the side or rear of the unit.

This is a high-voltage device and is capable of integrating a potentially dangerous charge onto a suitable capacitor. Exercise caution and discretion whenever using any electrical project of this nature.

CIRCUIT THEORY

The unit (Fig. 26-1 and Table 26-1) is powered by a 9-volt ni-cad battery or a 12-volt ni-cad (higher powered unit). These cells are easily replaced by removal of the rear plastic cover and carefully sliding out the batteries. Actuation is accomplished by simply pushing on the switch and making contact with the electrodes to the target object. Fur or hair are poor conductors, consequently requiring a good pressure contact. The circuit is nothing more than a miniature switching inverter circuit using the latest available components and techniques. A high-frequency inverter circuit consisting of transistors (Q1) and (Q2) alternately switches current flow in the primary of T1, and induces a high-voltage square wave at its secondary. Diodes (D1) and (D2) are the return path for the base current flow in the

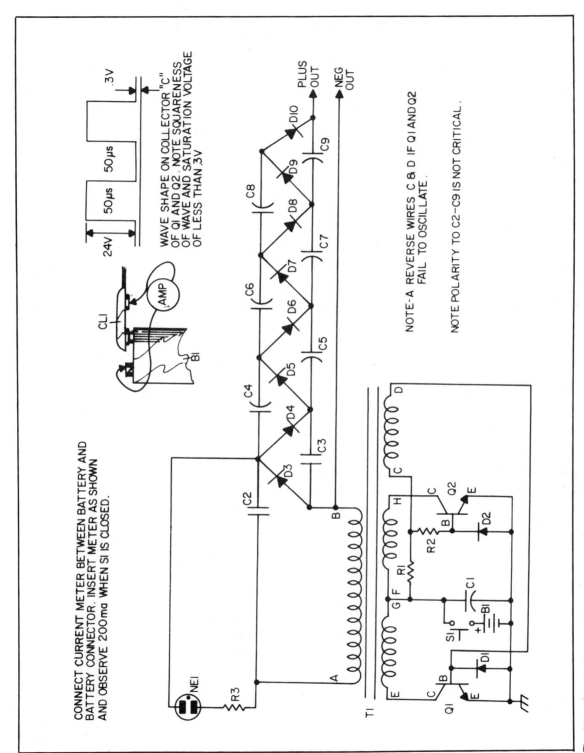

Fig. 26-1. Circuit schematic.

Table 26-1. PSW3 Phaser Burning Protection Device/Portable 3000V/4 mA Supply Parts List.

R1	(1)	2.2 k ¼ watt resistor
R2	(1)	220 ohm 1 watt resistor
R3	(1)	390 k ¼ watt resistor
C1	(1)	10 μF/25 V elect
C2-9	(8)	.1 μF/400 V paper or .01 μF/1600 V discs
TE1,2	(2)	Terminal lugs or use desired electrodes contacts
Q1,2	(2)	D4OD5 npn power transistor
D1,2	(2)	1N4002 diodes 100 V
D3-10	(8)	1N4007 diodes 1000 V
S1	(1)	Push-button switch
NE1	(1)	Neon lamp NE51 with leads
BU1	(1)	⅜ bushing
CA1,2	(2)	#988 plastic caps
TU1	(1)	8″ × 1.5″ × .035″ al tube
PL1		Thin plastic formed into insulating cylinder
WR5	(24″)	#22 plastic stranded hook-up wire
PB1	(1)	1 ¼″ × 4″ perfboard
CL1	(1)	Battery clip
T1	(1)	Assembled Type I transformer, Figs. 2-2 and 2-3
B1	(1)	Optional 9-volt ni-cad battery
LAB1	(1)	HV warning label

Complete kit with optional PC board available from Information Unlimited, Inc., P.O. Box 716, Amherst, N.H. 03031. Write or call 1-603-673-4730 for pricing and delivery.

conducting transistor obtained from the feedback winding on T1. Resistor (R2) limits this base current to a value necessary for causing saturation of the transistors. Resistor (R1) causes a temporary positive unbalance condition to start the switching action. A voltage multiplier consisting of multiplier diodes (D3-D10) and capacitors (C2-C9) develops a voltage of 2000 to 4000 volts or more. An external capacitor can be selected by the builder but must be handled with extreme caution as a potentially dangerous condition will exist.

CONSTRUCTION STEPS

1. Layout and identify all components. Read instructions and all notes.

2. Assemble transformer (Figs. 2-2 and 2-3) Type I.

3. Cut perfboard as shown (Fig. 26-2) (1¼″ × 4″).

4. Assemble components to board as shown (Fig. 26-2). Watch polarity of Q1, Q2, C1 and all diodes (D1-D10). Note contact terminals (TE1,2). Use component leads when possible to eliminate bussing jumps. Watch dress of components for proximity of shorting. Remember there is high voltage

successively increasing to 3 to 4 kV on C9 and D10.

5. Carefully attach correctly identified wires from T1 to points on board. Be careful as wires break easily. Use RTV and secure T1 to board. Use wire identification chart and wire matching chart (Fig. 26-3).

6. Attach wires to S1, CL1, and NE1 as shown (Fig. 26-3).

7. Verify wire and soldering for accuracy shorts, etc., and perform the following tests:

8. Insert a 9-volt ni-cad battery as shown in Fig. 26-3. Connect a 500 mA current meter as shown in Fig. 26-1. Please note some 9-volt ni-cads are only 7.2 volts. These will work but will reduce output.

9. Press S1 and note current reading of 200 mA, lamp NE1 should be glowing. If not, refer to Note A on Fig. 26-1.

10. Obtain a high resistance voltmeter capable of reading up to 5000 Vdc, such as a Simpson 260. Using the high volt range, carefully measure the output across TE1 and 2. It should read about 3000 volts. Obtain a resistor of about 1.5 meg at 2 watts and connect across output. Observe voltage dropping to about 2400. Input current will be about

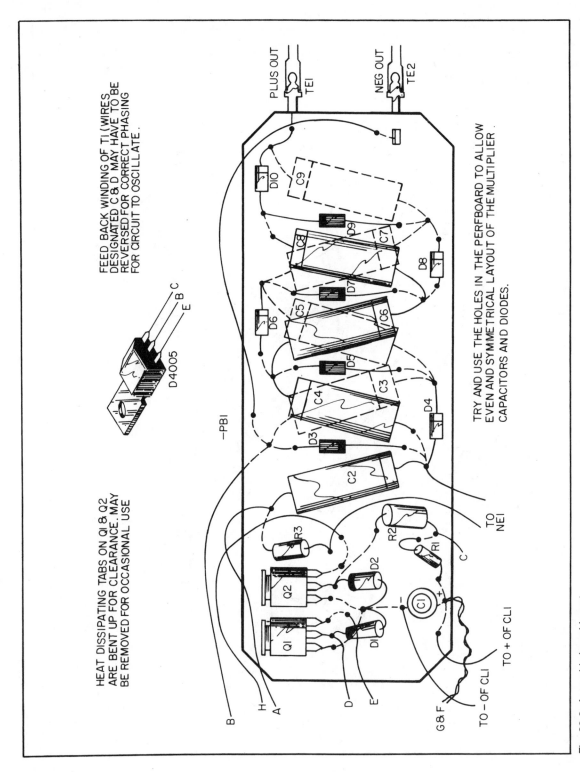

FEED BACK WINDING OF T1 (WIRES DESIGNATED C & D MAY HAVE TO BE REVERSED FOR CORRECT PHASING FOR CIRCUIT TO OSCILLATE.

HEAT DISSIPATING TABS ON Q1 & Q2 ARE BENT UP FOR CLEARANCE. MAY BE REMOVED FOR OCCASIONAL USE

TRY AND USE THE HOLES IN THE PERFBOARD TO ALLOW EVEN AND SYMMETRICAL LAYOUT OF THE MULTIPLIER CAPACITORS AND DIODES.

Fig. 26-2. Assembly board layout.

347

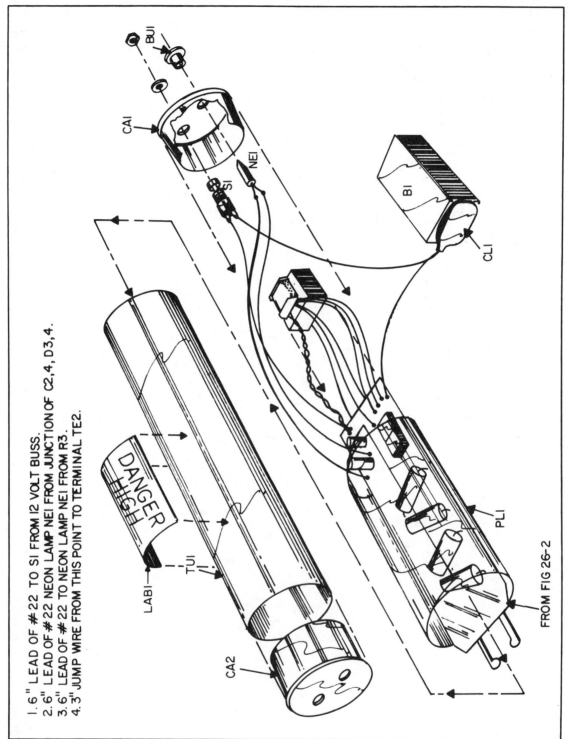

1. 6" LEAD OF #22 TO SI FROM I2 VOLT BUSS.
2. 6" LEAD OF #22 NEON LAMP NEI FROM JUNCTION OF C2,4, D3,4.
3. 6" LEAD OF #22 TO NEON LAMP NEI FROM R3.
4. 3" JUMP WIRE FROM THIS POINT TO TERMINAL TE2.

BUI

CAI

NEI

SI

BI

CLI

DANGER
HIGH

LABI

TUI

CA2

PLI

FROM FIG 26-2

Fig. 26-3. Final assembly.

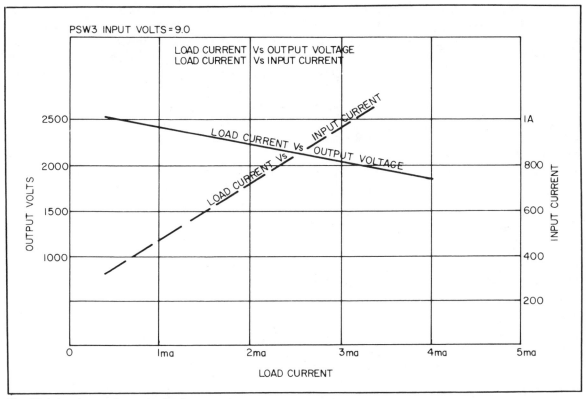

Fig. 26-4. Load current vs output voltage graph.

600 mA. This checks circuit loading. See Fig. 26-4 for operating parameters.

11. Allow unit to run for 1 minute and finger touch Q1 and Q2. These should not be hot to the touch.

12. Fabricate CA1, CA2, and TU1 as shown in Fig. 26-3. Note holes for BU1 (bushing for neon lamp NE1 and S1). Note holes for TE2, 2, and CA2.

13. Insert CA2 over TE1, 2 of assembly board so that they protrude through the holes. You may wish to solder extensions on the terminals to increase their length. Use second set of terminals if necessary. Liberally apply RTV or equivalent in CA2 to secure to PB1 board (Fig. 26-2).

14. Carefully wrap assembly board with a thin cylindrical sheet of plastic and hold together with scotch tape. This forms an insulating cylinder over the HV components and prevents any potential shorting to the metal enclosure TU1.

15. Sleeve wires to S1, NE1, and CL1 into

tube first. Proceed to insert assembled board via transformer end as shown in Fig. 26-3. Position CA2 into TU1 securing assembly board. Secure S1 and NE1 to CA1 via nut and RTV. Connect battery B1 to CL1 and insert into tube TU1. Place CA1 into rear of tube. Note wires to CL1, NE1, and S1 should be long enough to easily allow assembly via these steps. You may wish to also insert some foam rubber to position B1 from moving about in tube.

16. Push button and note NE1 glowing. Contact metal object and note healthy discharge at TE1 and 2. Roll a piece of paper and moisten. Place across contact and apply power. You will note paper smoking and sparking eventually catching on fire. This demonstrates the continuous output energy capable of not only shocking a subject but also of destroying living tissue.

OPERATION AND APPLICATIONS

To operate your unit, remove rear cover CA1

and insert batteries, observing polarity markings (Fig. 26-3). To verify operation, activate the operate button and place an *insulated* screwdriver momentarily across the electrodes and note a healthy spark sounding like a cap gun. An interesting point demonstrating the effectiveness of the unit is to dampen or moisten a piece of paper such as a match book cover and bridge the output contacts noticing a burning, smoking, and eventually flaming up. The dampened cardboard is similar to that of living tissue. When using against an attacking animal, you should attempt to contact a moist section of the body such as the nose and mouth area. This is a low resistance area, compared to fur or hair and allow a large jolt of current that can temporarily stun the subject. In most cases, a good jolt will discourage the animal from future attacks and only seeing the device the second time and associating it with the pain sometimes can be enough. A higher powered completed version of this device is available from Information Unlimited, Inc. P.O. Box 716, Amhurst NH. Order BLS1 Blaster Wand.

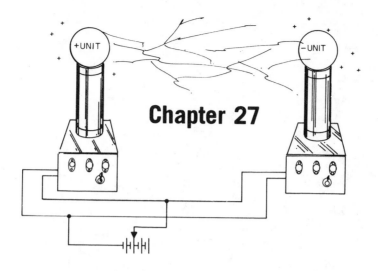

Chapter 27

Electronic Paralyzing Device (PG1W)

This project can be constructed in several different configurations. The circuit and electronic components are the same for these alternate construction approaches. It is suggested that the builder construct the "Simple Contact Device" as this requires less effort and is not as dangerous in the wrong hands as the second approach.

The Simple Contact Device requires physical contact to be made with the subject that causes him to retreat via extremely painful yet harmless electrical shocks. Contact is made via a fixed electrode and contact ring placed at the end of the device. The fixed electrode is at the center and can be a screw or other simple metallic object. This method is the least effective, however, it is the simplest to construct. See Figs. 27-1, 27-2, 27-3, and Table 27-1. The second approach is referred to as the "initial contact control method". This method consists of the contact head separating from the main power handle when contact is made. The contact head consists of barbed hooks or electrodes being connected via 5 to 10 feet of HV teflon insulated wire.

Contact is momentarily made once to the attacking subject by jamming him with the barbed electrode. The head now remains attached to the subject via his clothing or flesh depending on the structure and length of the contacts. The user now has total control via the connecting wires between the contact head and main power handle. The subject loses control via his central nervous system being jammed by the electrical impulses administered at the users discretion. The user is now in control of the attacking subject administering these pulses as needed until help arrives. Remember the subject will temporarily lose voluntary control as he is hit by these electrical pulses rendering him harmless (with the exception to himself via his own secondary reaction).

The third approach is similar to the initial contact control method with the exception of the contact electrical head being projected via a mechanical propulsion system such as a spring or other (non-chemical) means. (Please note that an explosive propellant places the device under Federal Law and

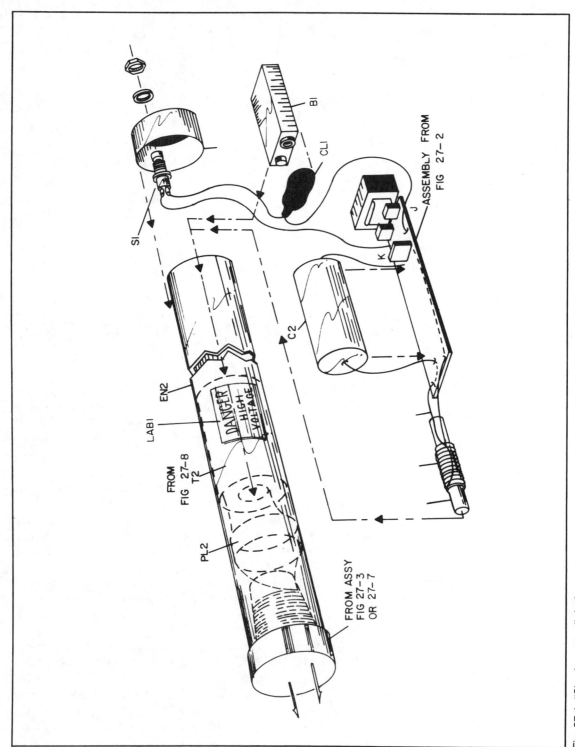

Fig. 27-1. "Simple contact" device.

352

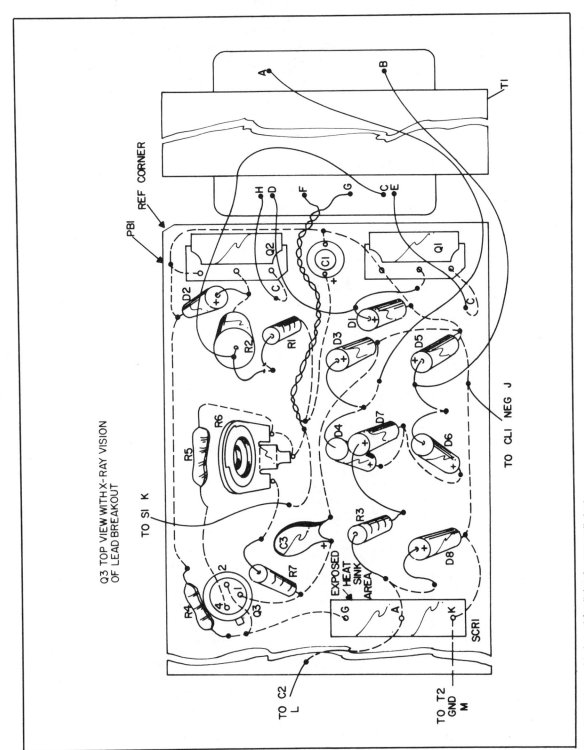

Fig. 27-2. Assembly board for "simple contact" device.

requires the same regulations as a hand gun.) This method allows the user the advantage of being able to subdue his attacker from a distance without contact, and is the most effective and consequently most difficult to construct owing to the propulsion technique. It is suggested that the builder obtain the Taser Information Pack offered by Information Unlimited, Inc. P.O. Box 716, Amherst, N.H. 03031 for $10.00, as this device utilizes an explosive propulsion system.

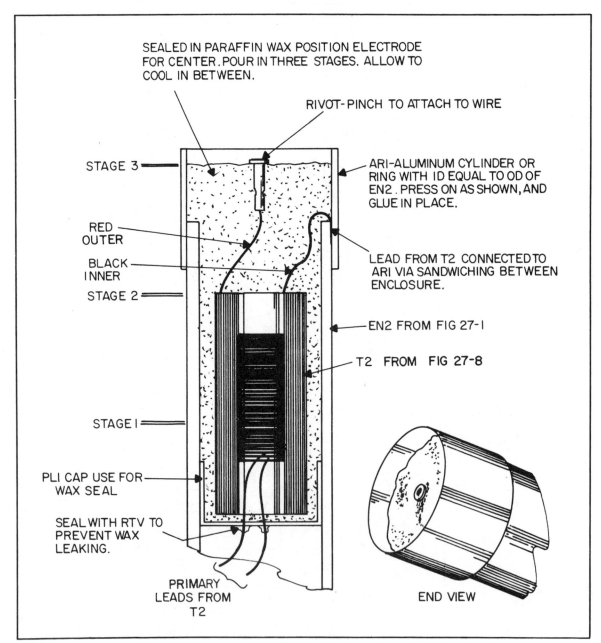

Fig. 27-3. X-ray view of contact wand head.

Table 27-1. PG-1 Paralyzer Wand Parts List.

R1	(1)	1 k ¼ watt resistor
R2	(1)	220 ohm 1 watt resistor
R3	(1)	220 ohm ¼ watt resistor
R4	(1)	10 k ¼ watt resistor
R5	(1)	100 ohm ¼ watt resistor
R6	(1)	500 k trimpot vertical resistor
R7	(1)	22 k ¼ watt resistor
C1	(1)	10 μF/25 V electrolytic cap
C2	(1)	1 3 μF at 400 V special pulse discharge
C3	(1)	1.0 μF/25 V electrolytic cap or tantalum
D1,2	(2)	1N4002 100 V 1 amp diode rectifier
D3,4,5,6 7,8	(6)	1N4007 1 kV 1 amp diode rectifier
Q1,2	(2)	D4OD5 npn power tab transistor
SCR1	(1)	2N4443 SCR
Q3	(1)	2N2646 UJT
PB1	(1)	Perfboard 3″ × 1.1″
T1 BOB	(1)	Small bobbin for T1
T1 CORE	(2)	Small cores "E" for T1
CL1	(1)	9-volt battery clip 6″
S1	(1)	Pushbutton switch
CA1,2	(2)	1 ⅞ plastic cap
CA3	(1)	1 ⅝ plastic cap
WN1,2	(2)	Small wire nuts
WR3	(6″)	6″ #24 plastic covered wire
WR2	(12″)	12″ #18 plastic covered wire
WR31	(60″)	60″ #18 enamel for pri of T2
WR40	(10′)	10′ #28 teflon .036 or #24 plastic stranded
EN2	(1)	10″ enclosure × 1.5 PVC
HA1	(1)	6″ × 1 ⅝″ al tube
PL1,2	(2)	Plastic cap 1 ½″ alliance A 1 ½ or #53
FER1	(1)	½ × 3″ ferrite rod for T2
WPL1	(1)	Special fab wood plug
T1		Type 1 400 volt inverter transformer, REF Fig. 2-2, 2-3
T2		Special HV pulse transformer, Fig. 27-8
LAB1		High voltage label
AR1	(1)	Aluminum ring 2″ OD × .049 wall cut 1.5″

OPTIONAL ITEMS

B1		9-volt ni-cad Varta
HK1	(2)	Small fish hooks

Complete kit with optional PC board available from Information Unlimited, Inc., P.O. Box 716, Amherst, N.H. 03031, write or call 1-603-673-4730 for pricing and delivery.

CIRCUIT THEORY

Your PG-1 circuit (Fig. 27-4) consists of two basic sections: (A) the inverter power supply and (B) the capacitor discharge section. The inverter section consists of switching transistors (Q1) and (Q2) that alternately switches the primary windings of a saturable core transformer (T1). A high-voltage square wave is induced in the secondary of T1 via the switching action and is rectified by diode bridge D3, 4, 5, and 6. Base current drive for Q1 and Q2 is obtained by a tertiary feedback winding on T1 and is applied in the correct phase to turn the appropriate transistor "on". This base current is limited by resistor R2. Diodes D1 and D2 provide a return path for the base current flowing in the opposite transistor respectively. R1 serves to unbalance the circuit to initiate switching. A voltage of approximately 400 volts is obtained in this circuit from the 9-volt ni-cad battery B1. Higher powered operation may be obtained by increasing B1 to a 12-volt battery pack, however, more space is required and care must be taken not to overrate the components if continued use is anticipated.

The capacitor discharge section consists of high-voltage pulse transformer T2 (Fig. 27-4) being current pulsed via SCR1 shorting a charged capacitor C2 across its primary. C2 and the primary inductance of T2 provides a ringing wave whose negative overshoot commutates SCR1 to turn "off." It is important that this primary inductance be sufficient so when combined with capacitor C2 allows a ringing frequency with a period considerably larger than the required commutation turn-off time of the SCR1. Diode D8 provides energy recovery of the negative overshoot component of this discharge pulse.

Transformer T2 now force-induces a very high-voltage pulse in its secondary with a high instantaneous peak current (this system is similar to a capacitor discharge ignition). Diode D7 and R3 limit the dc current to the SCR1 and prevents dc lock on, which also provides a high impedance to the negative turn "off" pulse.

SCR1 is triggered by the UJT pulse timing circuit consisting of Q3. Pulse repetition rate is determined by capacitor C3 and the charging resistor R6. SCR1 switch rate can be adjusted "from one to ten pps." Higher pulse repetition rates may have a tendency to overload the inverter power supply, where it will be unable to supply the current necessary to successfully charge C2, consequently with its charge voltage dropping off.

The voltage output of T2 is well over 25,000 volts at a peak current of 3-6 *amps*. The energy waveforms consists of a train of 30 microsecond pulses decaying exponentially. This produces a *very* painful electric shock and is what causes the temporary jamming of the nervous system but will not electrocute due to the *low average* current plus the fact that the current flow is only across a small section of the body, that is between the contact electrodes. The *peak power* of this device is 100,000 watts plus!

CONSTRUCTION STEPS

1. Select the mechanical configuration of the device you desire. You will note that the plans show two alternative methods of construction. Figure 27-1 shows the mechanical layout of the device built into a 12" long PVC cylinder that houses the entire assembly. Figure 27-2 shows the board layout for this configuration utilizing a 4" length of perfboard (PB1) also serving for the mounting of C2. Figure 27-5 shows the alternate mechanical layout of the device built into a pistol configuration where the handle of the device serves as the housing for the board layout and battery pack. Figure 27-6 shows the board layout built on a 3" piece of perfboard. The main housing consists of a piece of PVC 10" long.

2. Select the contact head desired. Figure 27-7 shows the use of small fishhooks straightened out and pressed into a wooden plug also serving as a dispensing spool for the wire. This configuration is intended to be used for "initial contact" control with the head remaining attached to the subject and separating from the main assembly being electrically attached via the dispensed wires. Figure 27-3 shows the less severe and suggested approach for the device.

3. Construct T1 transformer as shown in Chapter 2 Figs. 2-2 and 2-3.

4. Construct T2 HV pulse tranformer as shown in Fig. 27-8. Please note: T1 and T2 are available assembled through Information Unlimited, Inc. P.O. Box 716, Amherst, N.H. 03031.

5. Cut and fabricate a piece of perfboard 4" × 1⅛". Mark a reference corner for layout purposes as shown (Figs. 27-2 and 27-6).

6. Layout perfboard as shown. Observe positioning of SCR1, Q1, Q2, and Q3. Note polarity of capacitors and diodes D1-D8. Use component leads for interconnecting points whenever possible. Allow 1/16" to ⅛" leads for adequate heatsinking when wiring all semiconductors.

7. Carefully wire leads from T1 to respective points as shown. Note wires designated by letters in Figs. 27-2, 27-4 and 27-6 coinciding with T1 assembly sheet (Figs. 2-2 and 2-3). Secure to board

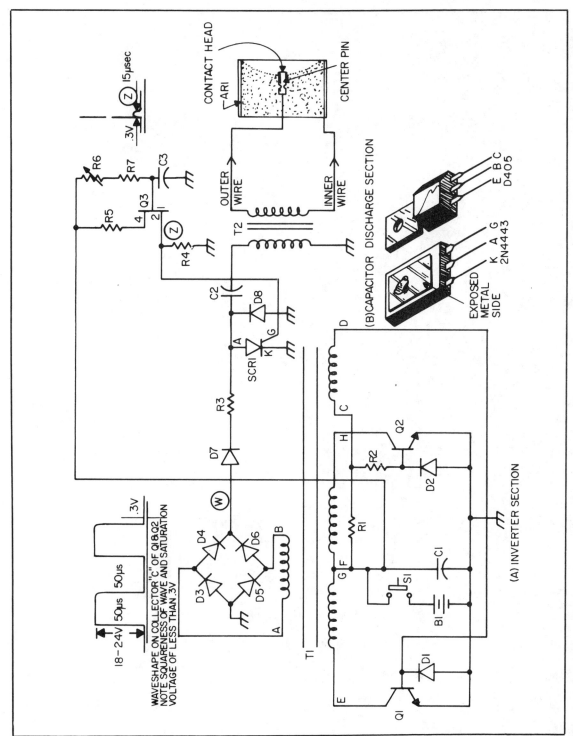

Fig. 27-4. Circuit schematic.

357

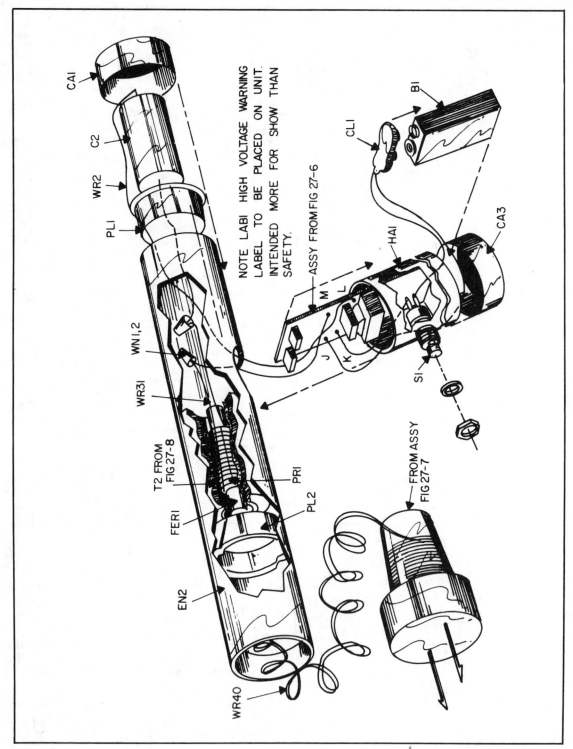

NOTE LAB1 HIGH VOLTAGE WARNING LABEL TO BE PLACED ON UNIT. INTENDED MORE FOR SHOW THAN SAFETY.

Fig. 27-5. "Initial contact" method.

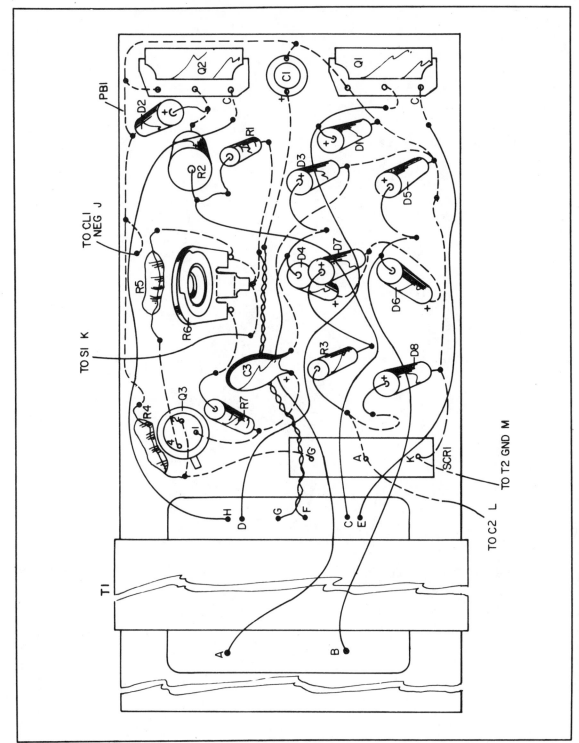

Fig. 27-6. Assembly board for "initial contact" method.

359

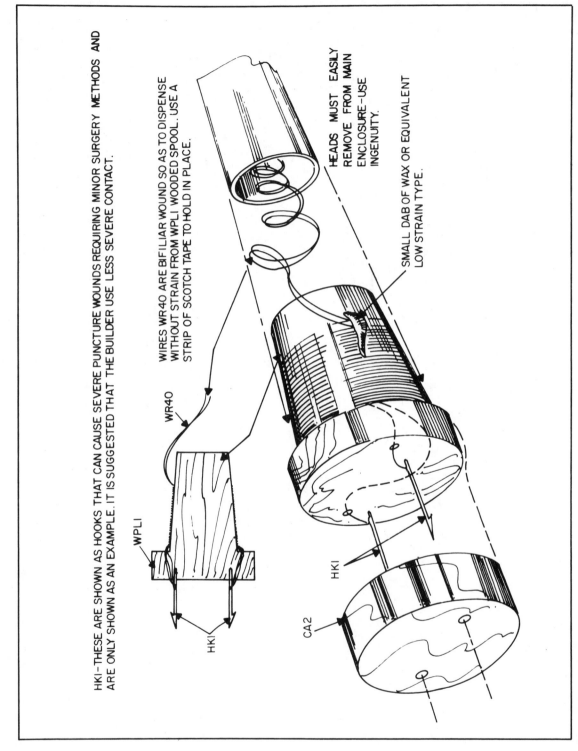

HKI – THESE ARE SHOWN AS HOOKS THAT CAN CAUSE SEVERE PUNCTURE WOUNDS REQUIRING MINOR SURGERY METHODS AND ARE ONLY SHOWN AS AN EXAMPLE. IT IS SUGGESTED THAT THE BUILDER USE LESS SEVERE CONTACT.

WIRES WR40 ARE BIFILIAR WOUND SO AS TO DISPENSE WITHOUT STRAIN FROM WPLI WOODED SPOOL. USE A STRIP OF SCOTCH TAPE TO HOLD IN PLACE.

HEADS MUST EASILY REMOVE FROM MAIN ENCLOSURE – USE INGENUITY.

SMALL DAB OF WAX OR EQUIVALENT LOW STRAIN TYPE.

WR40

WPLI

HKI

HKI

CA2

Fig. 27-7. Sample head for "initial contact" method.

THE FOLLOWING INSTRUCTIONS ARE FOR THOSE WHO DESIRE TO ASSEMBLE THEIR OWN HV OUTPUT PULSE TRANSFORMER T2. WINDING IS TEDIOUS AND MUST BE PROPERLY INSULATED FOR RELIABLE OPERATION. HOWEVER, FOR THOSE WHO HAVE THE PATIENCE AND THE PERSEVERENCE TO ATTEMPT THEIR OWN, WE OFFER THE FOLLOWING CONSTRUCTION STEPS.

1. OBTAIN A CARDBOARD OR PLASTIC FORM, 3" LONG WITH AN INNER DIAMETER OF 5/8".

2. WIND 75 TURNS OF CLOSE WOUND #34 ENAMELED MAGNET WIRE. THIS SHOULD OCCUPY ABOUT 1" OF LENGTH AND SHOULD BE CENTERED AS SHOWN IN FIG 27-8A. YOU WILL NOTE THE "START" IS SECURED VIA A SMALL PIECE OF TAPE. VARNISH OR SHELLAC WINDING TO HOLD IN PLACE. NOTE EXITING OF INNER OUTPUT LEAD.

3. OBTAIN SOME INSULATING KRAFT PAPER (MYLAR TAPE MAY ALSO BE USED) FROM AN OLD TRANSFORMER AND WRAP AROUND FIRST WINDING AS SHOWN IN FIG 27-8B. SECURE WITH A SMALL PIECE OF TAPE AND OVERLAP APPROXIMATELY 1/4". TAPE TO HOLD IN PLACE. VARNISH OR SHELLAC. PAPER SHOULD BE APPROXIMATELY 10 MILS THICK AND COVER ENTIRE FORM WITH SOLID 1/4" OVERLAP.

A

4. START SECOND LAYER AS SHOWN IN FIG 27-8C AND WIND 75 MORE TURNS SECURING AS FIRST LAYER. INSULATED SECOND LAYER SIMILAR TO FIRST AND REPEAT ASSEMBLY STEP FOR 20 LAYERS OF APPROXIMATELY 1500 TURNS IN TOTAL. SHELLAC OR VARNISH EACH LAYER MAKING SURE OF ADEQUATE "BETWEEN LAYER" INSULATION.

5. THE FINISHED HV COIL SHOULD CONSIST OF 20 LAYERS WITH SEVERAL LAYERS OF INSULATING PAPER OVER THE LAST ONE. THE OUTER "OUTPUT" LEAD MUST EXIT AT THE SAME END BUT DIRECTLY OPPOSITE OF THE INNER OUTPUT LEAD, NOTE FIG 27-8D.

6. TIGHTLY WIND 20 TURNS OF #18 ENAMELED WIRE AROUND A 1/2" DIAMETER FERRITE CORE AND INSERT INTO CENTER OF HV COIL ASSEMBLY. EXIT LEADS OF THIS COIL ARE AT OPPOSITE END OF OUTPUT LEADS. LENGTH OF THE WINDING SHOULD OCCUPY ABOUT 1 1/4" OF FERRITE CORE AND MUST BE CENTERED. LEAVE 4 TO 5" LEADS FOR CONNECTING INTO CIRCUIT.

7. ASSEMBLY NOW SHOULD BE POTTED. PARAFFIN WAX IS SIMPLE AND FORGIVING, ALLOWING REPAIR SHOULD THERE BE A CONSTRUCTION ERROR. THE ENTIRE ASSEMBLY SHOULD BE PLACED IN THE HOT WAX AND LEFT FOR SEVERAL HOURS. PARAFFIN WAX CAN EASILY BE MELTED BY PLACING IN A CONTAINER WITHIN A SECOND CONTAINER OF BOILING WATER. THIS LIMITS THE TEMPERATURE OF THE WAX TO THAT NOT MUCH ABOVE 100° C. LEAVING THE ASSEMBLY IN THE MELTED WAX WILL HELP REMOVE ANY AIR POCKETS AND ASSURE PROPER WAX IMPREGNATION. THE ASSEMBLY MAY THEN BE PLACED IN A THIN PLASTIC OR EQUIVALENT CYLINDER AND TOTALLY IMMERSED IN WAX. EPOXY SUCH AS CASTOLITE MAY BE USED, BUT IS UNFORGIVING ONCE SET AND REQUIRES VACUUM EQUIPMENT FOR PROPER IMPREGNATION. NOTE THE DRAWING EXAGGERATES THE SPACING BETWEEN TURNS FOR CLARITY. CLOSE WIND ALL LAYERS.

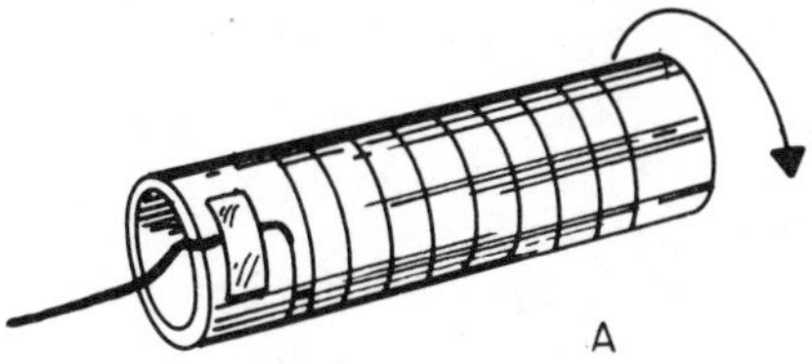

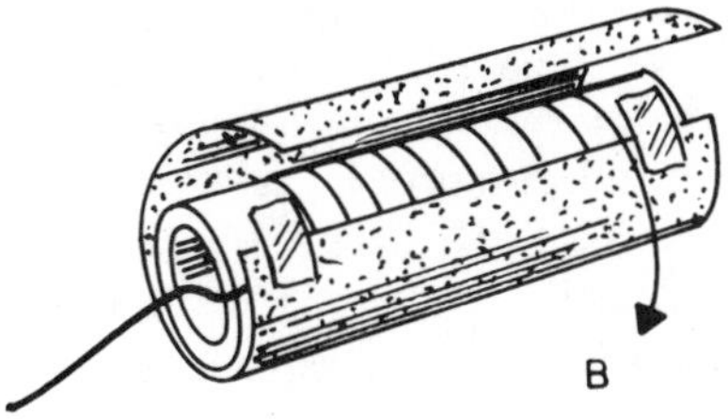

B

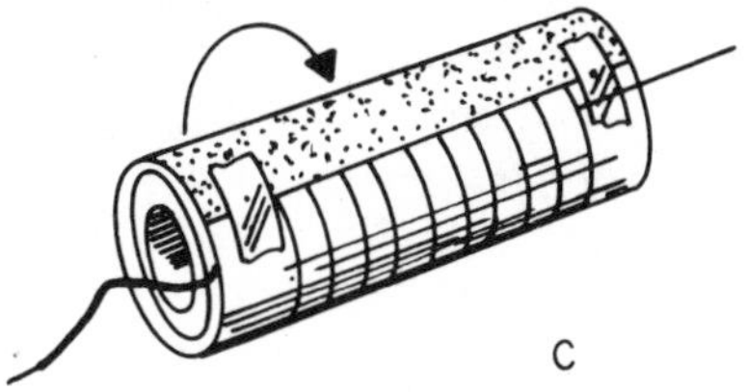

C

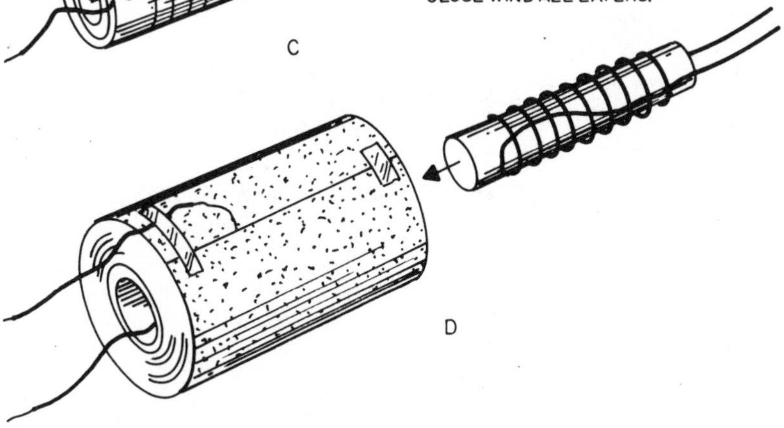

D

Fig. 27-8. Hv pulse coil assembly steps.

with a bead of RTV to eliminate stress on T1 wires from handling.

8. Connect black wire from battery clip (CL1) to point J. Connect 6″ length of wire from point K to switch S1. Connect red lead of CL1 to remaining contact of S1.

9. Connect a 10″ length of #18 wire (WR2) from C2 to point L on board. Connect free lead of C2 to heavy primary lead of T2. Join free lead of T2 with 6″ length of #18 wire (WR2). Connect other end of "M" on board. Use wire nuts where shown in Fig. 27-5 as these points must be disconnected when inserted into housing. Insert FER1 into primary winding of T2. *Do not insert into HV secondary section at this time.*

10. Insert a freshly charged 9-volt ni-cad into CL1 and monitor the dc voltage point W to ground. Momentarily press S1 and quickly note voltage immediately going to nearly 400 volts. Check input current by inserting millimeter across S1 contact and note kicking around 250 to 300 mA. If no voltage is observed, reverse wires C & D from T1 as this is the feedback voltage and must be phased properly relative to the other windings.

11 . When proper voltage and current measurements are verified, energize S1 and check pulsing action of Q3 and note point Z. Adjust R6 and note pulse repetition rate varying from 2 to 10 pps. Do not adjust higher. You should hear a faint ticking sound as the core FER1 is being magnetized by the dumping action of the SCR1 switching the energy from C2 to the primary of T2. Setting R6 for a pulse rate of 5 pps is a good compromise.

12. Place HV secondary coil cover primary and FER1 assembly. Position the thin output leads to about ½″ separation. Energize S1 and note a good spark discharge. It is not advisable to increase gap length any further as overstressing of T2 may occur.

MECHANICAL ASSEMBLY

13. Fabricate EN2 from a 10″ piece of 1.9″ OD, sked 40 PVC. Place a 1.625″ hole for handle using a circle saw 4″ from rear. Angle slightly for pistol grip effect (see Fig. 27-5).

14. Fabricate HA1 handle from a 5-6″ piece of 1⅝″ OD aluminum tubing. Drill hole for S1.

15. Finally assemble unit as shown in Figs. 27-1 and 27-5. Note the use of plastic cap covers PL1 and PL2 inserted and used for positioning C2 and strain-relieving of these wires from T2. Positive strain relieving can be accomplished by passing the heavier wires (WR40) from the contact head through PL2 and tying knots in this wire to prevent it from pulling through. The knotted ends are now attached to the thinner more fragile wires of T2. Further strengthening may be accomplished by pouring in some paraffin wax or RTV liquid (room temperature vulcanizing silicone). *Caution—make sure that secondary coil of T2 is well isolated electrically from any other part of the circuit. Wires must be clear of primary circuit.*

16. Construct contact head as shown in Fig. 27-3 or 27-7. Note contact electrodes shown can be straightened out fishhooks and are extremely severe when used. A more suitable electrode may be screws, nails, etc. This assembly is placed into enclosure EN2 and secured via weak glue, wax, etc., to allow easy separating when used. Fishhook type electrodes that are shortened for only adhering to outer clothing will be sufficient for effective use. Also note that the head may be securely attached to the device and strictly used as a contact device. Contact wires WR40 may be rewound on wooden plug if not damaged by prior use.

APPLICATION

This device can be a dangerous device both mechanically and electrically. It is not intended as a toy or joke and should never be used on a person unless no other alternative is available. It is intended for unruly, vicious animals and can produce puncture wounds depending on the type electrodes used.

Contact should be made simultaneously energizing S1 and noting convulsive, involuntary actions of the subject. Secondary reactions may also injure the subject from the above actions. However, electrocution is impossible due to the low average current and method of body contact. A complete study on devices of this type are available through Information Unlimited, Inc. P.O. Box 716, Amherst,

N.H. 03031, for $10.00 and fully describes the Taser Electronic Paralyzing Gun.

SPECIAL NOTES

Please note that this device is not made to operate in an open circuit condition, i.e., without some type of load for any period of time due to the insulation of the output wires and the head assembly. In actual use, any contact with living flesh would act as a resistance of no more than 25-100 kilohms. If open circuit testing is required to test the voltage waveform it is suggested that the HV secondary leads be disconnected from the head and separated by a gap of ½ to 1" to allow arcing to occur between the leads rather than the other points of T2.

Shocking or paralyzing power is achieved via the pulses of high peak current, characteristic in this type of circuit. The repetition rate of these pulses is obviously low or a tremendous amount of energy would be discharged across the contact electrode causing flesh to explode and burn. The pulse power is approaching 100 kW but the energy expended is only several joules owing to the low repetition rate and short pulse duration.

When testing the assembled unit it is suggested that all connections be made with the exception of the electrode wires from T2. These wires should be connected to a 100 ohm carbon or noninductive resistor. The current pulse can be determined by observing the voltage across the resistor and should be at least several hundred volts with a freshly charged 9-volt ni-cad battery. This corresponds to a peak current = to

$$\frac{voltage\ observed}{resistance\ \Omega\ ohms}.$$

Increased output may be obtained by using AA cell ni-cads in series for voltages up to 12 volts. This higher voltage may cause circuit stress and eventual failure if used continually.

Increasing the repetition rate of the pulses to a high value via R6 may start to reduce the charging voltage on C2 decreasing individual pulse energy. This is due to the power supply being unable to supply the necessary average current and limits the overall energy output. Without this characteristic output would approach lethal values.

Typical measurements taken in our lab are the following: Load 100 ohm carbon resistor, input 9 V charged ni-cad, repetition rate set at 2 pps, ringing frequency 30 microsec, voltage across 100 ohm resistor at 400 volts corresponding to a peak current of 4 amps.

For optimum performance it is suggested to pot the HV coil T2 (potting sketch Fig. 7-3) with paraffin wax. This is done by inserting the assembly and retaining cap as shown. The primary leads of T2 are dabbed with RTV to prevent the hot paraffin from leaking out. The wax is then poured into the enclosure until a ¼ to ½" layer is formed over T2 further securing the output leads.

Note on safety and common sense: Do not use on a person with a weak heart. Do not expose yourself to a hostile situation and fully depend on this device. Do not energize S1 for any more than several discharges unless in use as damage may result to head and coil.

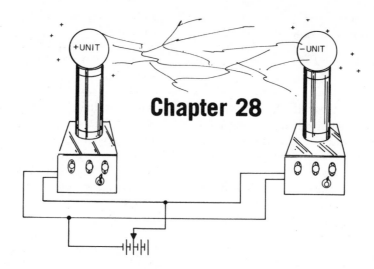

Chapter 28

Sound-operated
Switch and High-gain Amplifier (SOX2)

The following project shows how to construct an electronic device capable of controlling a tape recorder or other related items with sound or voices. It is intended for eliminating wasted dead tape time when recording conversation, nature studies, intrusion detection, automatic announcing of one's presence, etc. It can be made to interface with other systems such as hidden cameras, alarms, lights, etc., by activating them when triggered by sounds or voices.

The device is self-contained with all batteries and contains a built-in mike. A sensitivity adjustment is capable of presetting the activation threshold from the sound of a shot to a whisper. Circuitry also includes a high-gain low-noise amplifier for direct use with directional, parabolic mikes, etc., while at the same time being able to sound activate other circuits. The device uses no short cuts in circuitry and is a first-rate amplifier along with a reliable and sensitive sound-controlled switch.

CIRCUIT OPERATION

The circuit utilizes a transistor array type integrated circuit with a low-noise front end (Fig. 28-1 and Table 28-1). Input is a high-impedance crystal mike either mounted to the enclosure or can be an externally selected type with cable and plug connecting to jack (J1) on enclosure. (The latter method allows placement of the mike at focal points of parabolas, etc., for low-noise, high-sensitivity directional mikes.) The mike element is capacitively coupled to a Darlington pair for impedance transforming and is further fed to a capacitively-coupled cascaded pair of common-emitter amplifiers for further amplification. Sensitivity control (R7) controls base drive to the final transistor of the array and hence controls overall system sensitivity. Output of the amplifier is capacitively coupled to a one-shot consisting of Q1 and Q2 in turn integrating the output pulse of Q2 on to capacitor C8 through D1. This dc level now drives relay drivers Q3 and Q4 activating K1 consequently controlling external circuitry. You will also note that an auxiliary output, via P2 is adjusted by R20. This allows individual setting for proper recording signal to "aux" input of recorder and allows independent threshold of tape recorder activation via "remote" input.

CONSTRUCTION STEPS

1. Identify all components. Note that indicated layout should be followed for proper performance (Figs. 28-2, 28-3, and 28-4).

2. Fabricate aluminum minibox, as shown for J1, BU1, and R7/S1. Assemble these parts in place as shown. Note ground lug under J1.

3. Reference corner of assembly board and mark for identification.

4. Insert the CA3018 as shown and carefully position as shown. Bend over several leads to keep from falling out. This insertion of the CA3018 may require several attempts before aligning the 12 leads with the appropriate holes in the perfboard as shown. Note Fig. 28-4 showing pins 1, 4, 7, and 10 form the corners of a square.

5. Using standard audio frequency wiring techniques, proceed to wire and solder starting with C1 inserting the designated components and soldering point-by-point. Carefully check for accu-

Table 28-1. SOX2 Sound Operated Switch and Ultrahigh-Gain Amplifier Parts List.

R1	(1)	Resistor 470 k ¼ W
R2	(1)	Resistor 5.6 M ¼ W
R3,5,6	(3)	Resistor 6.8 k ¼ W
R4,8	(2)	Resistor 390 k ¼ W
R7,S1	(1)	5 k pot & switch
R10,16	(2)	Resistor 5.6 k ¼ W
R11,14,15,17	(4)	Resistor 100 k ¼ W
R13 AB	(2)	Resistor 10 ohm ¼ W (use two in parallel)
R18,19	(2)	Resistor 220 ¼ W
R20	(1)	Resistor 100 k trimpot horz
C1,9	(2)	.05 μF/25 V disc cap
C2,3,5,6,7	(5)	1. μF @ 25 V L or electrolytic cap (preferably non polarized)
C4	(1)	.01 μF/50 V disc cap
C8	(1)	4.7 μF @ 25 V electrolytic cap
A1	(1)	Semiconductor amp array CA 3018
Q1,2,3,4	(4)	PN2222 npn transistor
D1	(1)	1N914 signal diode
D2	(1)	600 V 1 amp rect 1N4007
K1	(1)	Dip 8 pin miniature
M1	(1)	High-impedance crystal mike
WR10	(12")	Shielded mike cable
CL1,2	(2)	Battery clips for 9 V
BU1	(1)	Plastic bushing
KN1	(1)	Small plastic knob
WR3	(36")	#24 hook-up wire
CA1	(1)	4 × 2 ⅛ × 1 ⅝ al mini box
J1	(1)	Phono jack
P1	(1)	Sub min 2.5 mm plug for remote of most recorders
P2	(1)	Mini 3.5 mm plug for "Aux" or "Mik" of most recorders
P3	(1)	Phono plug to mate with J1
PB1	(1)	Printed circuit board or use perfboard
B1,2	(2)	9 V batteries (not included in kit)

A complete kit of this device is available from Information Unlimited, Inc., P.O. Box 716, Amherst, N.H. 03031, write or call 1-603-673-4730 for price and availability.

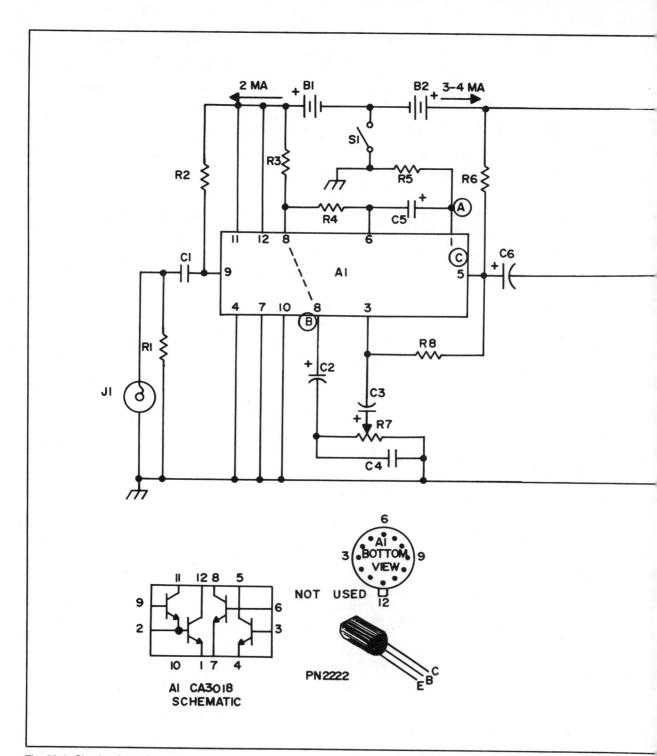

Fig. 28-1. Circuit schematic.

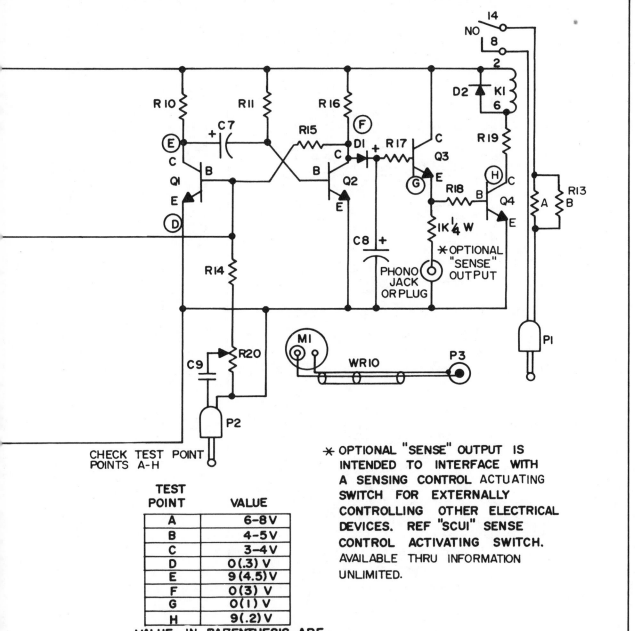

CHECK TEST POINT
POINTS A-H

TEST POINT	VALUE
A	6-8 V
B	4-5 V
C	3-4 V
D	0(.3) V
E	9(4.5)V
F	0(3) V
G	0(1) V
H	9(.2) V

VALUE IN PARENTHESIS ARE
IN THE "ACTUATED" STATE
AND ARE WITH A REALISTIC
#CTR-43 RECORDER

* OPTIONAL "SENSE" OUTPUT IS
INTENDED TO INTERFACE WITH
A SENSING CONTROL ACTUATING
SWITCH FOR EXTERNALLY
CONTROLLING OTHER ELECTRICAL
DEVICES. REF "SCUI" SENSE
CONTROL ACTIVATING SWITCH.
AVAILABLE THRU INFORMATION
UNLIMITED.

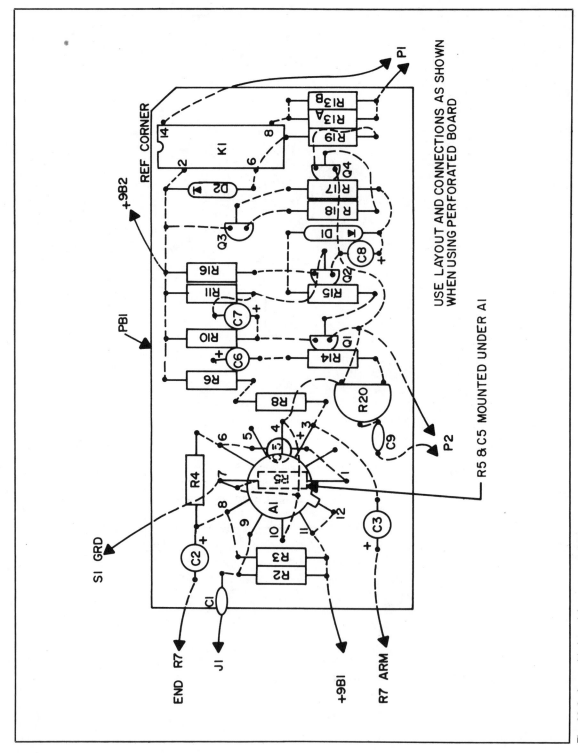

USE LAYOUT AND CONNECTIONS AS SHOWN
WHEN USING PERFORATED BOARD

R5 & C5 MOUNTED UNDER A1

Fig. 28-2. Assembly board (component side).

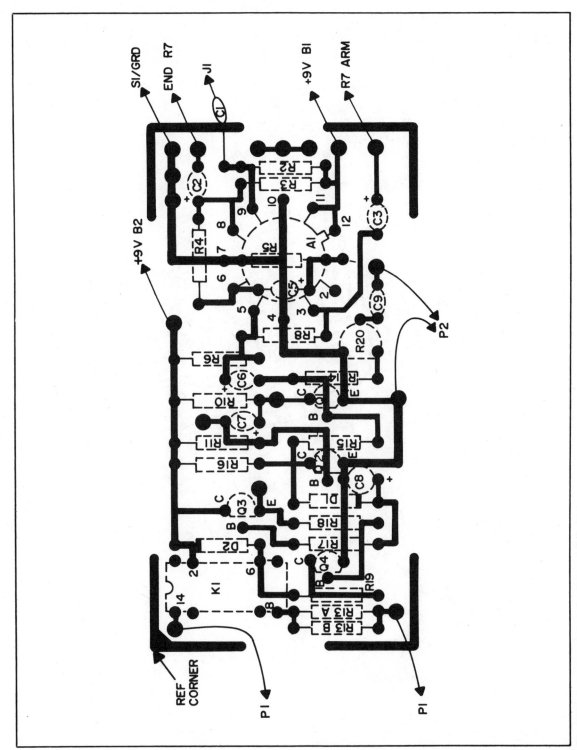

Fig. 28-3. Assembly printed circuit board (foil side).

369

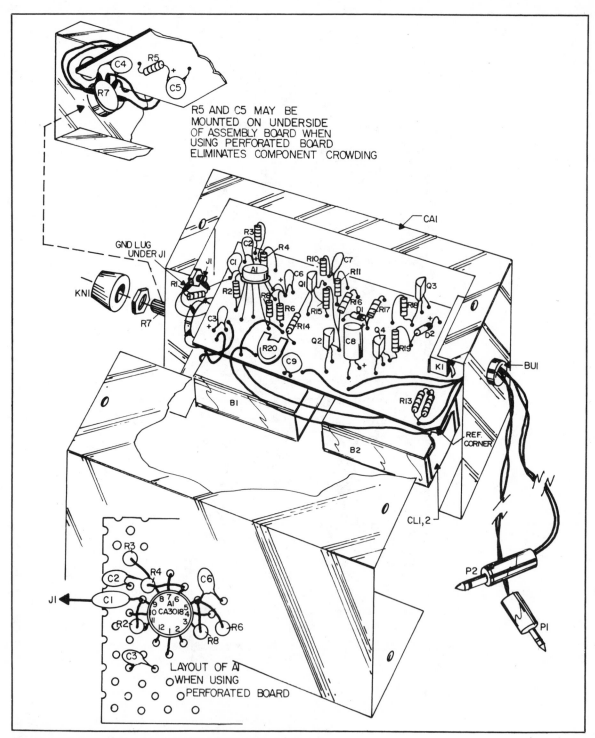

R5 AND C5 MAY BE
MOUNTED ON UNDERSIDE
OF ASSEMBLY BOARD WHEN
USING PERFORATED BOARD
ELIMINATES COMPONENT CROWDING

LAYOUT OF A1
WHEN USING
PERFORATED BOARD

Fig. 28-4. Final assembly wiring and CA3018 layout on perfboard.

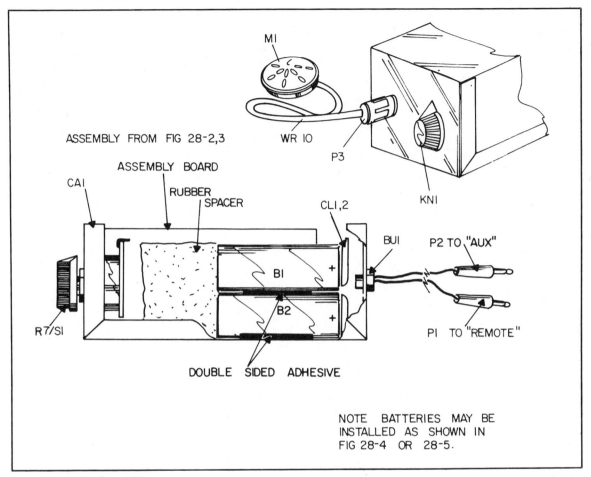

ASSEMBLY FROM FIG 28-2,3

ASSEMBLY BOARD

CA1

RUBBER
SPACER

M1

WR 10

P3

KN1

CL1,2

BU1

P2 TO "AUX"

B1

B2

R7/S1

DOUBLE SIDED ADHESIVE

P1 TO "REMOTE"

NOTE BATTERIES MAY BE
INSTALLED AS SHOWN IN
FIG 28-4 OR 28-5.

Fig. 28-5. Finished assembly.

racy and quality of solder joints. Remember mistakes can ruin the CA3018. *Observe polarity of tantalum and electrolytic capacitors and position of relay.*

6. Attach battery clips by inserting leads through holes in perfboard adjacent to their termination points as shown. This method strain-relieves these wires.

7. Connect C4 across R7 as shown in Fig. 28-4. Connect ground end of R7 to ground lug J1 via uncut lead from C4. Connect buss jump from this point to the two normally open lugs of S1 as shown. Connect R1 to J1 and ground lug of J1 with above and solder this point. Note these leads must be as short and direct as possible to prevent noise and hum.

8. Note that the assembled board should have the following leads for connection to the components in the aluminum minibox.

A. Input lead of C1 for connecting to J1 (short as possible).

B. Ground lead from pin 4, 7, 10, of CA3018 for connecting to grounded connections of R7/S1.

C. Two buss leads to R7 end and R7 arm (use uncut leads of components if possible). These leads must be as short as possible.

9. Visual check for solder, wiring errors, and shorts and install into CA1 box.

10. Preset R20 at midrange. Install mike into J1.

11. With S1/R7 "off" check for zero current through both B1 and B2. Turn switch "on" and note

371

B1 current at 2 mA. B2 at 3-4 mA. Turn up gain R7 and note B2 current increasing to 12 mA when sound is made. This is the relay current and indicates an "on" state. Current should drop back to 3 mA when relay turns "off" in several seconds. Note that R7 must not be set for too high gain or unit will not turn off due to ambient background noise.

12. Plug in P1 to "remote" jack on the recorder and set recorder to record mode. Plug in P2 to "aux" input of recorder. See Fig. 28-5.

13. Turn R7/S1 "on" and slowly advance until recorder starts. The setting of this adjustment is strongly related to the ambient background noise and probably cannot be run anywhere near maximum unless used in a recording studio or extremely low-noise environment. The best results are when R7 is set just below the threshold of activation. Take into account external noises such as appliances and subways that may falsely activate the recorder during the desired recording interim. Remember that proper positioning of the unit may allow the sensitivity to be increased minimizing activation by these other noises and sounds.

INSTRUCTION AND APPLICATIONS

This unit is equipped for use as an optional amplifier along with the sound controlling application. It contains a jack adjacent to the R7 adjust knob for an external mike, that is intended to be placed at the focal point of a parabolic reflector, etc. Also a second plug (auxiliary) is used for applying this amplified signal to the recorder greatly increasing overall system sensitivity. This system is a sensitive directional mike allowing automatic recording of desired sound levels.

For automatic sound control of a transmitter, the "remote" plug is removed and the wires connected as a switch in series with the battery supplying the transmitter. The device by itself can be adjusted to record sounds in an area for monitoring and surveillance functions.

Please note: That certain recorders take longer to start than others, therefore, the first syllable of a word may be missing. The unit can be demonstrated to activate at the very first sound and cannot be blamed for slow recorder start-up. Also, we have allowed a 6-10 second "turn-off" time allowing positive holding of most sounds and voices. Unit may be placed on premises to record certain levels of conversation or sounds. It is suggested to experiment with the proper settings of R7 and R20 to obtain the proper tape activation and recording levels.

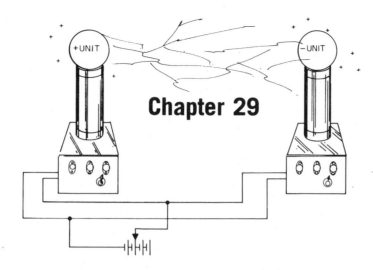

Chapter 29

Remote Wireless Repeater Transmitter (RWM-3)

The following project shows how to construct a wireless attachment capable of transmitting the audio output of a tape recorder, directional mike, another radio, etc., over a remotely located FM or aircraft radio for a distance of well over one mile. The unit when properly installed produces a crystal clear reproduction of the input. Unit is easily installed via a simple mating plug that fits the normal 8 Ω output of most modern electronic equipment. This device can also turn most video recorders into a small local TV transmitter allowing wireless operation of other sets without the cumbersome cables.

CIRCUIT DESCRIPTION

Transistor (Q1) forms a relatively stable rf oscillator whose frequency is determined by the value of L1 and variable capacitor (C2). C2 also determines the desired operating frequency. Frequency range is between 90 MHz on the standard FM broadcast band to 110 MHz in the aircraft band (Fig. 29-1 and Table 29-1).

The circuit with the components listed operates best at about 109 MHz. This is a clear spot without interference from FM radio stations. However, satisfactory performance is obtained between the above mentioned units of 90 to 110 MHz.

Capacitor (C3) supplies the necessary feedback voltage developed across R2 in the emitter circuit of Q1 sustaining an oscillating condition. Resistors R1 and R3 provide the necessary bias of the base-emitter junction for proper operation while capacitor (C4) bypasses any rf to ground fed through to the base circuit. C1 provides an rf return path for the tank circuit of L1 and C2 while blocking the dc supply voltage fed to the collector of Q1.

You will note that the junction of the base bias resistors (R1 and R3) is a feed point consisting of capacitor and resistor (C5 and R4). Because of the nature of the oscillator frequency being subject to change by varying the base bias condition, a varying ac voltage superimposed at this point causes a corresponding frequency shift (FM) along with an amplitude modulation (AM condition). It is this

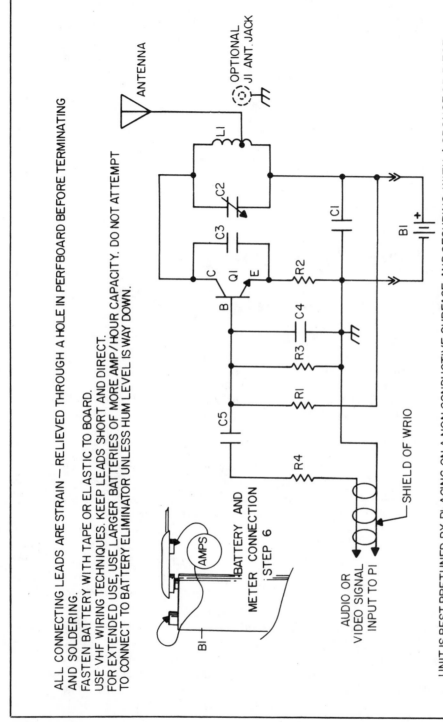

ALL CONNECTING LEADS ARE STRAIN – RELIEVED THROUGH A HOLE IN PERFBOARD BEFORE TERMINATING AND SOLDERING.

FASTEN BATTERY WITH TAPE OR ELASTIC TO BOARD.

USE VHF WIRING TECHNIQUES. KEEP LEADS SHORT AND DIRECT.

FOR EXTENDED USE, USE LARGER BATTERIES OF MORE AMP/HOUR CAPACITY. DO NOT ATTEMPT TO CONNECT TO BATTERY ELIMINATOR UNLESS HUM LEVEL IS WAY DOWN.

UNIT IS BEST PRETUNED BY PLACING ON A NON CONDUCTIVE SURFACE AND SECURING WITH A SPONGE OR PIECE OF RUBBER, ETC., WHILE MAKING ABOVE ADJUSTMENT TO C2 WITH TUNING WAND OR PLASTIC SCREW DRIVER. IT IS IMPORTANT TO KEEP HANDS AND ALL CONDUCTIVE OBJECTS AWAY FROM L1 COIL AND ASSOCIATED CIRCUIT.

PERFBOARD MAY BE REDUCED IN SIZE IF BATTERY OR POWER SUPPLY IS TO BE REMOTELY LOCATED.

R4 MAY BE DECREASED FOR MORE PICK-UP SENSITIVITY OR INCREASED FOR LESS SENSITIVITY.

Fig. 29-1. Circuit schematic.

Table 29-1. RWM-3 Remote Wireless Repeater Transmitter Parts List.

R1	(1)	15 k ¼ watt resistor
R2	(1)	220 ohm ¼ watt resistor
R3	(1)	3.9 k ¼ watt resistor
R4	(1)	180 k ¼ watt resistor
C1	(1)	.1 μF 50 V disc cap
C2	(1)	6/35 pF mini trimmer
C3	(1)	5 pF silver mica zero temp capacitor
C4,5	(2)	.01 μF/50 V disc cap
P1	(1)	Connector for interfacing, RCA phono plug and 3.5 mm mini plug included in kit
CL1	(1)	6″ snap clips for B1
L1	(1)	8 turns coil #16 wire
WR4	(12″)	#24 hook-up wire
WR10	(12″)	12″ shielded grey cable
PB1	(1)	1 × 2 ½″ perfboard .1 × .1
B1	(1)	9 V ni-cad battery (not included in kit)

A complete kit of this device is available from Information Unlimited, Inc., P.O. Box 716, Amherst, N.H. 03031, write or call 1-603-673-4730 for price and availability.

property that allows this circuit to clearly FM modulate intelligent speech that is detected in any FM receiver when properly tuned. C5 provides the necessary dc blocking component while R4 allows the proper attenuation for obtaining minimum distortion along with sufficient modulation signal.

CONSTRUCTION STEPS

1. Wind coil using the threads of a long #10 wood screw for a forming guide. Wind 8 turns of #16 wire, leaving 1 to 2″ leads. Coil should be approximately .6″ in length by .25″ in diameter (Fig. 29-2).

2. Fabricate and cut a piece of perfboard so that the holes are located as shown in Fig. 29-2. This will aid in layout of the components as shown.

3. Assemble the components to the perfboard starting with L1 and then C2. Note the leads of L1 should be used as connecting points. Proceed as shown using component leads wherever possible and avoiding wire bridges.

4. Attach battery clip leads via strain-relieving through holes in perfboard. Carefully solder 6″ antenna lead to tap of L1 and strain-relieve by passing through perfboard hole directly beneath connection.

5. Connect modulation input cable. This is an appropriate piece of shielded cable connected to a plug of your choice depending on the equipment the device is to be interfaced with.

6. Connect a 9-volt battery to clip. If a multimeter or 0-10 mA meter is available, connect in series with battery lead. This can be done by removing one of the fasteners of the clip and connecting the meter to the free contacts. Meter should read 5 to 7 mA. Pick up a short piece of the bare wire and touch the coil (L1) a turn at a time starting from the C1 end. Note that as you progress, turn-by-turn away from C1 that the indicated meter current will change—usually drop.

7. Slowly rotate C2 until the station being received by the radio at approximately 108 MHz is blocked out or disappears. It may be difficult at first to spot the signal as this adjustment is very touchy. Also, note that several spots in the adjustment *may* be found to produce a signal. Only two are valid (the rest being erroneous will be weak and unstable). If the circuit layout and component values are followed closely, it will be found that the correct setting of C2 for a signal at 108 MHz will be the position that is shown in Fig. 29-2.

8. Connect to desired device and adjust volume or output level for best results. For maximum sensitivity R4 may be reduced in value. The device should broadcast these outputs with very little distortion when adjusted correctly.

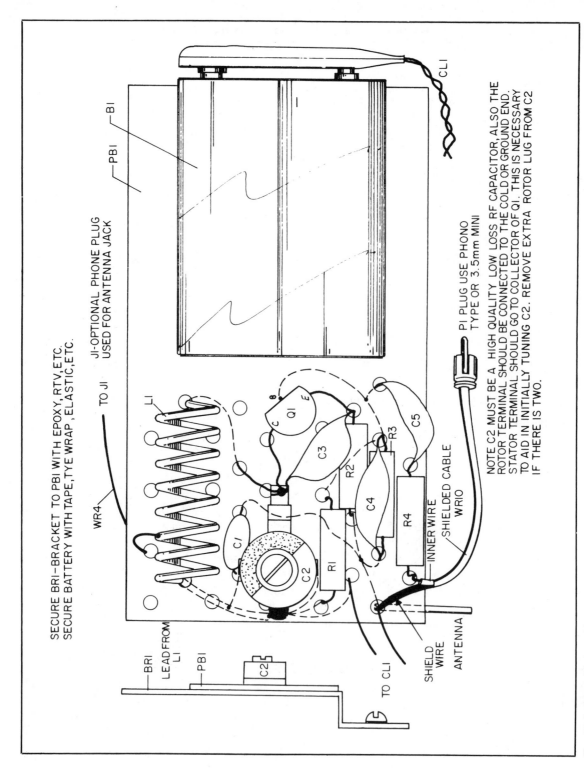

SECURE BRI–BRACKET TO PBI WITH EPOXY, RTV, ETC.
SECURE BATTERY WITH TAPE, TYE WRAP, ELASTIC, ETC.

JI–OPTIONAL PHONE PLUG USED FOR ANTENNA JACK

PI PLUG USE PHONO TYPE OR 3.5mm MINI

NOTE C2 MUST BE A HIGH QUALITY LOW LOSS RF CAPACITOR, ALSO THE ROTOR TERMINAL SHOULD BE CONNECTED TO THE COLD OR GROUND END. STATOR TERMINAL SHOULD GO TO COLLECTOR OF QI. THIS IS NECESSARY TO AID IN INITIALLY TUNING C2. REMOVE EXTRA ROTOR LUG FROM C2 IF THERE IS TWO.

Fig. 29-2. Assembly board layout.

APPLICATION AND TUNING YOUR
REMOTE WIRELESS REPEATER TRANSMITTER

Your RWM-3 Wireless Repeater Transmitter is a wireless attachment that connects to any device such as a phone, radio, tape recorder, dictating machine, or similar device and transmits with crystal-clear clarity, voices, sounds, etc., to any nearby FM radio tuned to pick up this signal. Applications are such as allowing one to listen to a tape or recording deck from a remote location over any portable FM radio while outside, etc. One very amusing use of this device was when I connected my unit to my own radio tuned to a classical station and transmitted this station over my daughters juvenile rock 'n' roll station on her radio whenever she had it turned up too loud.

Your unit is constructed on an oversized piece of perfboard for securing to brackets, etc. If desired, this excess may be removed. It is powered by a standard 9-volt transistor radio battery and is intended to be secured to the device via tape or an elastic. The unit is turned off by disconnecting the battery clip from the battery.

Units are designed to be operated on a clear spot around 109 MHz on or near this frequency. To fine tune the unit in the event that the present frequency falls near or on a local station, perform the following steps:

1. Tune to a station on an FM radio at the low end around 88. (A recorder or similar device can be used to apply radio to the RWM-3.)

2. Tune a second remote FM radio to the high end on a clear spot (no stations). Distance between these radios must be 50 feet for setup purposes.

3. Attach proper plug to RWM-3 (shown with phone plug attached) or use mini-plug as these are standard for most 8 ohm outputs of transistor radios.

4. Attach a standard 9-volt transistor radio battery to the clip. Power to unit is controlled by removal of this battery.

5. Plug in RWM-3 to 8 ohm ear or headphone jack to radio tuned to a station on low end.

6. Using a plastic tuning wand, very, very slightly tune frequency adjust screw until radio station on connected radio is heard on remote second radio at higher end. *Keep hands and objects away from coil and associated components.* Preferably use a tuning wand grasping unit with sponge or piece of foam rubber.

7. Adjust volume control of connected radio to proper level for clarity, etc. Note that a tape recorder or similar device can also be connected to the RWM-3. Please note that it may take several slight adjustments to properly tune the unit when in place. However, once tuned it should not require any readjustment until the battery starts weakening.

A device of this type is range-adjustable depending on the antenna system used with it. Care must be taken not to cause interference to radio stations. Adding an antenna on the unit may make it illegal. Maximum range of a device such as this will occur when the antenna is approximately 28" and the unit is set near or on a ground point such as the roof of a car, etc. If a ground point can't be found, a simulated ground can be obtained by attaching a similar length of wire to the common negative wire from the battery and stretching it out and directly away from the antenna. (Please note that any alterations in the antenna could result in a minor frequency adjustment.) *The unit is not intended to be handheld or moved during use.*

SPECIAL NOTE

If your FM radio is crowded with stations at around 108 MHz it may easily be detuned slightly to shift the dial reading down where 108 is 109. This is accomplished by carefully adjusting the osc padding trimmer located on the main tuning capacitor and "walking" a known station down the necessary megahertz or two. The antenna peaking trimmer should now be adjusted for maximum signal at the high end. See section on general use and helpful hints using wireless devices, antennas, etc.

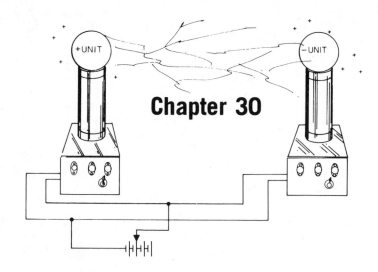

Battery Chargers and Eliminators (BLM3)

The following information shows how to construct several different circuits capable of charging or replacing the popular ni-cad (nickel-cadium battery) or gel cell (semi-dry wet cell). These batteries have several unique advantages. The most obvious one is that they may be recharged and discharged several hundred times thus eliminating costly one time use replacements. Even though the batteries are more expensive initially they quickly amortize themselves depending on usage.

Ni-cad and gel cells also have the capability of delivering high currents when necessary, therefore are required in equipment where high load currents and relatively short operating times are anticipated. Many products will not perform on standard batteries because of this feature. This asset also creates a safety problem as accidental short circuits can cause fire and quickly damage sensitive circuits. Some of the other disadvantages of ni-cads is their sensitivity to overcharging. This is indicated by the battery becoming warm and is sometimes used as an indication for sensing a charge state and

turning off the charging supply. Also, when ni-cads are connected in series, and are not equally charged, a polarity reversal on the low cell can result *causing* cell damage and obviously decreased voltage output. It should be emphasized that individual cell voltage is usually only 1.25 volts rather than 1.5 volts as with conventional batteries. This usually results in lower volts for the equivalent cell pack configuration. Some multicelled batteries compensate by adding more cells to make up for this deficiency in voltage.

Ni-cads also die fast, i.e., they do not give any warning of low charge like conventional batteries. This can be certainly a disadvantage when used in smoke detectors, ignition circuits, etc. The gel cell while being rechargeable is less prone to these negative points. Even though these cells produce high current the *total* ampere capacity is usually less than their conventional counterparts. The instructions show how to construct several charging circuits and battery eliminators for the devices described.

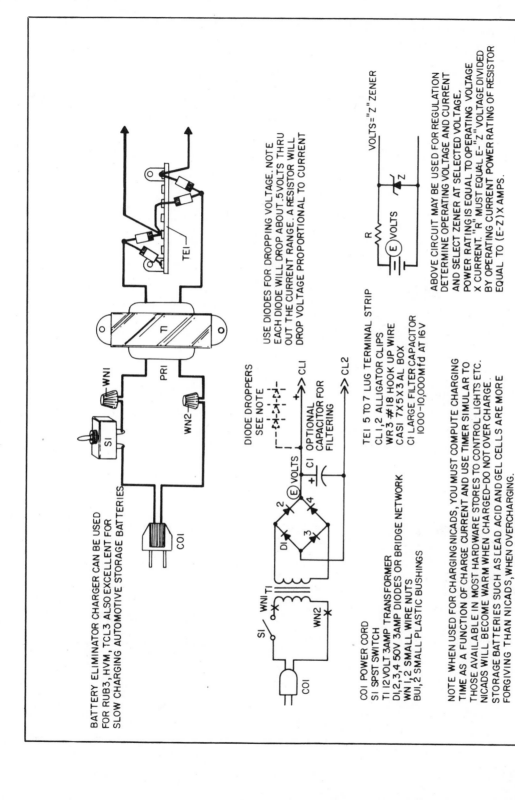

BATTERY ELIMINATOR CHARGER CAN BE USED
FOR RUB3, HVM, TCL3 ALSO EXCELLENT FOR
SLOW CHARGING AUTOMOTIVE STORAGE BATTERIES

USE DIODES FOR DROPPING VOLTAGE. NOTE
EACH DIODE WILL DROP ABOUT .5VOLTS THRU
OUT THE CURRENT RANGE. A RESISTOR WILL
DROP VOLTAGE PROPORTIONAL TO CURRENT

VOLTS="Z" ZENER

ABOVE CIRCUIT MAY BE USED FOR REGULATION
DETERMINE OPERATING VOLTAGE AND CURRENT
AND SELECT ZENER AT SELECTED VOLTAGE.
POWER RATING IS EQUAL TO OPERATING VOLTAGE
X CURRENT. "R" MUST EQUAL E – "Z" VOLTAGE DIVIDED
BY OPERATING CURRENT POWER RATING OF RESISTOR
EQUAL TO (E-Z) X AMPS.

DIODE DROPPERS
SEE NOTE

CLI

C1 OPTIONAL
CAPACITOR FOR
FILTERING

CL2

TEI 5 TO 7 LUG TERMINAL STRIP
CLI, 2 ALLIGATOR CLIPS
WR3 #18 HOOK UP WIRE
CASI 7X5X3 AL BOX
CI LARGE FILTER CAPACITOR
1000-10,000Mfd AT 16V

COI POWER CORD
SI SPST SWITCH
TI 12 VOLT 3AMP TRANSFORMER
DI,2,3,4 50V 3AMP DIODES OR BRIDGE NETWORK
WN I,2 SMALL WIRE NUTS
BUI,2 SMALL PLASTIC BUSHINGS

NOTE WHEN USED FOR CHARGING NICADS, YOU MUST COMPUTE CHARGING
TIME AS A FUNCTION OF CHARGE CURRENT AND USE TIMER SIMULAR TO
THOSE AVAILABLE IN MOST HARDWARE STORES TO CONTROL LIGHTS ETC.
NICADS WILL BECOME WARM WHEN CHARGED–DO NOT OVER CHARGE.
STORAGE BATTERIES SUCH AS LEAD ACID AND GEL CELLS ARE MORE
FORGIVING THAN NICADS, WHEN OVERCHARGING.

Fig. 30-1. BC12 12 V, 3 A charger/eliminator schematic.

379

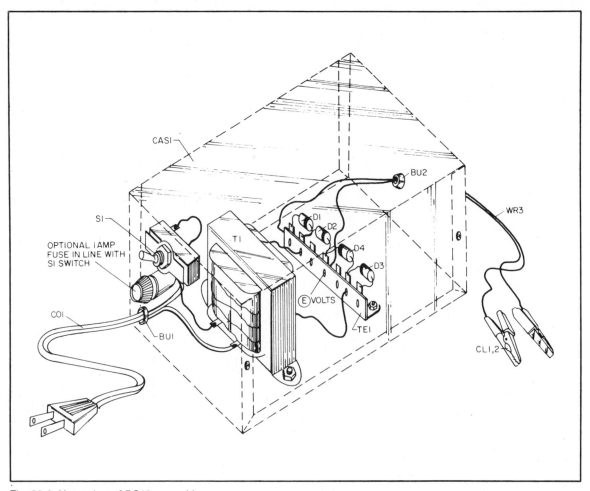

Fig. 30-2. X-ray view of BC12 assembly.

BC12 CHARGER AND COMBINATION BATTERY ELIMINATING POWER SUPPLY

Produces 12 V at 3 A dc unfiltered or filtered. Intended for fast charging ni-cad C- and D-cell packs, gel cells, or slow charging automotive storage batteries. The circuit can be built in any convenient metal or plastic box. See Figs. 30-1 and 30-2 (schematic and construction details).

BCM1 MULTIPLE NI-CAD CHARGER AND COMBINATION BATTERY ELIMINATING POWER SUPPLY

This power supply produces 15 V at 0.2 A dc. Intended for charging up to 12 ni-cad cells in series.

The circuit can be built in any convenient metal or plastic box. See Figs. 30-3 and 30-4 (schematic and construction details).

BC9 DUAL 9-VOLT CHARGER AND BATTERY ELIMINATOR

This unit is for use with wireless and other equipment using 9-volt batteries. The dual purpose charger and eliminator will charge 9-volt ni-cads at 10 mA and provide virtually hum-free operation when used as a battery eliminator with our listening and surveillance devices. Can be constructed on a small piece of perfboard. See Figs. 30-5 and 30-6.

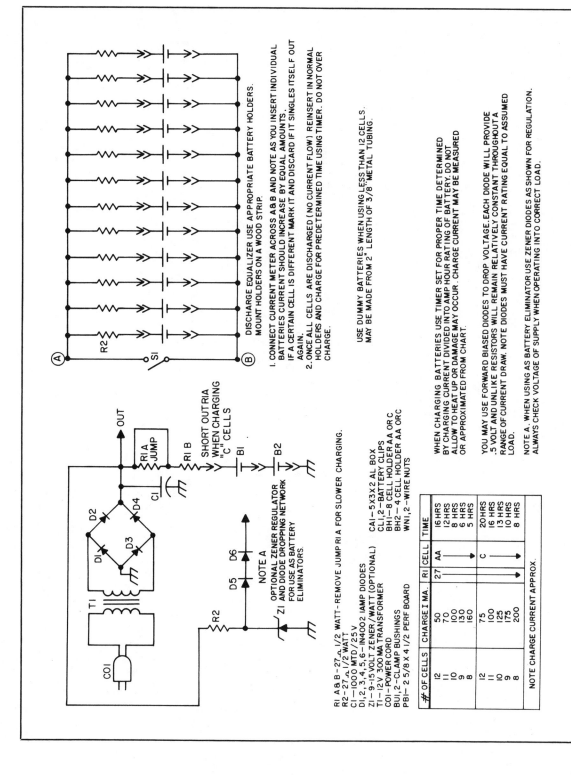

Fig. 30-3. BCM1 Multi-ni-cad battery charger/eliminator and discharge equalizer schematics (charges up to 12 cells).

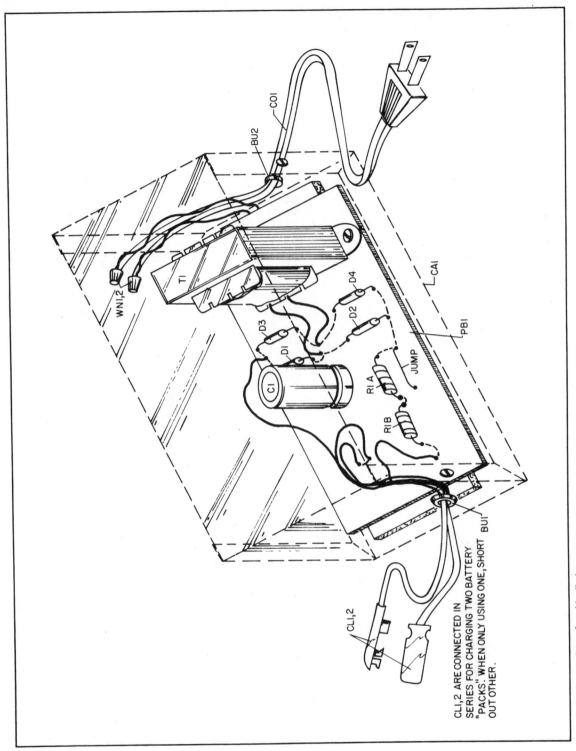

COI

BU2

WNI,2

TI

D4

D2

D3

D1

CAI

PBI

JUMP

RIA

CI

RIB

BUI

CLI,2

CLI, 2 ARE CONNECTED IN
SERIES FOR CHARGING TWO BATTERY
"PACKS." WHEN ONLY USING ONE, SHORT
OUT OTHER.

Fig. 30-4. X-ray view of multicell charger.

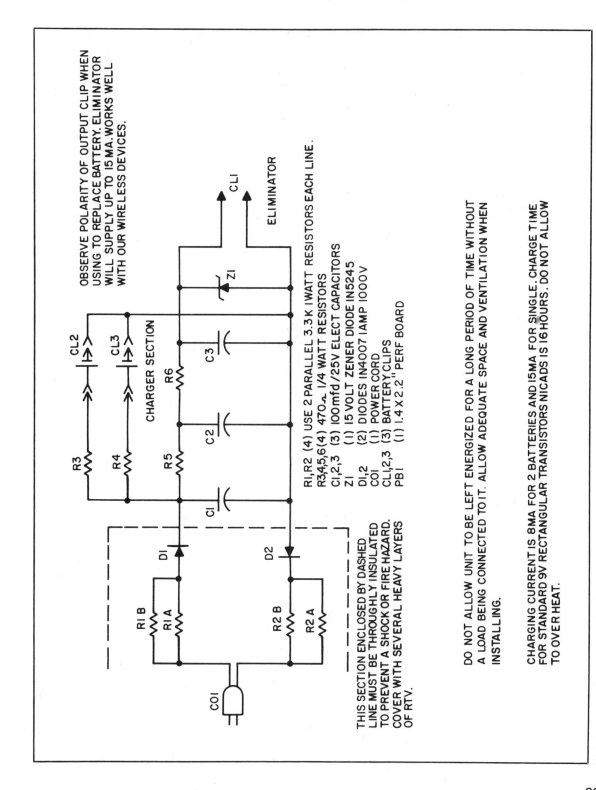

Fig. 30-5. BC9 dual 9 V battery charger/eliminator.

383

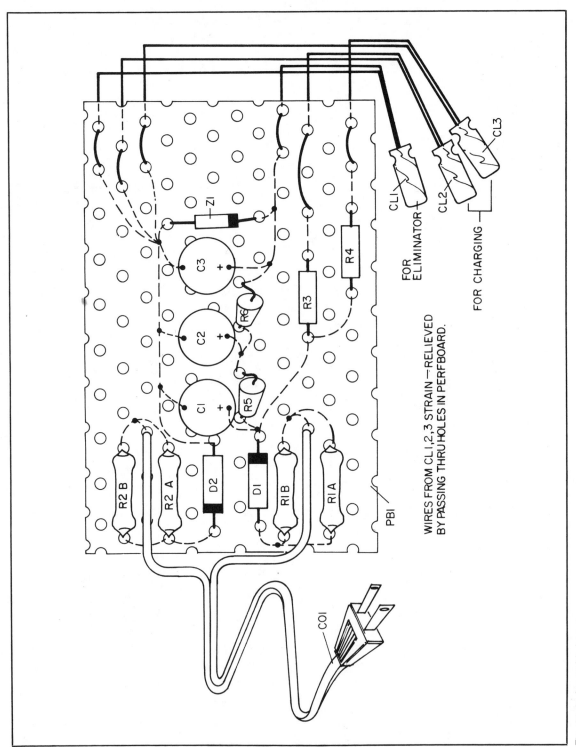

Fig. 30-6. Assembly board.

Index

Index